SpringerWienNewYork

Fortschritte der Chemie
organischer Naturstoffe

Progress in the Chemistry
of Organic Natural Products

Founded by L. Zechmeister

90

Fortschritte der Chemie
organischer Naturstoffe

Progress in the Chemistry
of Organic Natural Products

The Epothilones: An Outstanding Family of Anti-Tumor Agents

From Soil to the Clinic

Authors:
K.-H. Altmann, G. Höfle, R. Müller,
J. Mulzer, and K. Prantz

SpringerWienNewYork

Prof. A. Douglas Kinghorn, College of Pharmacy,
Ohio State University, Columbus, OH, USA

em. Univ.-Prof. Dr. H. Falk, Institut für Organische Chemie,
Johannes-Kepler-Universität, Linz, Austria

Prof. Dr. J. Kobayashi, Graduate School of Pharmaceutical Sciences,
Hokkaido University, Sapporo, Japan

Printed in Germany

SpringerWienNewYork is part of
Springer Science + Business Media
springer.at

Printed on acid-free and chlorine-free bleached paper

With 52 partly coloured Figures

ISSN 0071-7886
ISBN 978-3-211-99928-8
e-ISBN 978-3-211-78207-1

The Epothilones: An Outstanding Family of Anti-Tumor Agents
From Soil to the Clinic

Contents

List of Contributors

Altmann, Prof. Dr. K.-H., ETH Zürich, Department of Chemistry and Applied Biosciences, Institute of Pharmaceutical Sciences, ETH Hönggerberg. HCI H 405, Wolfgang-Pauli-Str. 10, 8093 Zürich, Switzerland
e-mail: karl-heinz.altmann@pharma.ethz.ch

Höfle, Prof. Dr. G., Helmholtz-Zentrum für Infektionsforschung, (formerly: *GBF*, Gesellschaft für Biotechnologische Forschung), Inhoffenstr. 7, 38124 Braunschweig, Germany
e-mail: G.Hofle@helmholtz-hzi.de

Müller, Prof. Dr. R., Institut für Pharmazeutische Biotechnologie, Universität des Saarlandes, Postfach 15 11 50, 66041 Saarbrücken, Germany
e-mail: rom@mx.uni-saarland.de

Mulzer, Prof. Dr. J., Institut für Organische Chemie, Fakultät für Chemie der Universität Wien, Währingerstr. 38, 1090 Wien, Austria
e-mail: johann.mulzer@univie.ac.at

Prantz, Mag. K., Institut für Organische Chemie, Fakultät für Chemie der Universität Wien, Währingerstr. 38, 1090 Wien, Austria
e-mail: kathrin.prantz@univie.ac.at

1. Preface

Johann Mulzer

Institut für Organische Chemie der Universtät Wien, Wien, Austria

This volume is dedicated to the epothilones, which have received unusual attention over the past ten years. Epothilones are novel antitumor drugs which very much like their predecessor paclitaxel (Taxol) act *via* microtubule stabilization. In comparison to paclitaxel and a number of alternative drugs with a similar mode of bioactivity (*e.g.* laulimalide, eleutherobin, peluroside, and discodermolide) the epothilones have significant advantages, above all ready availability, combined with high activity in the nanomolar range and low susceptibility towards multidrug resistance. Epothilone B and several derivatives thereof are in Phase I–III clinical trials; one of them (ixabepilone, BMS) is already on the market, others are supposed to appear on the market in the near future. All naturally occurring epothilones have been isolated from *Sorangium cellulosum* (Fig. 1); as mentioned above their antitumor action is traced back to the stabilization of microtubules. In consequence, the formation of the mitotic spindle is inhibited and the cell undergoes apoptosis (Fig. 2).

In the current publication, five experts in the field (*Karl-Heinz Altmann*, *Gerhard Höfle*, *Rolf Müller*, *Johann Mulzer*, and *Kathrin Prantz*) provide a comprehensive state of the art overview of history, discovery, biosynthesis, total synthesis, and industrial application of the epothilones.

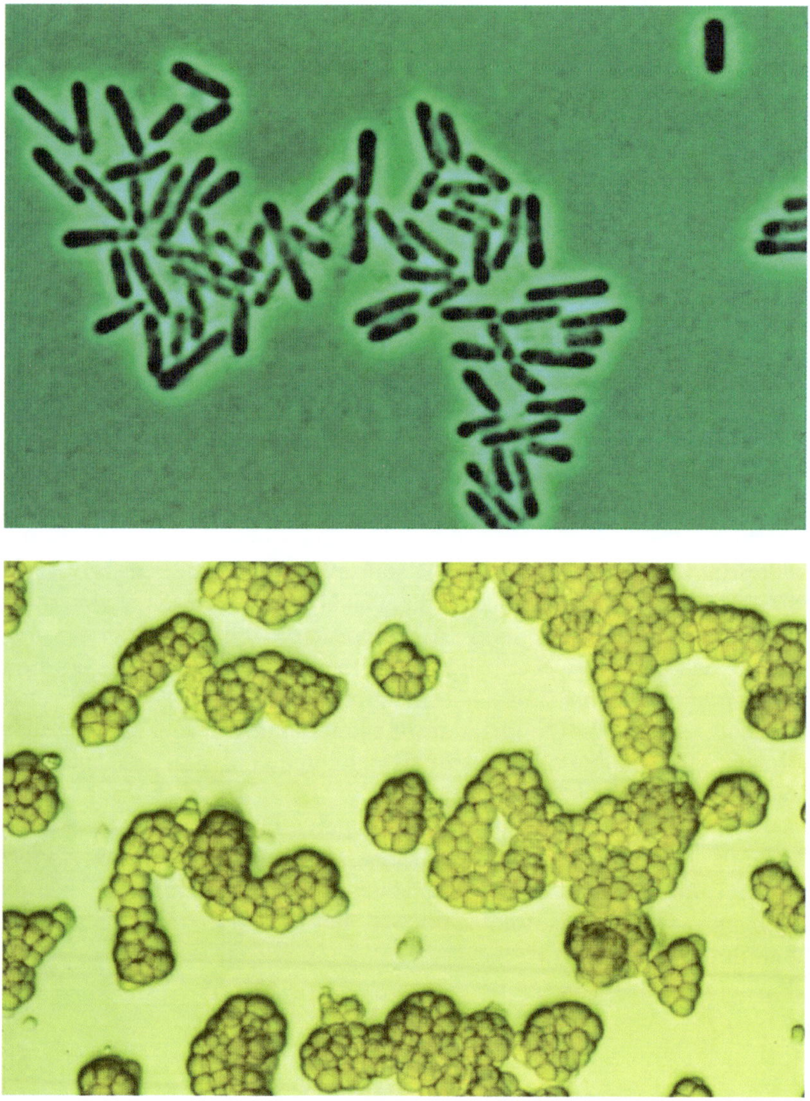

Fig. 1. Myxobacteria

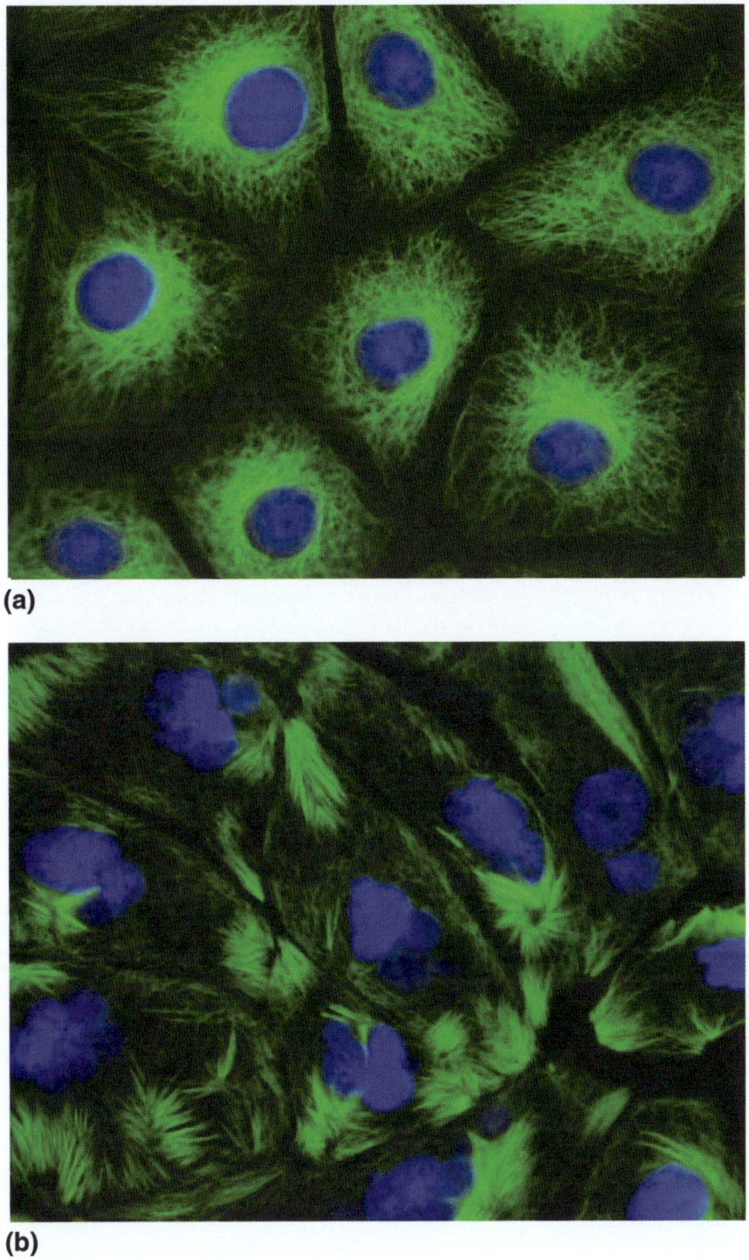

Fig. 2. Normal tubulin structure (a) extensive tubulin polymerization *e.g.* by taxol and epothilone (b)

2. General Aspects*

Gerhard Höfle

Helmholtz-Zentrum für Infektionsforschung (formerly: *GBF*, Gesellschaft für Biotechnologische Forschung), Braunschweig, Germany

Contents

2.1. History of Epothilone Discovery and Development

2.1.1. The Early Days

Epothilone is a microbial product, and thus its history may be traced back to the discovery of the respective microbe, *Sorangium cellulosum*, a bacterium belonging to the taxonomic group of myxobacteria, which originally has been described by *Roland Thaxter* in 1892 (*1*). Today this group of organisms comprises around 40 species, one of which is *Sorangium cellulosum*. For a long time, myxobacteria were only known

*Non-published papers and documents used for this article are in the possession of the Helmholtz-Center for Infection Research (formerly GBF, Gesellschaft für Biotechnologische Forschung).

for their gliding motility and sophisticated life cycle, although it had been occasionally speculated that they might produce secondary metabolites like actinomycetes or bacilli (*2*). In 1975 *Hans Reichenbach* and his group at the German Centre for Biotechnology (GBF; now called the *Helmholtz* Centre for Infection Research) set out to isolate strains of myxobacteria from soil samples collected all over the world, and to examine their secondary metabolism. In 1978, while work was already ongoing, I joined them and took over the chemistry part. In the same year the first structure of a myxobacterial metabolite, ambruticin, was published by a group from Warner-Lambert (*3*) making us very confident of being on the right track. Ambruticin had been isolated from a *Sorangium cellulosum* strain, and was identified as a unique cyclopropane polyketide structure exhibiting potentially useful antifungal properties. Ambruticin and its derivatives had been developed for medical application for some time, and recently gained new interest (*4*). Meanwhile we had been working at GBF quite successfully with the easily handled *Myxococcus*, *Corallococcus*, and *Stigmatella* strains, and only slowly shifted our focus to *Sorangium*. It took us considerable time to establish large-scale isolation and cultivation procedures of this slowly growing species. As soon as several hundred strains of *Sorangium cellulosum* had been accumulated by 1985, the screening for biological activity became productive, and a constant flow of unusual secondary metabolites came into our hands. Up to now, approximately 50 novel basic structures have been isolated from the various strains of this species, often with outstanding antifungal properties. The predominance of antifungal activity may be attributed to the fact that *Sorangium cellulosum* grows on cellulose as carbon source, and thus has to compete through chemical warfare (and other means) with fungi for its ecological niche.

In July 1985 *Sorangium cellulosum*, strain So ce90, the first producer of epothilone, was isolated by *Hans Reichenbach* from a soil sample collected at the banks of the river Zambesi in southern Africa in August 1980. Only two years after isolation, the strain was introduced in an antifungal screening of *Sorangium* strains by *Klaus Gerth* and identified as one of several hits in January 1987. Later, *Florenz Sasse*, responsible for cell culture tests, noticed high cytotoxicity of the culture extract. From these and other preliminary tests we were dealing with a new compound and *Norbert Bedorf* from the chemistry group immediately started to isolate the compound and elucidate its structure. Guided by biological activity, he isolated two closely related antifungal compounds later named epothilone A and B (*5*), and a structurally non-related family of polyene carboxylic acids, later named spirangiens in May 1987 (*5, 6*). This was not an easy task, as epothilones were only formed in the order of 1–2 mg/L;

however, spirangiens form in excess of 20 to 50-fold. After a preliminary structure had been deduced from spectroscopic data, an X-ray structural analysis of the nicely crystalline epothilone B was performed, which not only confirmed the structure but also revealed the configuration of the seven stereocenters in June 1987. Their absolute assignment was later corroborated for C-15 by degradation to (*S*)-malic acid (*7*, *8*).

Biological properties of pure compounds were re-examined, with the result that excess amounts of spirangiens were basically responsible for the toxicity of the culture extract. Epothilone A and B, in addition to antifungal activity, exhibited also significant toxicity in cell culture assays which seemed prohibitive for an application as a fungicide in medicine or agriculture. On the other hand, the cytotoxic activity did not seem to be of sufficient selectivity for an anticancer application. Nevertheless, *Johannes Gruner* from Ciba-Geigy, our industrial partner at that time, suggested checking this after receiving the first test samples in June 1989. Nothing happened, however, and the compound became dusty on the shelf and was forgotten. Moreover, at that time, we and Ciba-Geigy were extremely busy with the development of soraphen, another non-toxic macrolide from *Sorangium cellulosum*, to an agricultural fungicide (*9*).

2.1.2. Industry Becomes Interested in Epothilones – and Loses Interest Again

It was not before late 1990 that Ciba Geigy's plant protection department noticed a selective activity of epothilones against oomycetes, such as *Phytophthora*, still a threatening crop pest (*10*). At that time, the name epothilone was coined from the structural features epoxide, thiazole, and ketone. In spite of the apparent toxicity, greenhouse and small field trials were planned and carried out after the requested amount of 12 g of epothilones had been produced. This was, however, only possible after a clone had been selected from the genetically inhomogeneous wild strain by *Klaus Gerth*, increasing production more than 10-fold. Thus, with the help of the GBF Pilot Plant, the desired amount was produced with a single 1000 L fermentation batch by the end of 1991. Concurrently, in November 1991 a German patent, and a year later an international PCT application were filed, claiming matter and applications on the basis of the antifungal and cytotoxic properties (*11*). To our great disappointment, field trials revealed a phytotoxic side effect at the end of the growing season in 1992, and further development of epothilone A and B for plant protection was terminated.

At GBF we continued work with a limited semi-synthesis program to elucidate the structure/activity relationships in the hope of separating desired and undesired properties. *Michael Kiffe* synthesized around 30 derivatives of the functional groups around the macrocyclic ring of epothilone A and B in early 1993. Disappointingly, none of these derivatives showed appreciable biological activity. Nevertheless, investigations of the pharmacological properties of epothilones continued at Ciba-Geigy. Among others, the *in vitro* inhibition of lymphocyte proliferation at very low concentrations was investigated with regard to immunosuppression in animal models. However, compared with cyclosporine, epothilones only showed low selectivity and were therefore no longer investigated. At this time, at the latest, a mouse xenograft study would have been appropriate to discover the *in vivo* anticancer activity of the epothilones; it can only be speculated why this has not been done. In spite of the emerging clinical success of taxol, cytotoxic compounds had no future in anticancer therapy at that time. Various target-oriented approaches, by definition soft and without unwanted side effects, were *en vogue* and first promising successes were published. While we were waiting for possible further results from our industrial partner, the deadline for nationalizing the PCT patent application was approaching, and eventually forced us to abandon the international patent applications in May 1994. Only the German application was kept, for which a patent was granted October 17, 1995 (*11*).

Our very last action to rescue the compound was sending samples to the NCI (National Cancer Institute, USA) to be tested in the 60-cell line panel. First, the natural 4:1 mixture of epothilones A and B was submitted to the NCI *via* the European Liaison Office in January 1994. Five months later the NCI reported interesting activity, suggested repeated testing of the individual compounds in September 1994, and to refer the results to the NCI Biology Evaluation Committee in November. In December 1994, GBF also agreed to give the experts of the European Organisation for Research and Treatment of Cancer (EORTC) access to structural and biological information on the epothilones. However, as there was no follow-up action, interest in the epothilones faded. Moreover, because there was no visible advantage of epothilones over several other cytotoxic myxobacterial compounds supplied to the NCI (*12*), the epothilones have been ranked with low priority in our laboratory. Unfortunately, at that time the COMPARE program (*13*) for the analysis of inhibition profiles was not yet available from the NCI, and thus the close mechanistic relationship of epothilone and taxol was not uncovered.

2.1.3. Re-Discovery of Epothilone in the Nineties

It was serendipity that epothilone was rescued once more. In 1993 Merck Sharp and Dohme (MSD) decided, against the trend, to set up a

References, pp. 14–16

screening of natural products for taxane-like activity (*14*). At that time, taxol was the only compound known to inhibit cell division by stabilization of microtubules, and it seemed unlikely that taxanes would remain the only chemical class of compounds doing so. Moreover, taxol had just been registered for the treatment of ovarian cancer after a long-lasting development due to supply problems and unwanted side effects. Thus, more accessible and better tolerated alternatives were being sought for in several pharmaceutical companies in those days. Whereas 67,000 synthetic compounds and 70,000 extracts from natural sources were screened for taxol mimics with no direct hit by the Upjohn company, MSD had one confirmed hit for an extract from a *Sorangium cellulosum* strain coded SMP44. In contrast to Upjohn, MSD had a collection of only 7000 extracts from plants, and marine organisms, insects, and microorganisms including approximately 300 strains of myxobacteria from the strain collection of *John E. Peterson* from Emporia State University in Kansas (*15*, *16*). To their great surprise, the compounds responsible turned out to be epothilone A and B, known already from our published patent application. Epothilone B appeared to be even more active than taxol in the tubulin polymerization assay; it replaced bound taxol from microtubules, and, most remarkably, its activity against cancer cells was hardly impaired by the resistance to taxol and other cytostatics. These exciting results were published by *Bollag et al.* in July 1995 (*16*), and immediately triggered a variety of activities in pharmaceutical companies and academia.

At GBF the production of epothilones by fermentation was started again, and the derivatization program continued with more deep-seated transformations. In November 1995 a patent was filed for semisynthetic epothilones (*17*). At the same time the absolute configuration was passed on to interested colleagues who immediately started with the total synthesis of these apparently less complicated molecules. Parallel to that, we contacted pharmaceutical companies and distributed test samples in order to find a partner for joint development. While talks went on for more than one year, the first three total syntheses of epothilone A were completed and published at the end of 1996 and early 1997 by the groups of *S. J. Danishefsky* (*18*), *K. C. Nicolaou* (*19*), and *D. Schinzer* (*20*); those of epothilone B followed soon. These first papers created the basis for a broad variation of the natural structure independent of biological sources for the investigation of the structure/activity relationships and *in vivo* studies. *Danishefsky*'s and *Nicolaou*'s groups produced up to a thousand structural analogs, of which a large proportion was synthesized using a combinatorial approach. Meanwhile we had come to a license and joint development agreement with Bristol-Myers Squibb

(BMS) on May 12, 1997 and started to work on scale-up of fermentation and isolation of epothilones. *Klaus Gerth* and colleagues again improved the production strain, optimized nutrients and the fermentation process, while *Heinrich Steinmetz* in the chemistry department streamlined the extraction process and the large-scale purification by chromatography. From side fractions he and *Ingo Hardt* isolated epothilones C and D as major biosynthesis by-products and 36 minor epothilones (*21*). Also the basics of biosynthesis were investigated by feeding studies with labelled precursors and intermediates (*22*). By the end of 1997 more than 100 g of epothilone A and 50 g of epothilone B had been produced and supplied for the derivatization program and biological studies at BMS and GBF.

2.1.4. Development of Epothilones as Anticancer Drugs

Starting in fall 1995 the patent strain *Sorangium cellulosum* So ce90 was ordered from the German Strain Collection (DSMZ, Braunschweig) by pharmaceutical companies and academic groups, and used for production and genetic analysis by companies such as Novartis and Kosan Biosciences. This was in accordance with international patent laws, after our PCT application had been abandoned (now products derived from this strain may even be sold freely outside Germany). Meanwhile at GBF the collection of *Sorangium cellulosum* had expanded to more than 1700 strains, of which 40 were identified to produce epothilones. Thus, there is a good chance to find one's own producer in a small collection of, *e.g.*, 100 strains. Interestingly, most of these strains in addition to epothilones produce spirangiens and/or other secondary metabolites, often in large quantities. For that reason, a mutation program was established at the GBF in early 1996, yielding the first spirangien-negative mutant in September 1996. By the end of 1998 more than 30,000 HPLC runs from mutants had been carried out monitoring the stepwise improvement of the epothilone titer to more than 100-fold. Culture media, process, and isolation were concomitantly adapted, providing a technically useful production process. At this point it should be noted that epothilones have neither been observed in another myxobacterium other than *Sorangium cellulosum*, nor in any other organism.

Parallel to strain improvement by classical mutation we and others had been working on the identification of the biosynthesis genes, which, according to feeding studies, were expected to consist of a polyketide synthase/non-ribosomal peptide synthetase hybrid. During this time, Novartis and Kosan Biosciences published the cloning of the complete biosynthesis gene cluster from the GBF strain So ce90 (*23*) and the Merck/Peterson strain SMP44 (*24*), in August 2000. This opened the

References, pp. 14–16

door to an extensive manipulation of the epothilone structure by genetic engineering pursued primarily by Kosan Biosciences. However, in spite of great efforts, it was not possible to improve the productivity of strains by genetic engineering and to shut down formation of the nondesired epothilone A. The latter is by a factor of ten less active *in vitro* and, due to its chromatographic behavior, complicates the isolation of epothilone B considerably.

Initial reports on *in vivo* activity of epothilones in mouse xenograft studies came from the Sloan Kettering Cancer Research Center in October 1998. As soon as epothilones A and B were available from total synthesis, the behavior of sensitive and resistant human tumors was investigated by *Danishefsky*'s group. Only epothilone B was found to be effective, and only at 5 mg/kg, close to the maximum tolerated dose (*25*). Surprisingly, the significantly less active epothilone D showed a much broader therapeutic window, allowing extended treatments and observation of complete cures in mice (*26*). Later epothilone D was licensed to Kosan Biosciences, where it was produced with a genetically engineered strain, and introduced in clinical trials in July 2000. Hoffmann La Roche took a sub-license in August 2002 and since then has supported further development. *Danishefsky*'s group modified the structure of epothilone D by introducing an extra 9,10-double bond (dehydelone) and in addition replacing the 12-methyl by a trifluormethyl group (fludelone) (*27*). Both compounds show impressive *in vivo* activity and are going to replace epothilone D in clinical studies.

Similarly, Schering AG from the beginning relied on the total synthesis of epothilones, despite our offer to collaborate on the basis of the natural material. Based on their own total synthesis using *Schinzer*'s 6,7-aldol coupling, the structure-activity relationships with a variety of modifications were analyzed, focusing on activity in resistant tumor models. Impressive results have been reported since 2000 for the analog ZK-EPO whose structure was successfully kept secret until December 2006 (*28*). The superior properties result from two modifications of epothilone B – the 8-methyl replaced by allyl, and the C-15 side-chain replaced by a benzothiazole. ZK-EPO (Sagopilone) entered clinical trials in March 2002; the first results of a Phase I study were reported recently (*29*).

With a rich supply of epothilones from GBF from 1997 to 1999, and thereafter from their own production, Bristol-Myers Squibb relied exclusively on semi-synthesis to improve the pharmacological properties and therapeutic window of epothilone B. All functional groups including the epoxide were modified in a joint BMS/GBF project. From our early results on the rapid cleavage of the lactone by esterases (*17*) and the short half life in mouse serum, one of the primary goals was to improve metabolic stability. This was achieved by a change to the corresponding lactam, for which a short and practical synthesis from the lactone was developed (*30*). Unexpectedly, also remote modifications of the C-21-methyl group, such as aminomethyl, stabilized the lactone with the welcome improvement of water solubility and pharmacological properties (*31*). The epothilone B lactam, now called ixabepilone, was introduced in clinical trials in March 1999, and the 21-aminoepothilone B in the following year. Later, trials with 21-aminoepothilone B were terminated in favor of ixabepilone, which on October 16, 2007 received FDA approval for the treatment of metastatic or advanced breast cancer (*32*). It is marketed in the US by BMS under the trade name Ixempra® (*33*).

Novartis, the fourth big player in the game, was able to start its own production of epothilone as early as July 1996 with the GBF patent strain So ce90 and the in-house experience with large scale cultivation of *Sorangium cellulosum* gained during the soraphen project mentioned above. They stayed with epothilone B, renamed patupilone, as development candidate, and apparently managed the problems of unfavorable side effects by special formulation and application regimes. Next to ixabepilone, patupilone is the second most advanced epothilone and is in Phase III clinical trials. Nevertheless, also a backup candidate carrying a 2-methylsulfanyl substituted thiazole was developed by semi-synthesis and introduced in Phase I clinical trials in December 2004 (*34a*), which, however, were terminated recently (*34b*).

Apart from work directed toward the development of clinically useful drugs, the biochemistry of the epothilone/tubulin interaction and the resistance mechanisms of cancer cells against various epothilones have been extensively studied since 1995. The observation that epothilones can displace tubulin-bound taxol (*16*) immediately triggered speculations about a common pharmacophore for these two unrelated groups of compounds. Later, after the structure of the taxol-tubulin complex had been published, attempts were made to model epothilone within the taxol binding site; however, these results were controversial. A first more realistic picture was obtained by NMR experiments with a soluble form of the epothilone A-tubulin complex in 2003 (*35*). Soon after, this was questioned, however, by the electron-crystallographic structure of the complex on zinc-induced sheets (*36*), and recently by NMR experiments in solution (*37*). Because these results were obtained from artificial systems, the biological relevant conformation and binding site information may differ appreciably from both of these proposed forms (*38*).

References, pp. 14–16

Finally, a non-toxicity related activity of epothilones discovered recently should be mentioned. Based on the primary effect of epothilones to stabilize microtubules, experiments with a mouse model carrying inherent destabilized brain microtubules were performed in 2003. The phenotype of this mutant showed the typical signs of schizophrenia in humans which could be alleviated with a sub-toxic dose of epothilone D (*39*).

2.1.5. Epilogue

Today, exactly 20 years after the discovery of epothilones, we look back at a protracted and often interrupted development, reminiscent of paclitaxel in the seventies and eighties. In fact there are many similar cases, such as camptothecin, podophyllotoxin, conotoxins, and strobilurin, giving the false impression that retarded development is an inherent fate of natural products. However, other major products such as the *Vinca* alkaloids, the avermectins, and the statins were rapidly developed and brought to the marketplace soon after their discovery. So what is making the difference? Clearly, the supply issue is often a problem with natural products as we learned from the history of paclitaxel and more recently from numerous highly potent compounds from marine organisms. Even if supply is not the problem, recovery from the natural source and separation of complex mixtures make natural products less attractive than small synthetic compounds. Due to their complex structures peppered with stereocenters, total synthesis of natural products even on a small scale is often a time-consuming and expensive endeavor. The same applies for semisynthesis in lead optimization programs. Another disadvantage of natural products is their often untimely discovery and lack of information on the mode of action. In addition, while the compound is sitting around, intellectual property rights deteriorate which was exactly the case with the epothilones. The best chances of a speedy development occur when natural products are discovered *on demand* and *in-house* as was the case with the examples mentioned above. Thus, natural products, handicapped by their discovery outside the industry, should be as early as possible characterized with respect to their mode of action and simultaneously introduced in secondary screenings and application tests of different industrial partners.

Acknowledgements

I thank all members of the former groups of Natural Product Biology, Natural Product Chemistry, the Bio-Plant, and Instrumental Analytics at the GBF for their contributions in the epothilone project. I further thank Profs. *S. S. Crupper* and *J. E. Peterson* (Emporia State University, Kansas) for valuable information, and IneXel Medical Strategy and Communications in Lawrenceville, NJ, for proofreading. Financial support from Bristol-Myers Squibb and the Fonds der Chemischen Industrie is gratefully acknowledged.

References

1. Thaxter R (1892) On the Myxobacteriaceae, a New Order of Schizomycetes. Bot Gaz **14**: 389

2. Reichenbach H (1993) Biology of the Myxobacteria: Ecology and Taxonomy. In: Dworkin M, Kaiser D (eds.) Myxobacteria II, p. 13. American Society for Microbiology, Washington, DC; Reichenbach H, Höfle G (1993) Production of Bioactive Secondary Metabolites. In: Dworkin M, Kaiser D (eds.) Myxobacteria II, p. 347. American Society for Microbiology, Washington, DC

3. Connor DT, Greenough RC, von Strandtmann M (1977) W-7783, a Unique Antifungal Antibiotic. J Org Chem **42:** 3664; Ringel SM (1978) In vitro and in vivo Studies on Ambruticin (W7783): New Class of Antifungal Antibiotics. Antimicrob Agents Chemother **13**: 762

4. Shadomy S, Utz CJ, White S (1978) In vivo Studies with Ambruticin in Murine Histoplasmosis. Antimicrob Agents Chemother **14**: 99; Chiang LY, Ejzykowicz DE, Tian Z-Q, Katz L, Filler SG (2006) Antimicrob Agents Chemother **50**: 3464

5. Höfle G (1991) Biologically Active Secondary Metabolites from Myxobacteria – Isolation and Structure Elucidation. In: Walsdorff H-J (ed.) Scientific Annual Report of the GBF, p. 65

6. Niggemann J, Bedorf N, Flörke U, Steinmetz H, Gerth K, Reichenbach H, Höfle G (2005) Eur J Org Chem **2005**: 5013

7. Höfle G. *et al.* (1996) Epothilone A and B – Novel 16-membered Macrolides with Cytotoxic Activity: Isolation, Crystal Structure, and Conformation in Solution. Angew Chem Int Ed **35**: 1567

8. Gerth K, Bedorf N, Höfle G, Irschik H, Reichenbach H (1996) Epothilones A and B: Antifungal and Cytotoxic Compounds from *Sorangium cellulosum* (Myxobacteria). J Antibiot **49**: 560

9. Pridzun L, Sasse F, Reichenbach H (1995) Inhibition of Fungal Acetyl-CoA Carboxylase: A Novel Target Discovered with the Myxobacterial Compound Soraphen. In: Dixon GK, Copping LG, Hollomon DW (eds.) Antifungal Agents Discovery and Mode of Action, p. 99. BIOS Scientific Publishers Ltd, Oxford

10. At that time is was not yet known that taxol has very similar properties: Young DH, Michelotti EL, Swindell CS, Krauss NE (1992) Antifungal Properties of Taxol and Various Analogues. Experientia **48**: 882

11. Höfle G, Bedorf N, Gerth K, Reichenbach H (1993) (GBF) German Patent 4138042, filed Nov 11, 1991, granted Oct 14, 1993; World Patent 9310121 filed Nov 19 1992

12. These were, among others, chondramide, tartolon, icumazol, ratjadon, rhizopodin

13. Boyd MR, Paull KD (1995) Some Practical Considerations and Applications of the NCI *in vitro* Drug Discovery Screen. Drug Dev Res **34**: 91

14. Bollag DM (1997) Epothilones: Novel Microtubule-stabilising Agents. Exp Opin Invest Drugs **6**: 867

15. Lavelle F (1995) What's New About Tubulin/Microtubules Binding Agents. Exp Opin Invest Drugs **4**: 771; Crupper SS, Peterson JE, personal information

16. Bollag DM, McQuenay PA, Zhu J, Lazarides E, Hensens O, Koupal L, Liesch J, Goetz M, Woods C (1995) Epothilones, A New Class of Microtubule-Stabilizing Agents with a Taxol-like Mechanism of Action. Cancer Res **55**: 2325
17. Höfle G, Kiffe M (GBF) German Patent Application DE 19542986 A1, filed Nov 11, 1995
18. Balog A, Meng D, Kamenecka T, Bertinato P, Su D-S, Sorensen EJ, Danishefsky SJ (1996) Total Synthesis of (−)-Epothilone A. Angew Chem Int Ed **35**: 2801
19. Yang Z, He Y, Vourloumis D, Vallberg H, Nicolaou KC (1997) Total Synthesis of Epothilone A: The Olefin Metathesis Approach. Angew Chem Int Ed **35**: 166
20. Schinzer D, Limberg A, Bauer A, Böhm OM, Cordes M (1997) Total Synthesis of (−)-Epothilone A. Angew Chem Int Ed **36**: 523
21. Hardt IH, Steinmetz H, Gerth K, Sasse F, Reichenbach H, Höfle G (2001) New Natural Epothilones from *Sorangium cellulosum*, Strains So ce90/B2 and So ce90/D13: Isolation, Structure Elucidation, and SAR Studies. J Nat Prod **64**: 847
22. Gerth K, Steinmetz H, Höfle G, Reichenbach H (2000) Studies on the Biosynthesis of Epothilones: The Biosynthetic Origin of the Carbon Skeleton. J Antibiot **53**: 1373
23. Molnár I, Schupp T, Ono M, Zirkle R, Milnamow M, Nowak-Thompson B, Engel N, Toupet C, Stratmann A, Cyr DD, Gorlach J, Mayo JM, Hu A, Goff S, Schmid J, Ligon JM (2000) The Biosynthetic Gene Cluster for the Microtubule-stabilizing Agents Epothilones A and B from *Sorangium cellulosum* So ce90. Chem Biol **7**: 97
24. Julien B, Shah S, Ziermann R, Goldman R, Katz L, Khosla C (2000) Isolation and Characterization of the Epothilone Biosynthetic Gene Cluster from *Sorangium cellulosum*. Gene **249**: 153
25. Su D-S, Balog A, Meng D, Bertinato P, Danishefsky SJ, Zheng Y-H, Chou T-C, He L, Horwitz SB (1997) Structure-Activity Relationships of the Epothilones and the First in vivo Comparison with Paclitaxel. Angew Chem Int Ed **36**: 2093
26. Chou T-C, Zhang X-G, Balog A, Su D-S, Meng D, Savin K, Bertino JR, Danishefsky SJ (1989) Desoxyepothilone B: An Efficacious Microtubule-targeted Antitumor Agent with a Promising *in vivo* Profile Relative to Epothilone B. Proc Natl Acad Sci **95**: 9642
27. Chou T-C, Dong H, Rivkin A, Yoshimura F, Gabarda AE, Cho YS, Tong WP, Danishefsky SJ (2003) Design and Total Synthesis of a Superior Family of Epothilone Analogs, which Eliminate Xenograft Tumors to a Nonrelapsable State. Angew Chem Int Ed **42**: 4762
28. Klar U, Buchmann B, Schwede W, Skuballa W, Hoffmann J, Lichtner RB ((2006) Total Synthesis and Antitumor Activity of ZK-EPO (SAGOPILONE): The First Fully Synthetic Epothilone in Clinical Development. Angew Chem Int Ed Engl **45**: 7942
29. Schmid P, Kiewe P, Kuehnhardt D, Korfel A, Lindemann S, Giurescu M, Reif S, Thiel E, Possinger K (2005) A Phase I Study of the Novel Third-Generation Epothilone ZK-EPO (SAGOPILONE) in Patients with Advanced Solid Tumors. J Clin Oncol **23**(16S): Abstract 2051
30. Borzilleri RM, Zheng X, Schmidt RJ, Johnson JA, Kim S-H, DiMarko JD, Fairchild CR, Gougoutas JZ, Lee FYF, Long BH, Vite GD (2000) A Novel Application of a Pd(0)-Catalyzed Nucleophilic Substitution Reaction to the Regio- and Stereoselective Synthesis of Lactam Analogs of the Epothilone Natural Products. J Am Chem Soc **122**: 8890
31. Glaser N (2001) Semisynthese Seitenketten-modifizierter Epothilone, Doctoral Thesis, Technical University of Braunschweig
32. http://www.cancer.gov/cancertopics/druginfo/fda-ixabepilone
33. http://www.ixempra.com

34. (a) Wartmann M, Loretan J, Reuter R, Hattenberger M, Muller M, Vaxelaire J, Maira S-M, Flörsheimer A, O'Reilly T, Nicolaou KC, Altmann K-H (2004) Preclinical Pharmacological Profile of ABJ879, a Novel Epothilone B Analog with Potent and Protracted Anti-tumor Activity. AACR Ann Meet, Orlando, Abstr No 5440; (b) http://sec.edgar-online.com/2004/01/30/0001047469-04-002525/Section5.asp
35. (a) Carlomagno T, Blommers MJJ, Meiler J, Jahnke W, Schupp T, Petersen F, Schinzer D, Altmann K-H, Griesinger C (2003) The High-resolution Solution Structure of Epothilone A Bound to Tubulin: An Understanding of the Stucture-activity Relationships for a Powerful Class of Antitumor Agents. Angew Chem Int Ed **42**: 2511
36. Nettles JH, Li H, Cornett B, Kran JM, Snyder JP, Downing KH (2004) The Binding Mode of Epothilone A on $\alpha\beta$-Tubulin by Electron Crystallography. Science **305**: 866
37. Reese M, Sánches-Pedregal VM, Kunicek K, Meiler J, Blommers MJJ, Griesinger C, Carlomagno T (2007) Structural Basis of the Activity of the Microtubule-Stabilizing Agent Epothilone A Studied by NMR Spectroscopy in Solution. Angew Chem Int Ed **46**: 1864
38. Heinz DH, Schubert W-D, Höfle G (2005) Much Anticipated – The Bioactive Conformation of Epothilone and its Binding to Tubulin. Angew Chem Int Ed **44**: 1298
39. Andieux A, Salin P, Schweitzer A, Begou M, Pachoud B, Brun P, Gory-Fauré S, Kujala P, Suaud-Chagny M-F, Höfle G, Job D (2006) Microtubule Stabilizer Ameliorates Synaptic Function and Behaviour in a Mouse Model for Schizophrenia. Biol Psychiatry **60**: 1224

2.2. Natural Epothilones

2.2.1. Isolation and Large Scale Production

The first producer of epothilones, *Sorangium cellulosum*, strain So ce90 has been isolated by *Hans Reichenbach* from a soil sample collected at the banks of the Zambesi River in southern Africa, and was deposited at the German Culture Collection (Deutsche Sammlung von Microorganismen und Zellkulturen, DSMZ) under number DSM 6773 (*1*, *2*). A second strain, SMP44, isolated from a cotton field in Southern Missouri, came from the strain collection of *John E. Peterson* from Emporia State University, Kansas (*3*, *4*). Contrary to So ce90 it has never been available to the public; however, was passed on to certain US pharmaceutical companies (*4*, *5*). Both strains exhibit the characteristic nutrient requirements of *Sorangium cellulosum*, and grow even on cellulose and mineral salts, the preferred conditions for the isolation of this species (*6*). For production of epothilones, So ce90 was cultivated in a complex carbohydrate/protein medium in the presence of 2% of the neutral adsorber resin XAD-16 (*1*, *2*). During cultivation, excreted epothilones and other reasonably lipophilic metabolites are adsorbed on the resin (inner surface 800 m^2/g) and thus kept from re-absorption and metabolism by the organism. For harvest the resin is separated from the culture broth by filtration with a sieve and eluted with methanol to give a crude extract. Under these conditions combined yields of 1–2 mg/L of epothilone A (**1**) and B (**2**) were obtained with the wild strain at the variable A/B ratio of 3:1 to 2:1. In addition, the spirangiens, a structurally nonrelated group of polyketides, are formed by the strain So ce90 in excess of 20- to 50-fold, which seriously complicated separation by chromatography (*7*). In contrast, at the Merck Research Laboratories, a solid agar medium in large Petri dishes was used for production of epothilones with strain SMP 44 (*4*). After 10 days, cell mass and residual agar from 80 plates (1.4 m^2) were extracted with methyl ethyl ketone from which 2.7 mg of epothilone A and 0.9 mg of epothilone B could be recovered by chromatography.

Remarkably, the ratio of epothilone A and B was approximately the same as had been observed with strain So ce90; however, there was no indication on the formation of other metabolites, such as the spirangiens.

Optimization of strain So ce90 was first achieved by cloning. Wild strains and originally homogeneous laboratory strains, after longer periods of cultivation, often consist of a genetic mosaic of high and low producer cells. By picking colonies from So ce90 plates, cultures were established producing 22 mg/L and 11 mg/L of epothilone A and B on the 350 L scale (*1*, *8*). Further substantial improvements were achieved by mutation with either UV light or nitrosoguanidine (NTG) followed by high-performance liquid chromatographic (HPLC) analysis. Several of these mutants not only showed strongly increased production of epothilones, but were also spirangien-negative, which greatly simplified isolation of the epothilones (*9*). Other mutants produced only epothilone A; however, none of the more than 24,000 mutants that were analyzed produced epothilone B only, nor had the ratio A:B shifted significantly in favor of B (*10*, *11*). With the high-producing mutant So ce90/B2 on the 700 L scale, 30 g of epothilone A (**1**), 10 g of B (**2**), several grams of the epothilone congeners C (**3**) and D (**4**) (deoxyepothilones A and B) and small-to-trace amounts of 31 other structural variants **5–36** were isolated (see Fig. 1). In addition, open chain epothilone fragments **37–42** were recovered from So ce90/B2 and So ce90/D13 mutants (*12*). From the P450 defective mutant So ce90/D48, up 100 mg/L of epothilones C and D were obtained (*10*, *13*).

Isolation of pure epothilones from crude extracts usually requires consecutive chromatography on various sorbents with different separation characteristics like silica gel, Florisil, Sephadex LH20 and RP silica gels (*2*, *3*, *8*, *12*). Good separation of the homologous pairs like epothilone A/B or C/D is only achieved by reversed-phase chromatography. However, on a larger scale, epothilones A and B may be partially separated by fractional crystallization using either ethyl acetate/petroleum

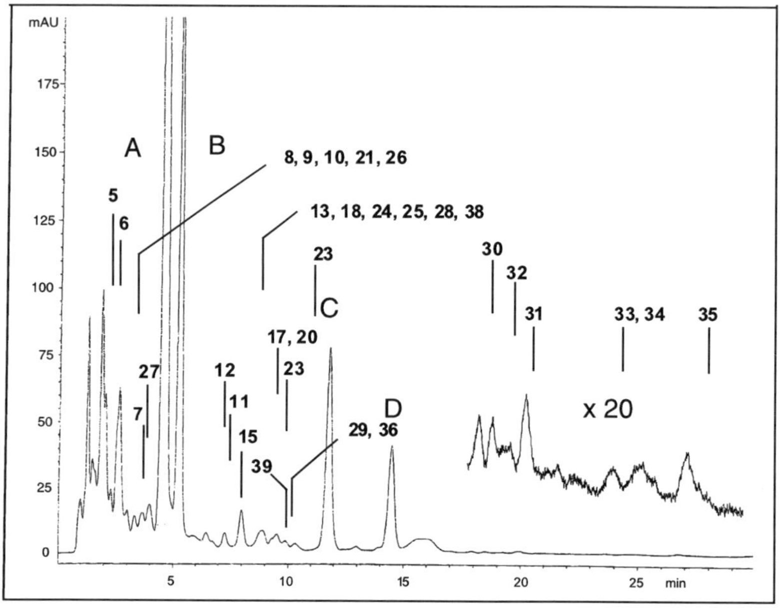

Fig. 1. HPLC analysis of a crude extract of *Sorangium cellulosum* strain So ce90/B2. In addition to the signals for epothilones A–D (**1**–**4**) retention times for structural variants **5**–**36** are marked (Nucleosil C_{18} (column 125 × 2 mm, 5 μm, 0.3 ml/min acetonitrile/ammonium acetate buffer *pH* 6.5, gradient 50:50 to 60:40 in 15 min, then isocratic, diode array detection) (*12*)

ether (*13*) or ethyl acetate/toluene (*14*), whereby the less soluble epothilone B is enriched in the first batch, and epothilone A is enriched in the mother liquor.

Epothilones B and D or their derivatives emerged as clinical candidates for the development of an anticancer drug during early preclinical studies (*15*, *16*). Thus, kilogram amounts of these compounds had to be made available, and in case one of the epothilones should reach the market, hundreds of kilograms over the long term would be required. Even though total synthesis has made great advances it is not competitive with a fermentation process for the natural epothilones. Nevertheless, improvements of strain, fermentation process, and product recovery were necessary to achieve a cost-effective production of such quantities.

In 1998 Bristol-Myers Squibb (BMS) started out with the GBF mutant So ce90/B2, which in their hands produced approximately 50 mg/L of epothilone B at an epothilone B/A ratio of 0.6:1 (*14*). By NTG mutagenesis, strains SC16408 and SC16449 (deposited as ATCC Nos. PTA-3880 and PTA-3881) were generated, which produced at least 100 mg/L of epothilone B. Further improvements were achieved by

References, pp. 27–28

modifications of the culture medium and feeding strategies, which, most importantly, includes feeding of sodium propionate during the production phase to stimulate epothilone B formation. In this way B/A ratios of 1.5:1 or better were achieved allowing direct crystallization of the crude extract obtained from the adsorber resin by elution with methyl-*t*-butyl ether. Scale-up to a 40,000 L production stage bioreactor in the presence of 550 kg of XAD-16 yielded 3.09 kg of epothilone B after final purification by RP-18 chromatography. From this material the semisynthetic lactam ixabepilone, the 21-amino derivative BMS-310705, and epothilone D by reduction of the epoxide, were prepared (*14*).

Similarly, for process optimization, Novartis started out with the GBF strain So ce90 obtained from the DSMZ culture collection. By a 3-step UV mutation program, strain So ce90 BCE33/10 (deposited at the DSMZ under No. DSM 11999) was generated, which is claimed to produce more epothilone B (*17*). For large-scale production with this strain, the XAD-16 adsorber resin was abandoned despite having great advantages for high yields and simple recovery procedures. Too many problems have been encountered with sterilization and breakdown of the resin beads, clogging tubes, valves, and stirrer bearings. Instead, a water-soluble adsorber was discovered from the cyclodextrins; its lipophilic molecular cavity accommodates a single epothilone molecule. During harvest, the culture is cleared by centrifugation and the epothilones are extracted from the cyclodextrin in the water phase by the stronger adsorbent XAD-16, which then is further processed following established lines. From a 450 L culture containing 10 g/L of 2-hydroxypropyl-β-cyclodextrin, more than 85% of the epothilones formed were recovered yielding 1.44 g of crystalline epothilone B after separation. This represents certainly a model study and not a later stage of optimization.

Kosan Biosciences, after cloning the epothilone biosynthesis gene cluster from *Sorangium cellulosum* SMP44, focused on the heterologous expression of natural and genetically modified epothilone variants in *Myxococcus xanthus* (see Chapter 3). The primary goal was the production of epothilone D, which, according to animal studies from *Danishefsky*'s group, had not only superior antitumor properties, but also a convenient therapeutic window (*16*). To this end a deletion mutant was constructed with an inactive epoK P450 epoxidase, resulting in a recombinant *Myxococcus xanthus* strain K111-40-1, incapable of converting epothilone C and D to the corresponding epoxides A and B (*18*). Even though the initial production was below 1 mg/L, as with the *Sorangium cellulosum* wild strain, the *Myxococcus xanthus* strain responded remarkably well to media and fermentation process optimization. Starting at a very favorable epothilone D to C ratio of 6:1 (*19*), addition of trace elements and intermittent feeding of casitone and methyl oleate as nitrogen and carbon sources increased production of epothilone D by 140-fold (*20*). A 1000 L fermentation batch in the presence of XAD-16 with modified work-up to remove residual methyl oleate yielded 63 g of crystalline epothilone D (*21*).

2.2.2. Structure of Epothilones and Related Compounds

Natural epothilones, in a strict sense, are 16-membered lactones with a thiazolylethylene side chain both spiked with methyl and oxygen groups in positions typical for polyketides. In principle, all these epothilones appear as pairs of homologues carrying on C-12 either a proton or a methyl group due to indiscriminate incorporation of an acetate or propionate unit in biosynthesis. However, both homologues could not be isolated in all cases because of low abundance and overlap with other strong peaks during chromatography (*12*). In general, the higher homologues are more active by a factor of 5–20, and thus of the most abundant epothilones only B (**2**) and D (**4**) are of practical interest.

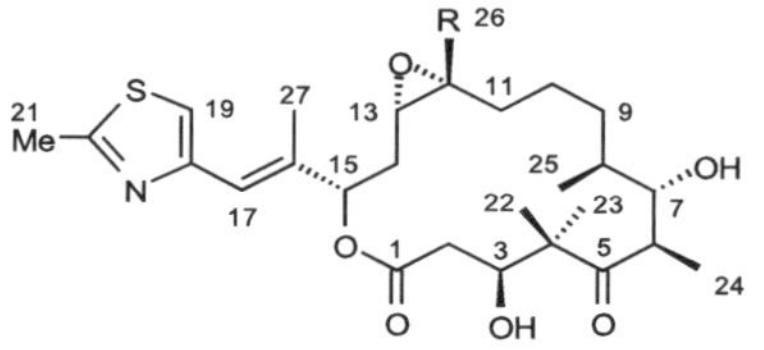

Epothilone A R = H (**1**)
B R = Me (**2**)

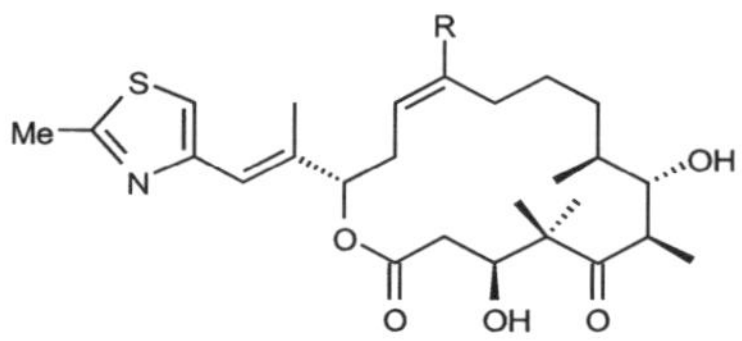

Epothilone C R = H (**3**)
D R = Me (**4**)

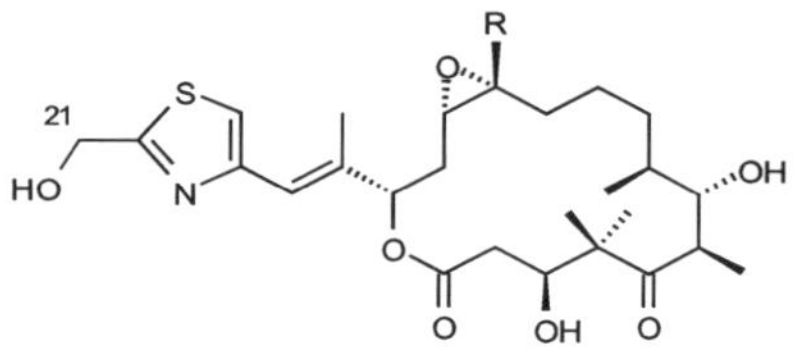

Epothilone E R = H (**5**)
F R = Me (**6**)

Characteristic structural features are two β-hydroxy carbonyl systems located in the south-eastern ring-segment, isolated from each other by a geminal dimethyl group. Probably for conformational reasons these groups are remarkably resistant to elimination of water and retroaldol reaction. Whereas four stereocenters are clustered in this lower part of the molecule, three others are located in the northwestern ring segment as β-acyloxyepoxide in epothilone A and B. In epothilones C and D the position of the epoxide is taken by a (*Z*) configured double bond, its biosynthetic precursor. The double bond linker in the side chain is in the

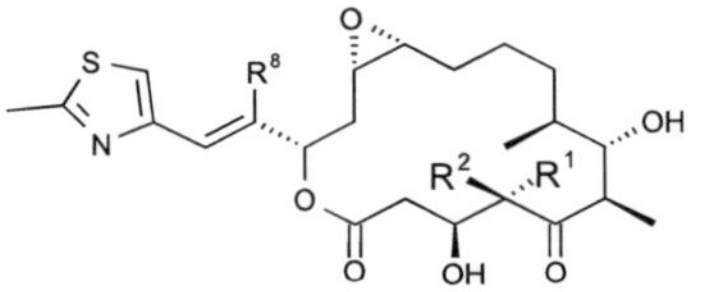

Epothilone A_1 (**7**) R^1 = H; R^2, R^8 = Me
Epothilone A_2 (**8**) R^2 = H; R^1, R^8 = Me
Epothilone A_8 (**9**) R^2 = H; R^1, R^2 = Me
Epothilone A_9 (**10**) R^1 = CH_2OH; R^2, R^8 = Me

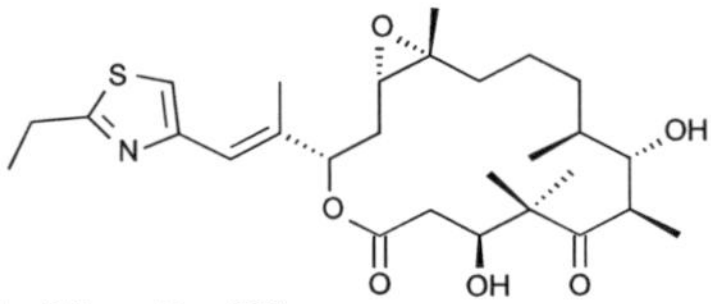

Epothilone B_{10} (**11**)

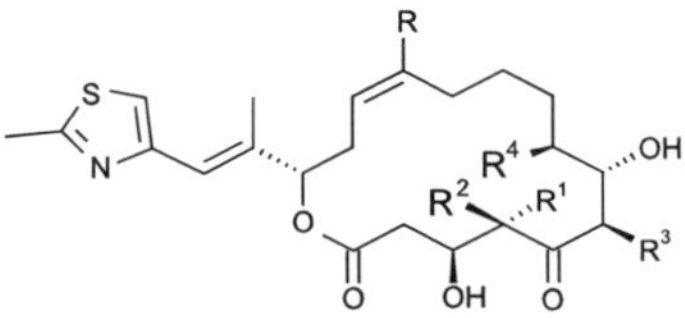

Epothilone C_1 (**12**) R^1 = H; R^2, R^3, R^4 = Me; R = H
Epothilone D_1 (**13**) R^1 = H; R^2, R^3, R^4 = Me; R = Me
Epothilone C_2 (**14**) R^2 = H; R^1, R^3, R^4 = Me; R = H
Epothilone D_2 (**15**) R^2 = H; R^1, R^3, R^4 = Me; R = Me
Epothilone C_3 (**16**) R^3 = H; R^1, R^2, R^4 = Me; R = H
Epothilone C_4 (**17**) R^4 = H; R^1, R^2, R^3 = Me; R = H

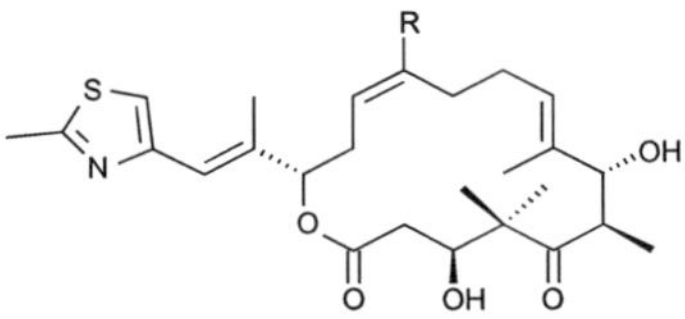

Epothilone C_5 (**18**) R = H
Epothilone D_5 (**19**) R = Me

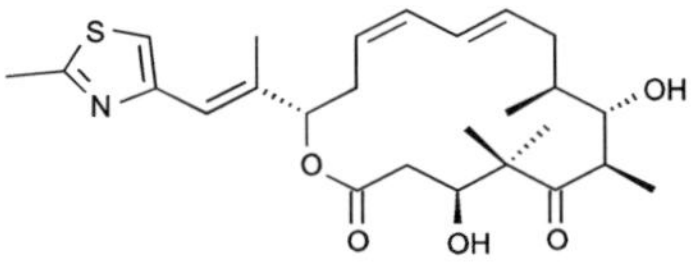

Epothilone C_6 (**20**)

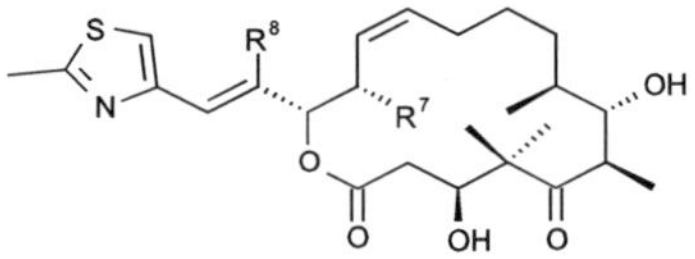

Epothilone C_7 (**21**) R^7 = OH; R^8 = Me
Epothilone C_8 (**22**) R^8, R^7 = H
Epothilone C_9 (**23**) R^8 = CH_2OH; R^7 = H

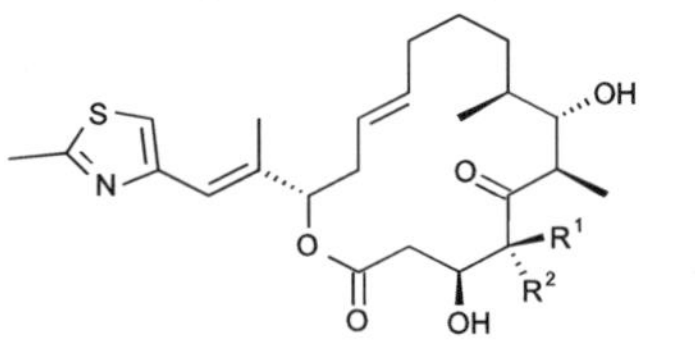

trans-Epothilone C_1 (**24**) R^1 = H; R^2 = Me
trans-Epothilone C_2 (**25**) R^2 = H; R^1 = Me

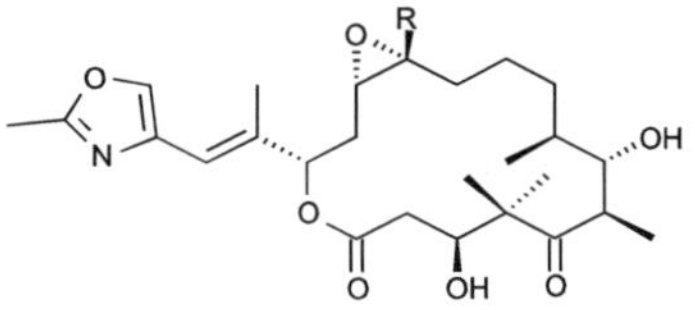

Epothilone G_1 (**26**) R = H
Epothilone G_2 (**27**) R = Me

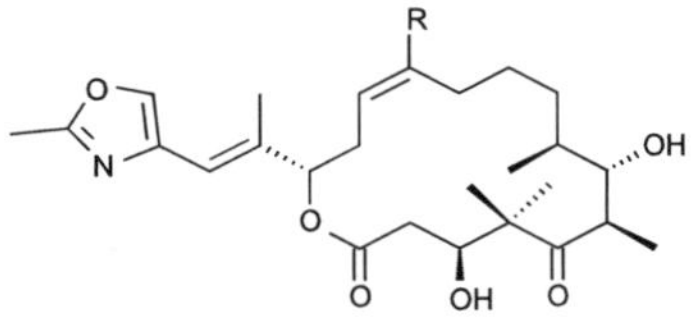

Epothilone H_1 (**28**) R = H
Epothilone H_2 (**29**) R = Me

(*E*) configuration in all epothilones. Hydroxylation, a very common secondary modification in natural product families, is also observed in the epothilone series. Thus C-21 is hydroxylated in epothilones E (**5**) and F (**6**), C-23 in A_9 (**10**), C-14 in C_7 (**21**), and C-27 in C_9 (**23**). All other structural variations may be considered as the result of infidelity of the biosynthetic machinery, which, however, is observed only at the very low level of 0.01 to 0.1%. Frequently, methyl groups are missing (**7–9**, **12–17**, **22**, **24**, **25**), or an additional one is present (**11**). An additional double bond is found in the 8- and 10-position (**18/19**, and **20**, respectively). Additional acetate or propionate units, incorporated in the northeastern ring segment, result in ring-enlarged 18-membered lactones (**30–35**), and, on the other hand, skipping of the C-3/C-4 building block leads to a ring-contracted 14-membered lactone (**36**).

Replacement of the cysteine in the construction of the thiazole ring for serine leads to the oxazole analogs **26–29**. Finally six open-chain fragments **37–42** were identified which presumably are derived from intermediates escaping at different stages from the biosynthetic enzyme complex. Most remarkably, not even traces of stereoisomers at tetragonal carbon centers have been observed.

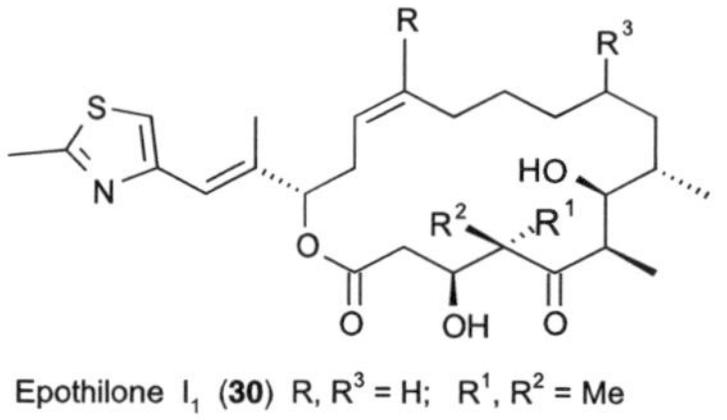

Epothilone I_1 (**30**) R, R^3 = H; R^1, R^2 = Me
Epothilone I_2 (**31**) R = H; R^1, R^2, R^3 = Me
Epothilone I_3 (**32**) R^1, R^2, R^3, R = Me
Epothilone I_4 (**33**) R^2, R = H; R^1, R^3 = Me
Epothilone I_5 (**34**) R^2 = H; R^1, R^3, R = Me
Epothilone I_6 (**35**) R^1 = H; R^2, R^3, R = Me

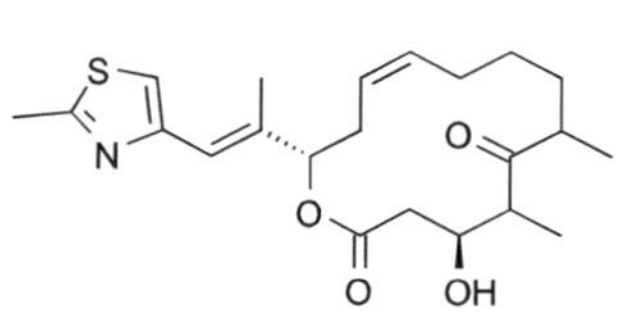

Epothilone K (**36**)

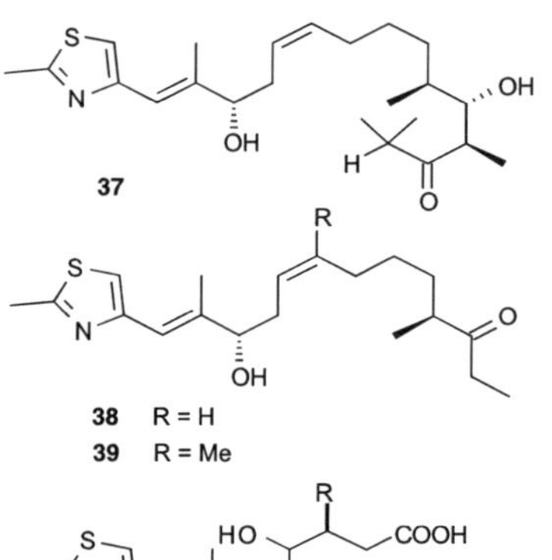

37

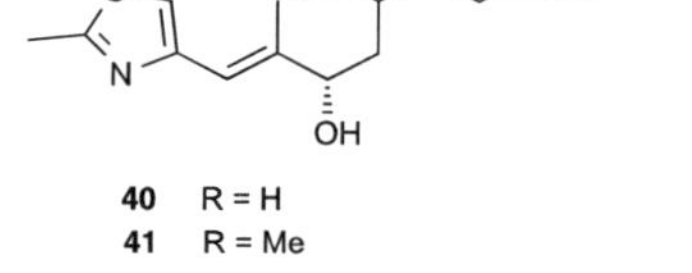

38 R = H
39 R = Me

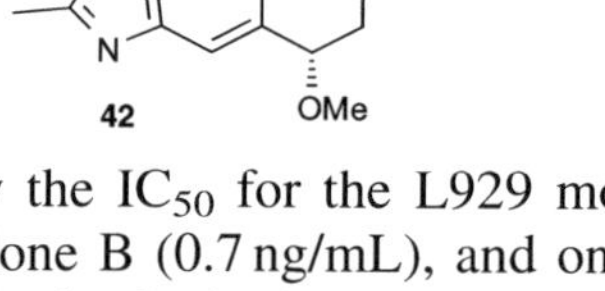

40 R = H
41 R = Me

42

The biological activity, as judged by the IC_{50} for the L929 mouse fibroblast cell line, is highest for epothilone B (0.7 ng/mL), and only a little less for the oxazole analog G_2 (**27**), the 21-hydroxy and 20-ethyl analogs F (**6**) and B_{10} (**11**). Replacement of one of the geminal methyl groups or the 27-methyl group by hydrogen (A_2 **8**, D_2 **15** and A_8 **9**) reduces activity 2–5-fold. All other analogs and partial structures **37–42** are of low activity or essentially inactive.

2.2.3. Physical and Chemical Properties

In pure state the epothilones are colorless oils; only epothilones A, B, and D have been crystallized. Epothilone A (**1**) is well soluble in polar organic solvents like methanol, ethyl acetate, acetone, diethyl ether and DMSO, whereas epothilone B is distinctly less soluble. Both are sparingly soluble in benzene, toluene and petroleum ether. Aqueous solubility at room temperature ranges from 722 mg/L for epothilone A (**1**) down to 24 mg/L for epothilone D (**4**) which is thus approximately as soluble as taxol (Table 1) (*1*, *8*, *11*, *13*). Generally, solubilities in D_2O are even lower making NMR experiments in this solvent very time-consuming. A variety of melting points have been reported ranging from 76 to 128°C depending on crystal form and nature of the solvate (*8*, *14*, *15*, *21*). Thus, epothilone A crystallized from ethyl acetate/toluene melts at 76–78°C, and, after more rigorous drying, at 95°C. Epothilone B crystallized from ethyl acetate melts at 93–94°C, from methanol/water at 85–87°C, and from methanol at 126–128°C (*13*, *15*). The latter was characterized by the X-ray diffraction pattern as crystal form A, another

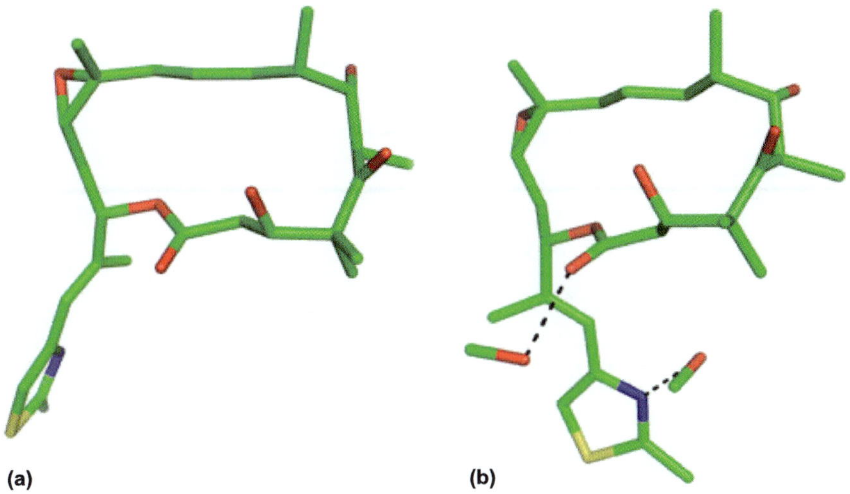

Fig. 2. (a) X-ray crystal structures of epothilone B (**2**) from dichloromethane/petroleum ether (*8*) and (b) methanol/water (*22*) (modelling by *W.-D. Schubert*)

Table 1. Selected physical and spectroscopic data of epothilones A–D (**1–4**) (*1, 8, 12, 13*)

	1	**2**	**3**	**4**
Elemental composition	$C_{26}H_{39}NO_6S$	$C_{27}H_{41}NO_6S$	$C_{26}H_{39}NO_5S$	$C_{27}H_{41}NO_5S$
EIMS *m/z* [M$^+$]	493	507	477	491
UV (MeOH) λ_{max}/nm (ε/mol^{-1}dm^{3}cm^{-1})	211 (17800)	211 (18600)	213 (16200)	210 (18400)
	249 (12500)	249 (14100)	248 (12500)	248 (13200)
$\alpha_D{}^{22}$(MeOH)/10^{-1} deg cm^2 g^{-1}	−47.1	−35.0	−74.4	−61.3
IR (KBr) ν_{max}/cm^{-1}	1738 1692	1738 1690	1734 1688	1734 1688
DC[a] R_f solvent[b]	0.63	0.65	0.72	0.75
R_f solvent[c]	0.36	0.40	0.70	0.74
HPLC[d] R_t/min	4.6	5.3	11.8	14.4
Solubility, H_2O, 21°C, mg/L	722	345	42	24[e]
D_2O, 21°C, mg/L	575	200		
Hydrolysis,[f] *pH* 1, RT	τ/2 60 min	τ/2 10 min		
pH 7, 37°C	τ/2 ~ 15 d	τ/2 ~ 15 d		
pH 11, RT	τ/2 6 h	τ/2 6 h		

[a] Silica gel Si60 aluminum sheets; [b] dichloromethane/methanol 95:5; [c] petroleum ether/ethyl acetate 6:4; [d] for conditions see Fig. 1; [e] Taxol 25 mg/L; [f] H_2O/MeOH 95:5 (*13*).

sample, obtained by treatment of form A with isopropanol, as crystal form B, for which unfortunately no melting point is given (*17*). Similarly, no melting points are given for solvates of epothilone B with ethyl acetate, toluene, isopropanol, and acetonitrile. However, in addition to X-ray diffraction patterns, X-ray crystal structures have been obtained showing very similar unit cells with a lipophilic solvent channel occupied by 2 solvent molecules (*14*). The original X-ray crystal structures of epothilone B from dichloromethane/petroleum ether and methanol/water were 2:1 solvates with dichloromethane and methanol (Fig. 2) (*8*, *22*).

Remarkably, in these structures the thiazole side-chain adopts different conformations, whereas the macrocycle is nearly identical. Epothilone D has been crystallized only from ethanol/water, melting point 120 to 121°C, but no crystal structure has been reported at this time (*21*). Mass spectrometry has not made a major contribution to structure elucidation, except of high-resolution electron ionization measurements of M^+ ions for the determination of molecular compositions (*8*, *12*). However, HPLC/electrospray ionization mass spectra were extremely important for the detection and quantitation of trace components in complex matrices like fermentation broths and biological fluids. From ammonium acetate buffered HPLC solvents, positive molecular ions and negative cluster ions with acetate are observed with a sensitivity two orders of magnitude higher than UV detection at 210 nm (*13*). Proton and carbon NMR spectra of epothilones in methanol and chloroform (Fig. 3) (*12*), or DMSO (*8*) are well resolved at 600/150 MHz. Signal assignments have been made by 2D techniques, and, in case of ambiguity, spectra from ^{13}C labelled samples (*23*) were consulted. Occasionally, signals in the vicinity of the thiazole ring are more or less broadened due to partial protonation of the nitrogen (*13*). Therefore, care should be taken to use acid-free NMR solvents – in particular chloroform – and eliminate acids from chromatography fractions and reaction products during work-up.

The preferred solution conformation of epothilone B in DMSO – analyzed by Nuclear *Overhauser* Effects and proton coupling constants – was found to be very similar to that in crystal structure from dichloromethane/petroleum ether (*8*). Only the thiazole side-chain rotates around both its single bonds, and the vicinal coupling constants in the ABX system at C-2/C-3 are significantly different from those expected, indicating some conformational freedom.

In a detailed conformational analysis, *Taylor* and *Zajicek* (*24*) defined a second minor conformer B in which the 3-OH is rotated from an axial to an equatorial position. Concomitantly, the torsion angle around C-6/C-7 is changed by 120°. Whereas conformer A dominates in DMSO/D_2O, conformer B is more prevalent in CD_2Cl_2.

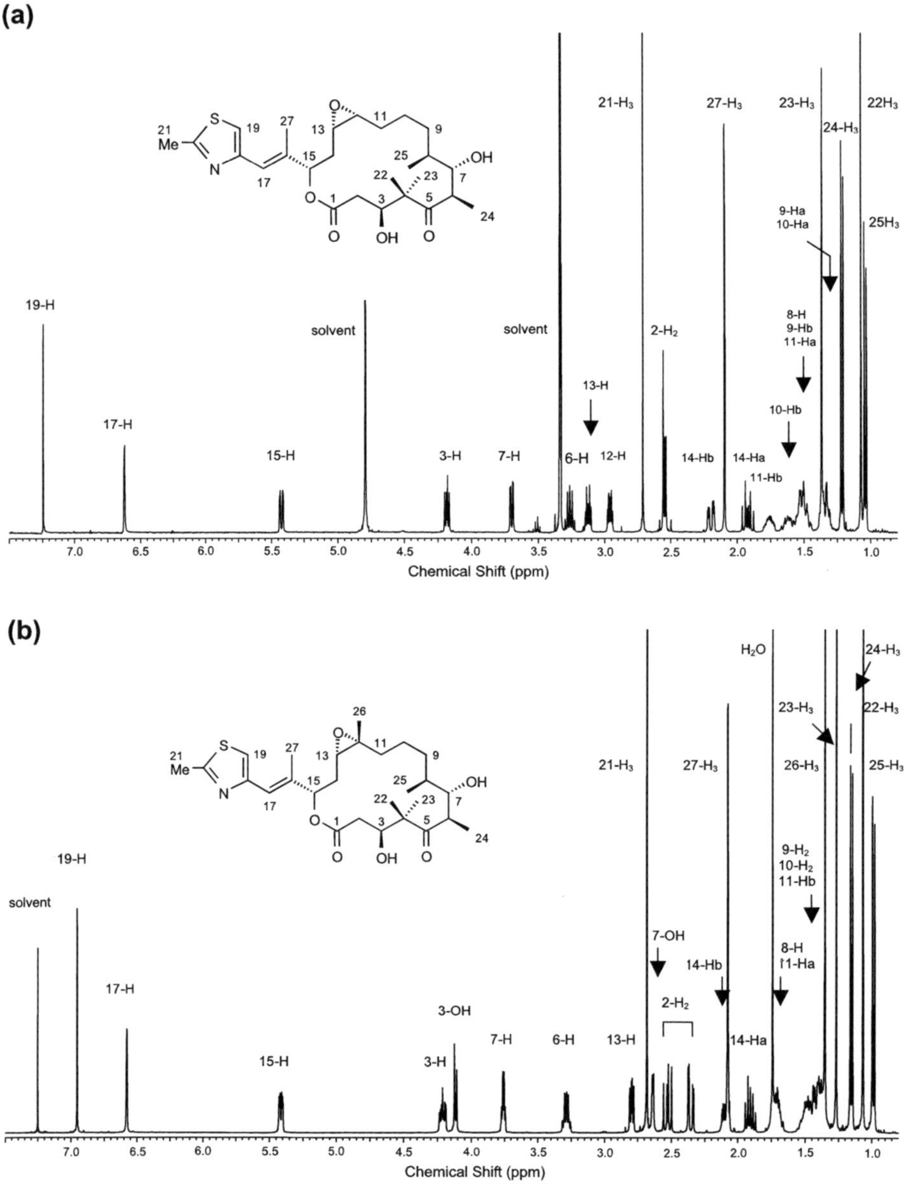

Fig. 3. ^{1}H NMR spectra of (a) epothilone A (**1**) in $CDCl_3$ and (b) epothilone B (**2**) in CD_3OD (for better presentation of coupling patterns 400 MHz spectra are depicted)

References, pp. 27–28

Lactone and epoxide are the most sensitive groups in the chemical reactivity of epothilones; however, in *pH* 7 buffered aqueous/methanolic solution, they have been found stable at room temperature for weeks (*13*). Whereas the lactone in epothilones A and B is rapidly hydrolyzed in aqueous systems above *pH* 11, the epoxide is acid sensitive and hydrolysed or rearranged below *pH* 3 (Table 1) (*25*). At neutral *pH*, pig liver esterase rapidly cleaves the lactone, whereas the epoxide is only attacked very slowly (*26*, *13*). In pure state, epothilones are perfectly stable at room temperature and in the presence of air; crystalline epothilone B after melting turns slightly yellow at 140°C. However, only traces of decomposition products are formed after one hour at this temperature.

Acknowledgements

I thank all members of the former groups of Natural Product Biology, Natural Product Chemistry, the Bio-Plant and Instrumental Analytics at the GBF for their contributions in the epothilone project. I further thank Dr. *N. Glaser*, Professor *H. Reichenbach*, *H. Steinmetz*, Dr. *W. D. Schubert* and Profs. *S. S. Crupper* and *J. E. Peterson* (Emporia State University, Kansas) for valuable information and help in the preparation of this article, and IneXel Medical Strategy and Communications in Lawrenceville, NJ, for proofreading. Financial support from Bristol-Myers Squibb and the Fonds der Chemischen Industrie is gratefully acknowledged.

References

1. Gerth K, Bedorf N, Höfle G, Irschik H, Reichenbach H (1996) Epothilones A and B: Antifungal and Cytotoxic Compounds from *Sorangium cellulosum* (Myxobacteria). J Antibiot **49**: 560
2. Höfle G, Bedorf N, Gerth K, Reichenbach H (GBF) German Patent 4138042, filed Nov 11, 1991, granted Oct 14, 1993; World Patent 9310121 filed Nov 19, 1992
3. Crupper SS, Emporia State University (Kansas), personal information
4. Bollag DM, McQuenay PA, Zhu J, Lazarides E, Hensens O, Koupal L, Liesch J, Goetz M, Woods C (1995) Epothilones, a New Class of Microtubule-stabilizing Agents with a Taxol-like Mechanism of Action. Cancer Res **55**: 2325
5. Tang L, Shah S, Chung L, Carney J, Katz L, Khosla, Julien B (2000) Cloning and Heterologous Expression of the Epothilone Gene Cluster. Science **287**: 640
6. Reichenbach H (1993) Biology of the Myxobacteria: Ecology and Taxonomy. In: Dworkin M, Kaiser D (eds.) Myxobacteria II, Chapter 2, p. 13. American Society for Microbiology, Washington, DC
7. Höfle G, Bedorf N, Gerth K, Reichenbach H (GBF), German Patent Application DE 4211056 C1 (1993) Niggemann J, Bedorf N, Flörke U, Steinmetz H, Gerth K, Reichenbach H, Höfle G (2005) Eur J Org Chem **2005**: 5013
8. Höfle G, Bedorf N, Steinmetz H, Schomburg D, Gerth K, Reichenbach H (1996) Epothilone A and B – Novel 16-membered Macrolides with Cytotoxic Activity: Isolation, Crystal Structure, and Conformation in Solution. Angew Chem Int Ed **35**: 1567

9. The first spirangien-negative mutant was obtained in Oct 1996 (Gerth K, personal communication)
10. Gerth K, Steinmetz H, Höfle G, Reichenbach H (2001) Studies on the Biosynthesis of Epothilones: The PKS and Epothilone C/D Monooxygenase. J Antibiot **54**: 144
11. Höfle G, Reichenbach H (2005) Epothilone, a Myxobacterial Metabolite with Promising Antitumor Activity. In: Cragg GM, Kingston DGI, Newman DJ (eds.) Anticancer Agents from Natural Products, p. 413. Taylor & Francis, London
12. Hardt IH, Steinmetz H, Gerth K, Sasse F, Reichenbach H, Höfle G (2001) New Natural Epothilones from *Sorangium cellulosum*, Strains So ce90/B2 and So ce90/D13: Isolation, Structure Elucidation, and SAR Studies. J Nat Prod **64**: 847
13. Steinmetz H, Höfle G, unpublished work
14. Benigni D *et al* (BMS) (2007) US Patent No. 7,172,884 B2
15. Su D-S, Balog A, Meng D, Bertinato P, Danishefsky SJ, Zheng Y-H, Chou T-C, He L, Horwitz SB (1997) Structure-activity Relationships of the Epothilones and the First in vivo Comparison with Paclitaxel. Angew Chem Int Ed **36**: 2093
16. Chou T-C, Zhang X-G, Balog A, Su D-S, Meng D, Savin K, Bertino JR, Danishefsky SJ (1989) Desoxyepothilone B: An Efficacious Microtubule-targeted Antitumor Agent with a Promising *in vivo* Profile Relative to Epothilone B. Proc Natl Acad Sci **95**: 9642
17. Hofmann H *et al* (Novartis) (2006) US Patent No. 7,101,702 B2
18. Julien B, Katz L, Khosla C, Tang L (Kosan Biosciences) (2001) US Patent No 6303342
19. Frykman S, Tsuruta H, Lau J, Regentin R, Ou S, Reeves C, Santi D, Licari P (2002) Modulation of Epothilone Analog Production Through Media Design. J Ind Microbiol Biotchnol **28**: 17
20. Lau J, Frykman S, Regentin R, Ou S, Tsuruta H, Licari P (2001) Optimizing the Heterologous Production of Epothilone D in *Myxococcus xanthus*. Biotechn Bioeng **78**: 280
21. Arslanian RL, Parker CD, Wang PK, McIntire JR, Lau J, Starks C, Licari PJ (2001) Large-Scale Isolation and Crystallization of Epothilone D from *Myxococcus xanthus* Cultures. J Nat Prod **65**: 570
22. Hecht H-J and Höfle G unpublished results; Crystallographic data of the structure have been deposited with the Cambridge Crystallographic Data Centre as supplementary publication no CCDC-241333 CCDC-241334. Copies of the data can be obtained free of charge on application to CCDC, 12 Union Road, Cambridge CB2 1EZ, UK
23. Gerth K, Steinmetz H, Höfle G, Reichenbach H (2000) Studies on the Biosynthesis of Epothilones: The Biosynthetic Origin of the Carbon Skeleton. J. Antibiot **53**: 1373
24. Taylor RE and Zajicek J (1999) Conformational Properties of Epothilone. J Org Chem **64**: 7224
25. Sefkow M, Kiffe M, and Höfle G (1998) Derivatization of the C12-C13 Functional Groups of Epothilones A, B and C. Bioorg Med Chem Lett **8**: 3031
26. Höfle G, Kiffe M (1997) Epothilon-Derivate und deren Verwendung, DE 19542986A1; Vite GD, Kim S-H, Höfle G (BMS) (2004) Preparation of Epothilone Intermediates US 2004/0023345A1

3. Biosynthesis and Heterologous Production of Epothilones

Rolf Müller

Institut für Pharmazeutische Biotechnologie,
Universität des Saarlandes, 66041 Saarbrücken, Germany

Contents

3.1. Introduction

Although a variety of chemical syntheses for the epothilones and various derivatives have been described, modifying the backbone of those natural products remains a major challenge. One alternative to chemical alteration is the elucidation and subsequent manipulation of the biosynthetic pathway *via* genetic engineering in the producing organism. This type of approach is known as "combinatorial biosynthesis" and holds great promise, especially in conjunction with semi-synthesis methods to alter the structure of the natural product. In parallel, production can be optimized in the natural producer if the regulatory mechanisms governing the biosynthesis are understood. Alternatively, the entire gene cluster can be transferred into a heterologous host, more amenable both to genetic alteration and overexpression.

In order to modify the structure of a metabolite in a rational way, the biosynthetic pathway within the producing organism – in this case, the

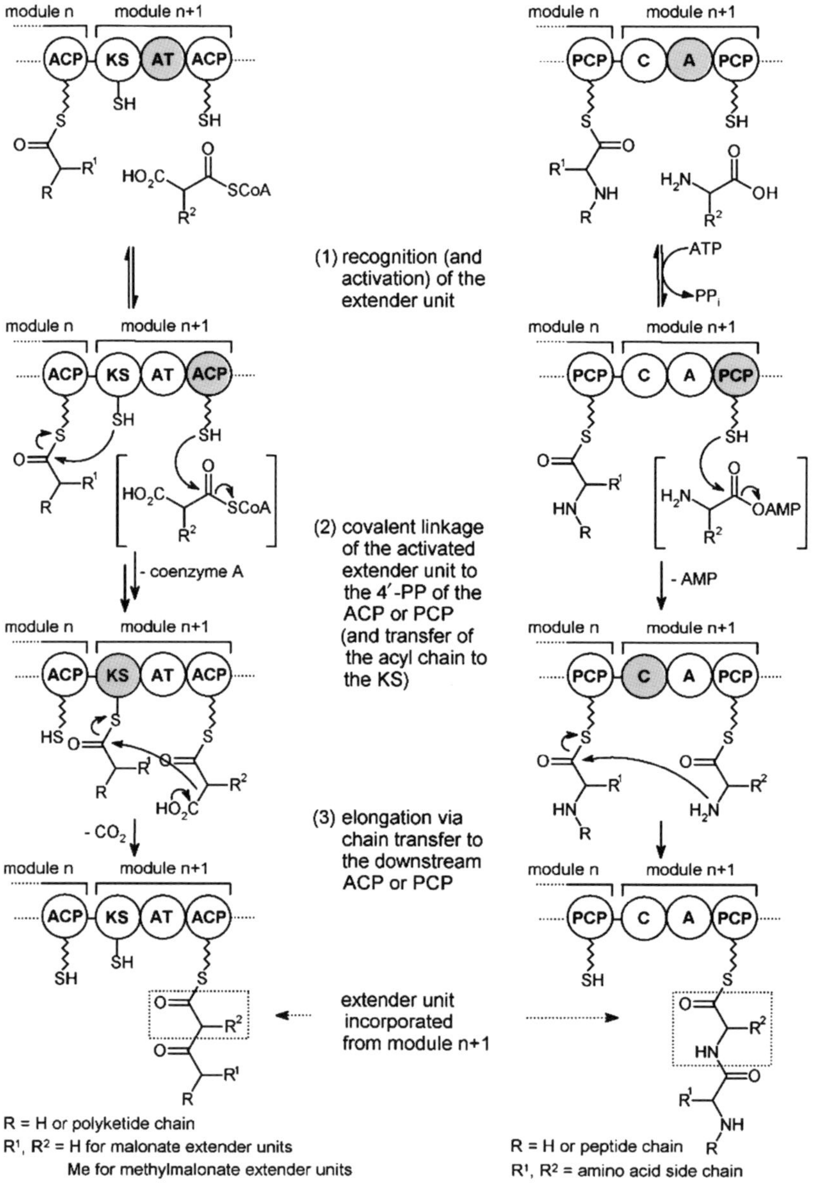

Fig. 1. Comparison of basic steps during assembly directed by polyketide synthases (left) and nonribosmal peptide synthetases (right). Active domains in each step are shown in grey. 4′-PP, 4′-phosphopantetheinyl; ACP, acyl carrier protein; KS, β-keto synthase domain; AT, acyl transferase domain; PCP, peptidyl carrier protein; C, condensation domain; A, adenylation domain

myxobacterium *Sorangium cellulosum* – must be known in detail. One essential step is to identify the building blocks used as precursors to assemble the compound. Together with the compound's structure, this information can be used to postulate a reasonable biosynthetic pathway. From such experiments it was originally speculated that the epothilones are assembled jointly by polyketide synthase (PKS) and non-ribosomal peptide synthetase (NRPS) multienzymes. These complex protein assembly lines catalyze the condensation of activated short chain carboxylic acids and amino acids *via* a thiotemplated mechanism (illustrated in Figs. 1 and 2). Natural product assembly on such multimodular megasynthetases has been

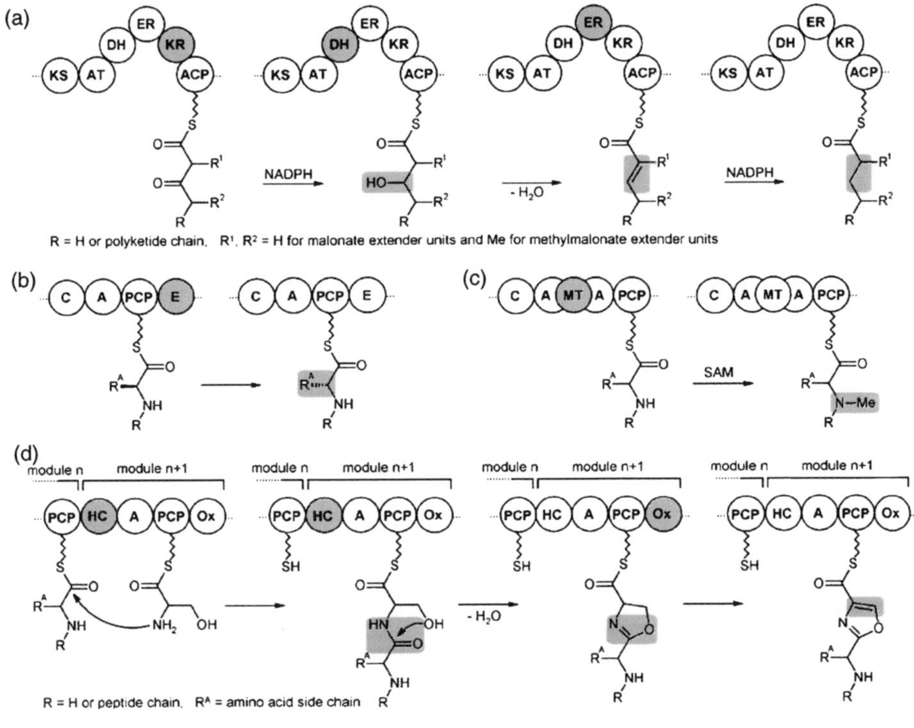

Fig. 2. Biochemistry performed by optional domains in PKS and NRPS assembly lines a) reduction of the β-keto intermediate by domains of the reductive loop. Active domains in each step are shown in grey. KR, β-keto reductase domain; DH, dehydratase domain; ER; enoylreductase domain. b) epimerization of L-amino acids to D-amino acids by epimerization (E) domains c) *N*-methylation on NRPS performed by methyltransferase (MT) domains. SAM, *S*-adenosyl methionine. d) Formation of heterocycles (oxazole, thiazole) from serine, threonine or cysteine exemplified by oxazole formation starting from serine in a hybrid PKS/NRPS. HC, heterocyclization domain; Ox, oxidation domain. For the description of minimal domains, see Fig. 1

described in a number of excellent reviews (*1*, *2*, *3*). Basically, the activated monomeric building blocks are selected as CoA esters in PKS by acyltransferase (AT) domains or selected plus activated as adenylates in NRPS by adenylation (A) domains. The activated acids are then covalently tethered to carrier proteins (acyl carrier proteins – ACPs – in PKS and peptidyl carrier proteins – PCPs – in NRPS) and then condensed with each other *via* the action of ketosynthase (KS) domains in a *Claisen* type reaction (PKS) or by condensation (C) domains in NRPS forming peptide bonds. A minimal module responsible for one chain elongation is thus comprised of KS-AT-ACP in PKS and of C-A-PCP domains in NRPS. Additional domains are frequently used in both types of enzymatic systems. Examples are ketoreductase (KR) domains, dehydratase (DH) domains, enoylreductase (ER) domains, and *O*-methyltransferase (O-MT) domains in PKS. In NRPS, heterocyclization (HC) domains forming thiazoline and oxazoline rings from cysteine and serine, respectively, can be employed. Additional chemical diversity is generated by using oxidation (Ox) domains generating the respective thiazole and oxazole structures. Additionally, *N*-methyltransferase (N-MT) domains or epimerization (E) domains forming D-amino acids from the natural L-forms are found frequently. Typically, the fully extended intermediate is released from the final carrier protein by the action of a thioesterase (TE) domain forming free acids, lactones, or lactams.

From the sequencing of many biosynthetic gene clusters, it has become increasingly clear that the co-linearity between gene sequence and product structure, which once seemed characteristic of these systems, is not the rule. Such "deviations from textbook logic" (*4*, *5*) are particularly prevalent among myxobacterial biosynthetic systems, including that responsible for epothilone assembly (*6*).

It must be mentioned at this stage, that although hybrid systems incorporating both PKS and NRPS elements are now commonplace, at the time that the epothilone biosynthetic genes were discovered, only a few examples of such mixed assembly lines were known (*e.g.* for rapamycin (*7*), mycosubtilin (*8*), and myxothiazol (*9*)). It only became clear through later sequencing efforts that many microorganisms assemble their secondary metabolites using hybrid biosynthetic machineries (*10*). Myxobacteria, in particular, are rich in such systems (*6*).

3.2. Feeding Studies and the Discovery of Natural Epothilone Variants

Based on the structure of the natural epothilones, the compounds were expected to arise from polymerization of short chain carboxylic acid monomers, as well as the amino acid cysteine. These expectations were confirmed through feeding studies carried out by *Gerth et al.* (*11*) in the natural host *Sorangium cellulosum* So ce90, using labelled precursors. These experiments revealed that the carbon atoms in the epothilone backbone are derived from acetate (from malonyl-CoA; mCoA),

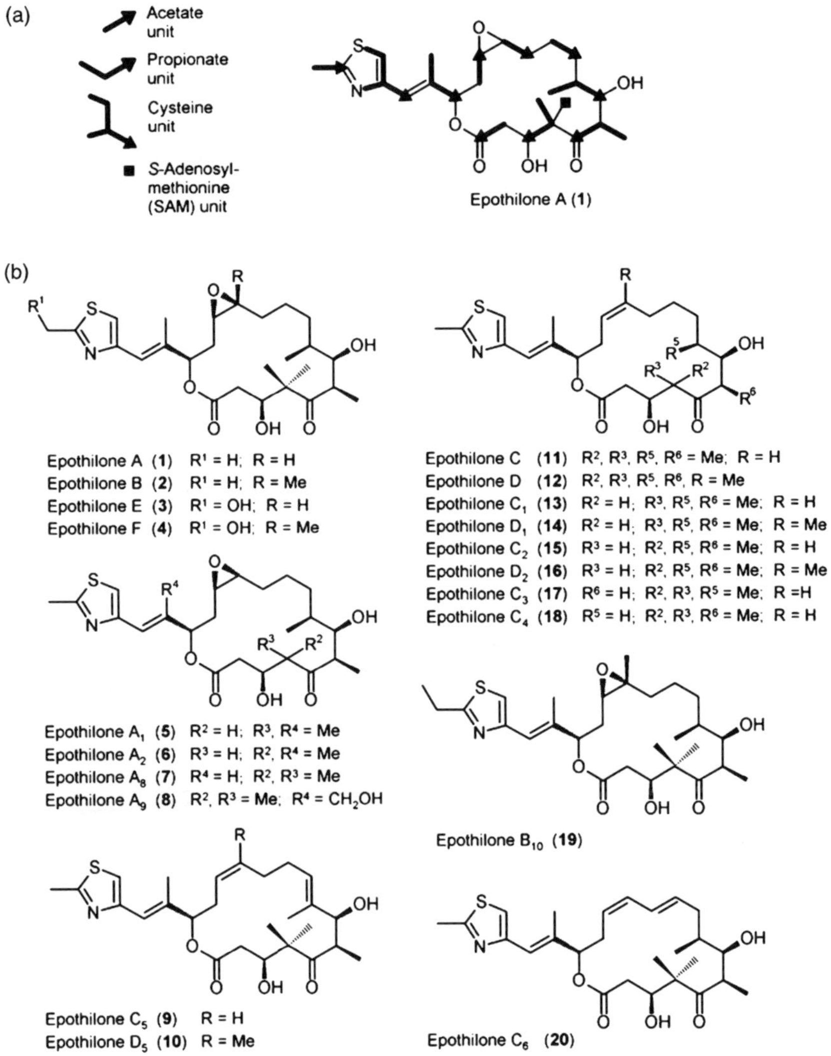

Fig. 3. a) Incorporation of isotope labelled precursors into epothilone A. b) Naturally occurring epothilones isolated from *S. cellulosum* So ce90 fermentation broth

(b)

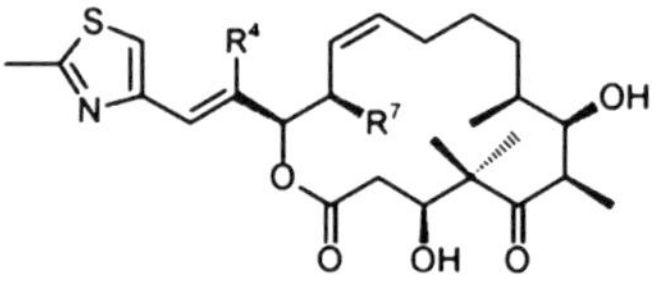

Epothilone C_7 (**21**) R^4 = Me; R^7 = OH
Epothilone C_8 (**22**) R^4, R^7 = H
Epothilone C_9 (**23**) R^4 = CH_2OH; R^7 = H

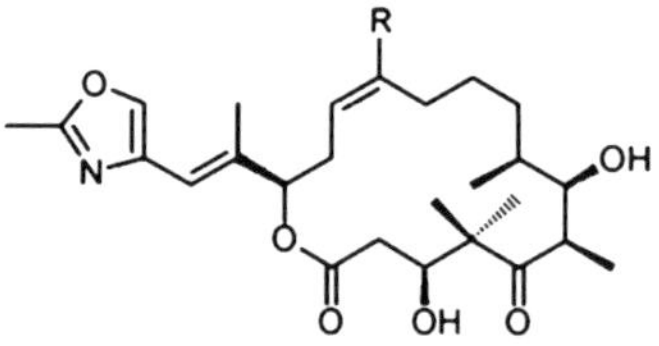

Epothilone H_1 (**28**) R = H
Epothilone H_2 (**29**) R = Me

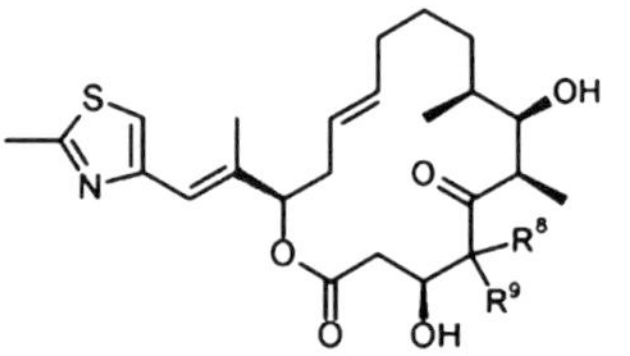

trans-Epothilone C_1 (**24**) R^8 = H; R^9 = Me
trans-Epothilone C_2 (**25**) R^8 = Me; R^9 = H

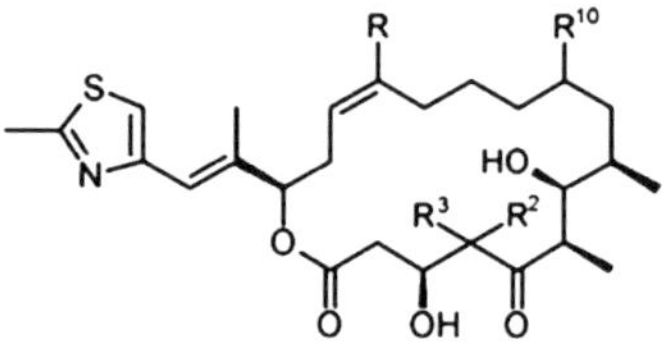

Epothilone I_1 (**30**) R, R^{10} = H; R^2, R^3 = Me
Epothilone I_2 (**31**) R = H; R^2, R^3, R^{10} = Me
Epothilone I_3 (**32**) R, R^2, R^3, R^{10} = Me
Epothilone I_4 (**33**) R, R^3 = H; R^2, R^{10} = Me
Epothilone I_5 (**34**) R, R^2, R^{10} = Me; R^3 = H
Epothilone I_6 (**35**) R, R^3, R^{10} = Me; R^2 = H

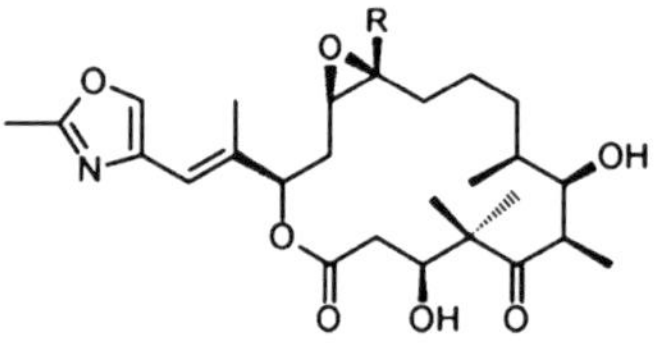

Epothilone G_1 (**26**) R = H
Epothilone G_2 (**27**) R = Me

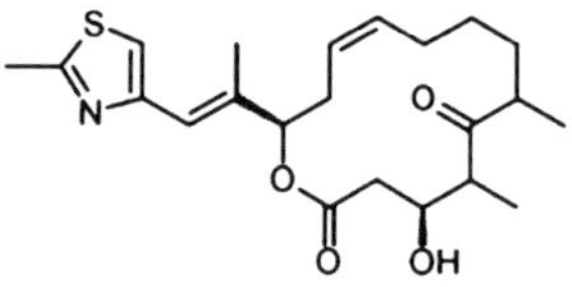

Epothilone K (**36**)

Fig. 3 (*continued*)

propionate (from methylmalonyl-CoA; mmCoA), the methyl group of *S*-adenosyl-methionine (SAM), and cysteine (which also introduces the sulfur and nitrogen atoms; see Fig. 3a). The authors also concluded that epothilone biosynthesis begins with the formation from acetate and cysteine of the thiazole portion of the molecule. Furthermore, they showed that epothilones A (**1**) and B (**2**) are formed from the alternative incorporation of mCoA or mmCoA at position C11–C12. Finally, these studies demonstrated that the epoxide functionality of the epothilones arises from molecular oxygen.

In a subsequent study, the same authors investigated non-producing mutants of the strain *S. cellulosum* So ce90, to evaluate whether epothilones A (**1**) and C (**11**) were generated by a different biosynthetic apparatus than that responsible for epothilones B (**2**) and D (**12**). As a single mutation was sufficient to abolish biosynthesis of both epothilones A (**1**) and B (**2**), this work showed that both metabolites are constructed by the same assembly line. In addition, the authors demonstrated that the first enzyme-free products of the PKS, epothilones C (**11**) and D (**12**), compete for the active site of a constitutively synthesized monooxygenase, which is regulated by product inhibition (*12*). The same group also provided evidence that the conversion of epothilones A (**1**) and B (**2**) to epothilones E (**3**) and F (**4**) by hydroxylation, is accomplished by a monooxygenase (*13*).

Taken together, this set of experiments showed that epothilone A (**1**) is assembled from one molecule of acetate, four molecules of malonyl-CoA, four units of methylmalonyl-CoA, one SAM-derived methyl carbon, and one cysteine. Epothilone B derives from the same precursors, with the exception that it incorporates an additional unit of methylmalonate in place of malonate, giving rise to a methyl branch at position 12. Based on the model for the biosynthesis of these metabolites, one would predict assembly of these compounds by multienzymes composed of a series of enzymatic modules, each of which carries out a single round of chain extension. Indeed, almost all of the expected modules (and their constituent domains) were found in the gigantic proteins encoded by the epothilone biosynthetic gene cluster (see below).

The biosynthesis of natural products in microorganisms is a multistep process, whose fidelity depends on the substrate specificity of every active site. If some of the activities exhibit promiscuity towards alternative substrates, and if subsequent modules tolerate "non-natural" chain extension intermediates, a mixture of compounds can result. In the case of epothilone biosynthesis, the result is the production of both epothilones A (**1**) and B (**2**), by alternative incorporation of mmCoA or mCoA at positions C11–C12. To assay for additional epothilone variants in *S. cellulosum* So ce90, *Hardt et al.* (*14*) performed a detailed analysis of a multitude of metabolites in the culture broth of a 700 L fermentation (a volume sufficient to isolate and characterize even compounds produced at very low levels). This experiment revealed an astonishing variety of epothilones: in addition to epothilones A (**1**) and B (**2**), 37 natural epothilone variants and epothilone-related compounds were isolated (see Fig. 3b). However, only the 12,13-deoxyepothilones and epothilones C (**11**) and D (**12**) were produced at significant levels (3–6 mg/L). Almost all of the other epothilones, including the C21-hydroxy derivatives and epothilones E (**3**) and F (**4**) were obtained in low yields (1–100 μg/L).

While it remains a formal possibility that some of these compounds resulted from degradation processes in the producing organism, others almost certainly derived from alternative operation of the epothilone assembly line. For example, among the variants (shown in Fig. 3b) were analogues incorporating an oxazole moiety in the side chain instead of a thiazole (most likely produced by activation, incorporation and cyclization of serine instead of cysteine), as well as a ring-expanded 18-membered macrolide (epothilone I (**30**)), and a ring-contracted 14-membered macrolide, epothilone K (**36**). The formation of the epothilones I (**30**) and K (**36**) can be explained by repeated use of a single module (termed "stuttering") and by module skipping, respectively, both processes which have been shown to happen in other multimodular assembly lines. For example, module skipping in the erythromycin PKS was shown to occur by direct transfer of the chain extension intermediate from an acyl carrier protein (ACP_n) to a downstream carrier protein (ACP_{n+1}), in the absence of chain extension by module $n+1$ (*15*). A similar process has been described for the NRPS responsible for myxochromide biosynthesis, although in this case, an active carrier protein in the skipped module was not required (*16*).

One mutant strain analyzed in the study by *Hardt et al.* did not produce macrolactones, but instead short-chain carboxylic acids bearing the characteristic thiazole side chain, indicating that the mutagenesis had affected the assembly line in a module needed for chain extension after thiazole ring assembly. The yields of these compounds were surprisingly robust, given that the failure to process the intermediates was expected to block the biosynthesis at a particular step, limiting turnover to a single round. However, *Yu et al.* (*17*) have shown that strains of *Amycolatopsis mediterranei* in which release of fully-processed "pre-rifamycin" by lactamization has been disabled, nonetheless hydrolyse alternative chains bound to the PKS, yielding all the predicted intermediates as their free acids. Similar hydrolytic activities may be present in *S. cellulosum*, and therefore release of the short-chain carboxylic acids isolated by *Hardt et al.*, may have been enzyme-catalyzed.

Hardt et al. also tested the new epothilone variants in cytotoxicity assays, establishing structure-activity relationships. Several of the natural epothilone variants exhibited activity comparable to epothilone A (**1**) and B (**2**), but none exceeded that of epothilone B (**2**), the most active natural variant.

3.3. Identification of the Epothilone Biosynthesis Gene Cluster

The most common approach to identify biosynthesis genes in the natural producer organism is to carry out gene inactivation studies. Mutant strains are screened for the absence of the desired metabolite, and then analyzed to identify the site of mutagenesis, thereby revealing

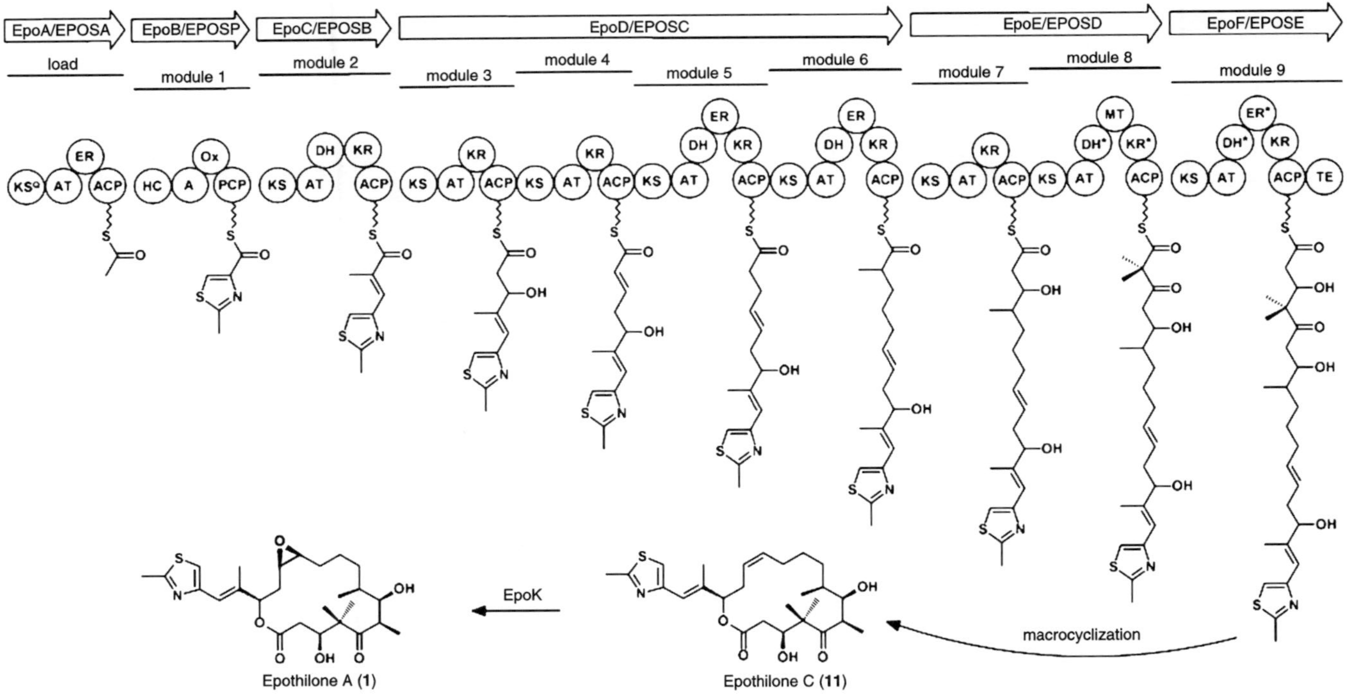

Fig. 4. Epothilone biosynthesis assembly line exemplified for the formation of epothilone A (protein names according to the Novartis and Kosan nomenclatures). For a description of the domains see Figs. 1 and 2

the associated biosynthesis gene cluster. The success of genome sequencing projects has revealed that numerous bacterial and fungal secondary metabolite producers harbor an enormous variety of biosynthesis gene clusters of both known and unknown function, typically including PKS, NRPS, and/or hybrid megasynthetases (*18*). Thus, searching for genes of mixed PKS/NRPS origin has a low likelihood of revealing the cluster responsible for a particular metabolite. Indeed, such an approach uncovered a number of hybrid systems in a cosmid library of *S. cellulosum* So ce90 (*19*), but the epothilone genes were not reported to be found among them.

Spurred on by the fact that total synthesis of natural epothilones could not provide an economically viable alternative to fermentation, a group at the pharmaceutical giant Novartis again attempted to identify the epothilone cluster in *S. cellulosum* So ce90 (the strain was obtained from the GBF collection). After analysis of a number of mutants, they discovered a gene cluster spanning 68,750 bp of the genome, and incorporating 22 open reading frames (*20*). Sequencing of the cluster revealed nine PKS modules, one NRPS module, a cytochrome P450, and two putative antibiotic transport proteins. Disruptions in the genes encoding the PKS abolished epothilone production, providing strong evidence for the clusters identity (Fig. 4).

From knowledge of the sequence of biosynthetic transformations required to generate epothilone, the authors were able to hypothesize roles for the various components within the gene cluster. Thus, the first PKS module and the NRPS module were proposed to co-operate in forming the thiazole heterocycle of epothilone from an acetate and a cysteine by condensation, cyclodehydration and subsequent dehydrogenation (see Figs. 1, 2, and 4). The remaining eight PKS modules were suggested to elaborate the rest of the epothilone carbon skeleton, as well as forming the macrolactone through thioesterase-catalyzed cyclization.

Overall, the domain organization of the epothilone biosynthesis gene cluster is highly consistent with the set of reactions required to generate the molecule, except for the absence of a dehydratase function in module 4, which appears to be required for formation of the double bond between C12 and C13. In addition, some domains found in modules 8 and 9 seem to be inactive.

In a parallel study, *Tang et al.*, at the biotech company Kosan, cloned an almost identical gene cluster from *S. cellulosum* SMP44 (*21*). Unfortunately, a different enzyme and gene nomenclature were reported, which has generated some confusion in the literature. As both publications appeared essentially simultaneously, and the postulated biosynthesis routes are almost identical, reference is given to both

References, pp. 50–53

nomenclatures as shown in Fig. 4. Interestingly, the sequences reported by both groups show high identity on the nucleotide level (less than 1% of the nucleotides are different, a discrepancy which may even be explained by sequencing errors; *R. M.*, unpublished data). This level of identity is astonishing, because the *S. cellulosum* strains were isolated from different continents. As an alternative to disrupting the gene cluster in order to confirm its identity, the Kosan group reconstituted the complete gene set in the heterologous host *Streptomyces coelicolor*, where transcription was driven by an engineered promoter system. The resulting strain assembled epothilone at a very low yield (no exact data are available), suggesting (although not proving) that all of the genes associated with its biosynthesis were in the transferred cluster. The choice of alternative host strain was motivated by the fact that *S. coelicolor* is more amenable to strain improvement and grows about 10-fold more rapidly than the natural producer. Thus, it was claimed that expression in *S. coelicolor* would allow for industrial-scale production of epothilones by fermentation. However, the epothilones are toxic to *S. coelicolor* (*R. M.*, unpublished results) and therefore the minor amounts of epothilones identified in their study presumably represent the quantity needed to kill the heterologous host.

Shortly thereafter, the Kosan group reported how they actually identified the epothilone genes within a gene library of *S. cellulosum* SMP44 in an alternative, statistical approach prior to heterologous expression (*22*). In the same work, they also expressed the P450 enzyme EpoK in *Escherichia coli*, and demonstrated that the purified protein converts epothilone D to B *in vitro*. Thus EpoK is also likely to be the activity responsible for epoxidation of epothilone C (**11**) to generate epothilone A (**1**).

3.4. Studies *in Vitro* into the Biochemistry of Epothilone Assembly

The availability of the epothilone genes set the stage for studies to uncover the molecular details of the biosynthesis. Such experiments are typically performed *in vitro*, using intact multienzymes or protein fragments expressed in recombinant form in *E. coli*, and purified to homogeneity. Although this approach is not expected to yield significant amounts of epothilones and/or new analogues, these studies can help to shed light on biosynthetic mechanisms, and to indicate which alternative substrates might be used by the megasynthetase *in vivo*.

In an initial study, *Chen et al.* (*23*) generated the chain-initiating methylthiazole ring of epothilone *in vitro* as an acyl-*S*-enzyme intermediate, using five domains derived from two modules of the megasynthetase. As EpoA could not be expressed in soluble form, the acyl carrier protein domain from EpoA was expressed as a discrete protein in *E. coli*, purified in its *apo* form, and then post-translationally primed with acetyl-CoA using a broad spectrum phosphopantetheinyl transferase, Sfp (*24*). The four-domain 150 kDa EpoB subunit harboring cyclization, adenylation, oxidase and peptidyl carrier protein domains was also expressed and purified. Sfp was again used to introduce the prosthetic group onto the *apo*-PCP, enabling subsequent loading with L-cysteine to generate the Cys-*S*-PCP acyl enzyme intermediate.

When acetyl-*S*-ACP (EpoA) and cysteinyl-*S*-EpoB were mixed, the cyclization domain of EpoB catalyzed acetyl transfer from EpoA to the amino group of the Cys-*S*-EpoB, generating a transient *N*-acetyl-Cys-*S*-EpoB intermediate that was cyclized and dehydrated to the five-membered ring methylthiazolinyl-*S*-EpoB. Subsequently, the FMN-containing oxidation domain of EpoB oxidised the thiazolinyl ring to the heteroaromatic oxidation state. Intriguingly, other acyl-CoAs could be substituted for acetyl-CoA in the priming of the EpoA *apo*-ACP, resulting in the production of additional alkylthiazolylcarboxy-*S*-EpoB acyl enzymes (Fig. 5). Taken together, these experiments established that chain transfer occurs across the EpoA and EpoB subunits of the epothilone megasynthetase, and showed that a cysteine group is converted into a methyl-substituted heterocycle during chain growth.

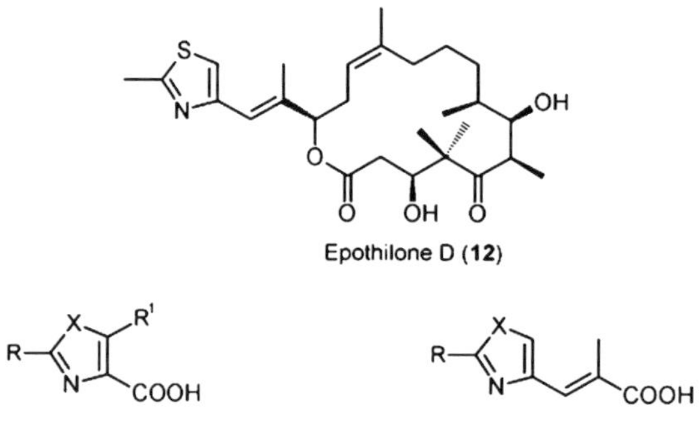

(**37**): X = O, R = Me, R^1 = H
(**38**): X = O, R = Me, R^1 = Me
(**39**): X = N, R = Me, R^1 = H
(**40**): X = S, R = phenyl, R^1 = H
(**41**): X = S, R = $CH(CH_3)_2$, R^1 = H
(**42**): X = S, R = C_5H_{11}, R^1 = H

(**43**): X = O, R = Me
(**44**): X = S, R = Me
(**45**): X = O, R = $CH(CH_3)_2$
(**46**): X = S, R = CH_2-NH_2
(**47**): X = O, R = CH_2-NH_2

(**48**): X = S, R = CH_2-NH-Boc
(**49**): X = S, R = Et
(**50**): X = S, R = $CH(CH_3)_2$
(**51**): X = S, R = phenyl
(**52**): X = S, R = methylpyrimidine

Fig. 5. Biosynthesis intermediates detected as free acids in *in vitro* studies with purified enzymes after loading of the carrier proteins with natural and unnatural precursors

At the time this report appeared in literature, the mechanism by which internal cysteine residues are converted into thiazolines and subsequently oxidized to heteroaromatic thiazole rings, was completely uncharacterized. *Schneider et al.* (*25*) addressed this question by excising the EpoB oxidase domain, EpoB-Ox, proposed to be responsible for thiazoline to thiazole oxidation in epothilone biosynthesis, and expressed it in soluble form in *E. coli*. The authors then showed that the purified domain represents an FMN-containing flavoprotein, which exhibits thiazoline to thiazole oxidase activity when incubated with thioester substrate mimics. The physiological electron acceptor in this reaction is not yet known, but molecular oxygen is needed in *in vitro* assays to enable reoxidation of reduced FMN.

In subsequent work reported from the *Walsh* group (*26*, *27*), the ACP of EpoA, along with the subunits EpoB and EpoC were expressed in recombinant form and characterized. In this way, the authors were able to reconstitute both the PKS/NRPS interface (EpoA-ACP/EpoB) and the NRPS/PKS interface (EpoB/EpoC) for epothilone biosynthesis *in vitro*. Additionally, a number of alternative substrates were shown to be incorporated into epothilone biosynthetic intermediates (Fig. 5). These experiments again raised the question as to whether these modified ACP bound precursors would also be processed by the complete megasynthetase, resulting in novel, biologically active epothilone analogues.

Another study by the same group addressed the question of the role of so called "linker" (*28*, *29*) or "docking domains" (*30*, *31*) in epothilone assembly. These domains were identified in other PKS systems and shown to play an essential role in protein-protein interactions, facilitating the transfer of biosynthetic intermediates between subunits of the megasynthetase (*29*, *32*). In light of their function in PKSs, the N- and C-terminal sequences of EpoB were examined for their roles in propagating the developing natural product. Eight amino acid residues at the EpoB C-terminus, of which six are positively charged, were found to be key components in chain transfer mediated by the C-terminal docking domain (*33*). A minimal sequence of 56 residues at the EpoB N-terminus was also required for transferring the acetyl group from the EpoA ACP to EpoB, to form methylthiazolyl-*S*-EpoB.

Based on the three enzymes discussed above (ACP of EpoA, EpoB and EpoC), *Schneider et al.* (*34*) showed that some oxazole and imidazole intermediates can be processed (Fig. 5). However, activated precursors of such derivatives have not yet been fed to the natural producer to prove the applicability of these findings.

In a follow-up study from the *Walsh* group (*34*, *35*), the docking domains were manipulated to enable chain transfer between proteins derived from the yersiniabactin, enterobactin, and rapamycin biosynthesis pathways, and subunits from the epothilone synthetase. These experiments generated not only the native thiazole intermediates of epothilone assembly, but additionally novel methylpyridine derivatives (Fig. 5).

However, although the feasibility of generating analogues has been demonstrated *in vitro*, it remains to be established whether such feeding experiments will work *in vivo* using the complete megasynthetases. One way to address this issue would be to attempt to reconstitute the entire biosynthesis *in vitro*, but the complexity of the pathway has so far precluded this approach.

Traditionally, thioester-templated reaction intermediates and products have been characterized by quantifying the radioactivity incorporated from biosynthetic precursors, either when the chains are bound to the multienzymes, or following release by chemical hydrolysis. More recently, analysis by mass spectrometry has emerged as a viable alternative methodology, generating data on the relative amounts of free enzyme compared to that bound to intermediates or products. This technology was applied to the multienyme EpoC, where direct detection of thioester intermediate mixtures bound to the enzyme was achieved using limited proteolysis coupled with *Fourier*-transform mass spectrometry (*F*TMS) (*36*). Incubation of EpoC with various *N*-acetylcysteamine thioester (NACS) substrate mimics produced mass increases on the ACP domain consistent with condensation of the introduced subsubtrate with an enzyme-bound carbanion produced by the decarboxylation of methylmalonyl-*S*-EpoC. Reconstitution of the assembly line including EpoA ACP, EpoB, and EpoC (see above) resulted in a further mass shift on EpoC consistent with the formation of the methylthiazolylmethacrylyl product by incorporation of acetyl-CoA, cysteine, and methylmalonyl-CoA.

The *Khosla* group (*37*) reported the production of the excised epothilone thioesterase as a recombinant protein in *E. coli*. The enzyme was characterized kinetically and employed to generate epothilone B from *seco*-epothilone activated as its NAC thioester. With the thioesterase at hand, it should now be possible to probe the substrate tolerance of the enzyme using artificial NACS-derivatized substrates, (*3*, *38*) with the aim of generating novel epothilone analogues from linear precursors. However, such studies have not been reported to date.

The first insights into the mechanism of formation of the epothilone epoxide came from experiments in which a P450 encoded in the cluster was deleted (*22*). As the resulting metabolite lacked the epoxide, this study identified the enzyme responsible for introducing this functionality. The enzyme, EpoK, was then expressed in recombinant form and characterized, and its crystal structure solved (*39*, *40*). Structures were reported for substrate-free EpoK and EpoK bound to epothilone D and B, at 2.10, 1.93, 2.65, and 2.1 Å resolution. Although the substrate for EpoK is the largest yet identified for a P450 enzyme whose X-ray structure has been solved, the changes to the structure upon substrate binding or product release were found to be minor, and the overall fold in all binding states was very similar to other P450s. Interestingly, the interactions between epothilone and EpoK appear to resemble those between paclitaxel and tubulin (*39*). Based on these structural similarities, *Nagano et al.* have proposed a possible binding mode for epothilone with tubulin.

The crystal structure of EpoK was reported before the enzyme was kinetically characterized. This delay may have been due to the difficulties in obtaining the natural interaction partners required for efficient catalysis by the enzyme. For example, P450 enzymes are known to often require specific ferredoxins as well as reductases, which would have to be isolated from *S. cellulosum* (*41*). This fact may explain the very low enzymatic activity reported for EpoK employing as surrogates, spinach ferredoxin and spinach ferredoxin reductase (*40*). In this study, *Ogura et al.* reported that the natural substrate can prevent and even reverse denaturation of the protein. Ligands and surrogate substrates were described, as well as a high throughput fluorescence assay (based on the oxidation of H_2O_2 dependent of 7-ethoxy-4-trifluoromethylcoumarin) to identify inhibitors of EpoK.

3.5. Heterologous Expression and Genetic Engineering of the Epothilone Biosynthesis Gene Cluster

The *in vitro* studies provided very significant insights into the molecular basis for epothilone assembly. However, exploiting this information to generate new analogues requires genetic manipulation of the producing organism, a task which remains challenging (*42*). In fact, to date no study has been published in which the genes for epothilone biosynthesis were manipulated in *S. cellulosum* to yield an active modified assembly line. *S. cellulosum* is also very slow growing (16 h doubling time), making the transfer of the complete biosynthesis system into a favorable heterologous host organism an even more desirable experiment. Ideally, this host would grow more quickly, and additionally allow for the efficient manipulation of the gene cluster.

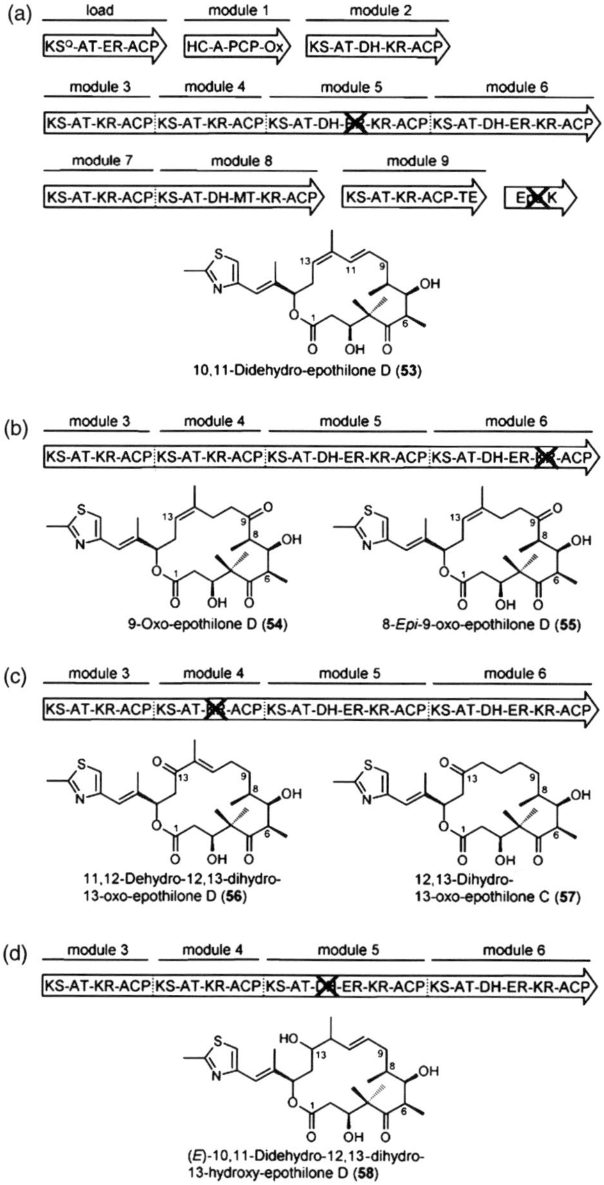

Fig. 6. Novel epothilone derivatives generated after heterologous production of epothilone in *M. xanthus* and genetic engineering of the epothilone megasynthetase. a) Inactivation of the ER domain of module 5. b) Inactivation of the KR domain of module 6. c) Inactivation of the KR domain of module 4. d) Inactivation of the DH domain of module 5. For a description of the domain abbreviations and the assembly line see Figs. 1, 2, and 4

A number of approaches have been used to heterologously express complex megasynthetases, including some from myxobacteria (*21*, *43*, *44*, *45*, *46*, *47*, *48*, *49*). Although the epothilone cluster was successfully relocated to *S. coelicolor* (see above), the resulting strain yielded only very small amounts of the metabolite, and so no follow-up work has been reported.

To address this shortcoming, the Kosan group also transferred the gene cluster into the well-studied myxobacterium *Myxococcus xanthus*, by splitting it into two parts (*48*). As genetic manipulation in this host is more difficult than in *Streptomyces* and no plasmids are available for myxobacteria, this was a significant achievement. Briefly, a 65.4 kb fragment of *S. cellulosum* DNA encompassing the entire epothilone gene cluster was inserted into the chromosome of *M. xanthus* by a series of homologous recombination events. The resulting strain produced epothilones A and B in yields of 0.16 mg/L (production from *S. coelicolor* was approximately 0.05 μg/L (*21*)). Construction of a strain that contained a mutation in *epoK*, encoding the P450 epoxidase, resulted in production of epothilones C (**11**) and D (**12**). Further studies to increase yields were only reported in *M. xanthus* (see below).

Heterologous expression in *M. xanthus* allowed for more straightforward genetic manipulation of the megasynthetase. Analysis of a strain engineered to produce epothilone D (**12**) by deletion of *epoK*, also revealed a new epothilone, 10,11-didehydro-epothilone D (**53**) (see Fig. 6) (*50*). The epothilone PKS was further modified by inactivating the ER domain of module 5 in a recombinant *M. xanthus* strain to produce 10,11-didehydro-epothilone D (**53**) as the major epothilone-related metabolite. The cytotoxicity of this compound against a panel of tumor cell lines and the effects on tubulin polymerization, were comparable to those of epothilone D (**12**).

Subsequently, the epothilone PKS genes were engineered to generate novel unnatural natural products that might be useful as new scaffolds for chemical modification (*51*). Inactivation of the KR domain in module 6 of the epothilone PKS resulted in accumulation of 9-oxo-epothilone D (**54**) and its isomer 8-*epi*-9-oxo-epothilone D (**55**) as the major products (see Fig. 6). Modification of the KR domain in module 4 yielded the expected compound 12,13-dihydro-13-oxo-epothilone C (**57**), in trace amounts. Interestingly, the major product of the fermentation was 11,12-dehydro-12,13-dihydro-13-oxo-epothilone D (**56**), whose biosynthesis was unanticipated. The other predicted compound, 12,13-dihydro-13-oxo-epothilone D, was not detected. The authors speculated that production of the unanticipated 13-oxo derivative indicates that the ER domain of module 5 has substrate specificity requirements, suggesting a second role for the domain (*51*).

As discussed earlier, the domain organization of the epothilone gene cluster is consistent with the chemical structure of the molecule, except for the absence of a dehydratase function in module 4 which would be expected to produce the *cis* double bond between carbons 12 and 13. To elucidate a possible function of the dehydratase of module 5 in this biosynthetic step, *Tang et al.* (*52*) performed three different deletions/ replacements of this domain. The three recombinant strains did not produce detectable amounts of 11-hydroxy epothilones (the compounds expected if the dehydratase of module 5 were not involved in forming the double bond between carbons 12 and 13). However, the resulting strains produced compounds lacking the double bond at the positions introduced by module 4, thus confirming the role of the module 5 DH domain in dehydration. Interestingly, the novel compound, (*E*)-10,11-didehydro-12,13-dihydro-13-hydroxy-epothilone D (**58**) (see Fig. 6) found in these mutants was also unanticipated because dehydration occurred forming the C10–C11 double bond in spite of the inactivation of the module 5 DH domain. These findings indicate an iterative function of the dehydratase in module 5 but the generation of the particular novel metabolite cannot be explained by iteration only. The iterative use of domains and modules has been proposed for type I PKS systems from both Streptomycetes and myxobacteria (*53*, *54*, *55*).

Recently, Kosan researchers took this work a step further, and engineered epothilone production in *E. coli* (*56*). To accomplish this, the genes encoding the entire gene cluster, *epoABCDEF*, were redesigned *in silico* and then synthesized to allow for improved expression. The motivation behind synthesizing more than 50,000 base pairs, was to allow for a change in codon usage from that preferred by *S. cellulosum*, to codons which were optimal for expression in *E. coli*. Expression of the largest of the proteins, EpoD, further required that the protein be separated into two subunits carrying compatible docking domains. By optimizing expression using low temperature and alternative promoters, as well as co-expressing chaperones, the researchers succeeded in producing soluble protein from all genes in the epothilone cluster. Expression of the cluster in a strain of *E. coli* modified to enable polyketide biosynthesis, resulted in the production of epothilones C (**11**) and D (**12**). In addition, feeding of an activated thioester of the normal substrate for EpoD to cells expressing only the *epoDEF* genes, also led to the generation of epothilones C (**11**) and D (**12**).

Based on these results, the authors argued that the design of the synthetic epothilone genes together with *E. coli* expression provides an ideal platform for both the biochemical investigation of the epothilone PKS and the generation of novel biosynthetic epothilone analogues. In

References, pp. 50–53

Table 1. Heterologously produced myxobacterial secondary metabolites

Compound	Producing organism	Host strain	Yield	References
Soraphen	*S. cellulosum*	*S. lividans*	0.3 mg/L	(*47*)
Epothilone	*S. cellulosum*	*S. coelicolor*	50–100 μg/L	(*21*)
		*E. coli**	10 μg/L	(*56*)
		M. xanthus	1–23 mg/L	(*48, 57*)
Flaviolin	*S. cellulosum*	*P. putida*	10 mg/L	(*49*)
Myxochromid S	*S. aurantiaca*	*P. putida*	40 mg/L	(*45*)
		M. xanthus	>1 g/L	*S. C. Wenzel* and *R. M.*, unpublished
Myxothiazol	*S. aurantiaca*	*P. putida**	0.8 mg/L	(*43*)
		M. xanthus	20 mg/L	(*44*)

* After genetic engineering of the host strain.

light of the yields reported (0.01 μg/L; see Table 1), however, it remains to be seen whether the *E. coli* host can be used effectively in such experiments.

3.6. Nutrient Regulation in *S. cellulosum* and *M. xanthus*

Nutrient regulation has been studied in the heterologous host *M. xanthus* as well as in *S. cellulosum* with the aim of increasing secondary metabolite production (*58*). It is known that *S. cellulosum* species producing the same secondary metabolite can exhibit striking differences in their physiology, and therefore optimal production conditions have to be established for each strain (*59*, *60*). Differences in physiology were also detected with the epothilone producer strains So ce90 and So ce1198. Production of epothilone by So ce90 is stimulated by increasing concentrations of glucose, while the same compound inhibits biosynthesis by strain So ce1198 (*59*). The initial amount of epothilones produced by the two strains vary from 0.3 to 25 mg/L.

As a result of a strain improvement programme at the German Centre for Biotechnology (GBF) involving conventional methods, production of epothilones was increased approximately 25-fold (*K. Gerth*, personal communication). Recently, it has become possible to study the regulation of secondary metabolite biosynthesis in *S. cellulosum* in a more rigorous way as a result of a functional genome project with strain So ce56. In an initial study, a specific regulator of chivosazol biosynthesis – ChiR – was identified and overexpressed, resulting in 5-fold overproduction of chivosazol (*61*). It remains to be seen, however, whether information gleaned from the genome of strain So ce56 can be applied to study and optimize the epothilone producer So ce90.

Research has also addressed the effects of media components on epothilone yields. For example, scientists at Kosan investigated epothilone production by *S. cellulosum* and *M. xanthus* in fermentation media containing various initial concentrations of ammonium and phosphate salts, in order to identify possible inhibitory effects of these additives (*58*). *S. cellulosum* was more sensitive to ammonium and phosphate than the heterologous producer in terms of epothilone yields. The effect of iron supplementation in the fermentation medium was also investigated. These studies employed an oil-based medium, which would obviously cause problems in industrial large-scale work up procedures. Nevertheless, a production process was developed based on this work, resulting in the crystallization of epothilone D (**12**) from the crude product in 97% purity (*62*).

The heterologous production of epothilone D (**12**) in *M. xanthus* was improved significantly from an initial titer of 0.16 mg/L, by the addition of adsorber resin XAD into the fermentation medium, the identification of a suitable carbon source, and the implementation of a fed-batch process (*57*). XAD was added to bind epothilone D (**12**) and thus reduce the degradation in the medium *via* stabilization of the compound (*13*, *59*). In general, this procedure facilitates the recovery of secondary metabolites, and enhances overall production levels in myxobacteria. In addition, the potential of using oils as a carbon source for cell growth and product formation was evaluated. This study revealed that among the oils tested (*e.g.* methyl oleate, coconut oil, sesame oil) methyl oleate had the most significant effects: at a concentration of 7 ml/L in a batch process, the maximum cell density was increased 5-fold, to 2 g dry cell weight/L. Metal ions were also shown to play a significant role in cell metabolism and in epothilone biosynthesis. To further increase the product yield, a continuous fed-batch process was developed to generate a higher cell density and to maintain an extended production period. All of these improvements resulted in an optimized fed-batch culture process, consistently yielding an average production titer of 23 mg/L.

Addition of biosynthesis precursors of epothilone, such as acetate and propionate to the fermentation media, resulted in alterations of the epothilone D:C ratio (*63*). Interestingly, addition of L-serine along with either acetate or propionate, yielded the oxazole containing epothilones H_1 (**28**) and H_2 (**29**) (see Fig. 2) as the major products, indicating that oxazole formation in *M. xanthus* is substrate limited and not caused by a preference of the enzyme complex for L-cysteine over L-serine.

A further study addressed the effects of oxygen limitation on epothilone production (*64*). In general, limiting the concentration of dissolved oxygen during cultivation of microbial strains decreases the activity of cytochrome P450 monooxygenases required for the processing of pathway intermediates into their final forms, which can result in the accumulation of these intermediates as the primary products. Cultivation of the recombinant *M. xanthus* strain with excess oxygenation resulted in the production of epothilones A (**1**) and B (**2**) as the primary products, while the same strain cultivated under oxygen depleted conditions yielded epothilones C (**11**) and D (**12**) as the primary products. In addition, the peak cell density in the oxygen-depleted cultures was found to be 60% higher than that observed in cultivations in which oxygen was present in excess. The same strategy was applied to the strain of *M. xanthus* harboring a genetically altered epothilone megasynthetase (see above) which generates 10,11-didehydro-epothilone D (**53**) (see Fig. 6). Fermentation of this strain under excess oxygenation revealed that EpoK catalysed production of a novel epothilone, 10,11-didehydro-epothilone B.

3.7. Conclusions

Over the past decade, significant progress has been made towards elucidating many aspects of epothilone biosynthesis, as well as producing the metabolite in heterologous hosts. In future, these advancements may allow alterations to the epothilone backbone, which would be difficult, if not impossible, to achieve by traditional synthetic methods. Such modified backbones could subsequently be employed as scaffolds for elaboration by medicinal chemists, towards the generation of new drug leads. However, a number of challenges must be overcome to achieve this goal. As production of engineered epothilones currently relies on heterologous expression hosts, yield improvements will be absolutely essential. Such gains may be achieved using alternative host organisms and/or by metabolic engineering of existing host strains. Rational construction of novel epothilone analogues should also be enabled by further deepening our understanding of the biosynthesis, which may result from studies of the pathway proteins *in vitro*, coupled with analysis by high-resolution mass spectrometry. Alternatively, methods for sophisticated genetic engineering in *S. cellulosum* may be developed, allowing productive pathway manipulation in the original host where production titers are much higher. Given the promise of epothilone as an anticancer agent, the next few years should see exciting developments in this already productive area.

Acknowledgements

I would like to acknowledge *Kira Weissman* and *Silke Wenzel* for helpful comments on this manuscript. Research in my laboratory is funded by the Deutsche Forschungsgemeinschaft (DFG) and the Bundesministerium für Bildung und Forschung (BMB + F).

References

1. Fischbach MA, Walsh CT (2006) Assembly-line Enzymology for Polyketide and Nonribosomal Peptide Antibiotics: Logic, Machinery, and Mechanisms. Chem Rev **106**: 3468
2. Finking R, Marahiel MA (2004) Biosynthesis of Nonribosomal Peptides. Annu Rev Microbiol **58**: 453
3. Weissman KJ (2005) Polyketide Synthases: Mechanisms and Models. Ernst Schering Res Found Workshop: 43
4. Müller R (2004) Don't Classify Polyketide Synthases. Chem Biol **11**: 4
5. Wenzel SC, Müller R (2005) Formation of Novel Secondary Metabolites by Bacterial Multimodular Assembly Lines: Deviations from Text Book Biosynthetic Logic. Curr Opin Chem Biol **9**: 447
6. Bode HB, Müller R (2006) Analysis of Myxobacterial Secondary Metabolism Goes Molecular. J Ind Microbiol Biotechnol **33**: 577
7. Aparicio JF, Molnar I, Schwecke T, Konig A, Haydock SF, Khaw LE, Staunton J, Leadlay PF (1996) Organization of the Biosynthetic Gene Cluster for Rapamycin in *Streptomyces hygroscopicus*: Analysis of the Enzymatic Domains in the Modular Polyketide Synthase. Gene **169**: 9
8. Duitman EH, Hamoen LW, Rembold M, Venema G, Seitz H, Saenger W, Bernhard F, Reinhardt R, Schmidt M, Ullrich C, Stein T, Leenders F, Vater J (1999) The Mycosubtilin Synthetase of *Bacillus subtilis* ATCC6633: A Multifunctional Hybrid Between a Peptide Synthetase, an Amino Transferase, and a Fatty Acid Synthase. P Natl Acad Sci USA **96**: 13294
9. Silakowski B, Nordsiek G, Kunze B, Blöcker H, Müller R (2001) Novel Features in a Combined Polyketide Synthase/Non-ribosomal Peptide Synthetase: The Myxalamid Biosynthetic Gene Cluster of the Myxobacterium *Stigmatella aurantiaca* Sga15. Chem Biol **8**: 59
10. Cane DE, Walsh CT (1999) The Parallel and Convergent Universes of Polyketide Synthases and Nonribosomal Peptide Synthetases. Chem Biol **6**: R319
11. Gerth K, Steinmetz H, Höfle G, Reichenbach H (2000) Studies on the Biosynthesis of Epothilones: The Biosynthetic Origin of the Carbon Skeleton. J Antibiot **53**: 1373
12. Gerth K, Steinmetz H, Höfle G, Reichenbach H (2001) Studies on the Biosynthesis of Epothilones: The PKS and Epothilone C/D Monooxygenase. J Antibiot **54**: 144
13. Gerth K, Steinmetz H, Höfle G, Reichenbach H (2002) Studies on the Biosynthesis of Epothilones: Hydroxylation of Epo A and B to Epothilones E and F. J Antibiot **55**: 41
14. Hardt IH, Steinmetz H, Gerth K, Sasse F, Reichenbach H, Höfle G (2001) New Natural Epothilones from *Sorangium cellulosum*, Strains So ce90/B2 and So ce90/D13: Isolation, Structure Elucidation, and SAR Studies. J Nat Prod **64**: 847
15. Thomas I, Martin CJ, Wilkinson CJ, Staunton J, Leadlay PF (2002) Skipping in a Hybrid Polyketide Synthase. Evidence for ACP-to-ACP Chain Transfer. Chem Biol **9**: 781

16. Wenzel SC, Meiser P, Binz T, Mahmud T, Müller R (2006) Nonribosomal Peptide Biosynthesis: Point Mutations and Module Skipping Lead to Chemical Diversity. Angew Chem Int Ed **45**: 2296
17. Yu TW, Shen Y, Doi-Katayama Y, Tang L, Park C, Moore BS, Hutchinson CR, Floss HG (1999) Direct Evidence That the Rifamycin Polyketide Synthase Assembles Polyketide Chains Processively. Proc Natl Acad Sci USA **96**: 9051
18. Bode HB, Müller R (2005) The Impact of Bacterial Genomics on Natural Product Research. Angew Chem Int Ed **44**: 6828
19. Beyer S, Kunze B, Silakowski B, Müller R (1999) Metabolic Diversity in Myxobacteria: Identification of the Myxalamid and the Stigmatellin Biosynthetic Gene cluster of *Stigmatella aurantiaca* Sg a15 and a Combined Polyketide-(poly)peptide Gene Cluster from the Epothilone Producing Strain *Sorangium cellulosum* So ce90. Biochim Biophys Acta **1445**: 185
20. Molnar I, Schupp T, Ono M, Zirkle R, Milnamow M, Nowak-Thompson B, Engel N, Toupet C, Stratmann A, Cyr DD, Gorlach J, Mayo JM, Hu A, Goff S, Schmid J, Ligon JM (2000) The Biosynthetic Gene Cluster for the Microtubule-stabilizing Agents Epothilones A and B from *Sorangium cellulosum* So ce90. Chem Biol **7**: 97
21. Tang L, Shah S, Chung L, Carney J, Katz L, Khosla C, Julien B (2000) Cloning and Heterologous Expression of the Epothilone Gene Cluster. Science **287**: 640
22. Julien B, Shah S, Ziermann R, Goldman R, Katz L, Khosla C (2000) Isolation and Characterization of the Epothilone Biosynthetic Gene Cluster from *Sorangium cellulosum*. Gene **249**: 153
23. Chen HW, O'Connor S, Cane DE, Walsh CT (2001) Epothilone Biosynthesis: Assembly of the Methylthiazolylcarboxy Starter Unit on the EpoB Subunit. Chem Biol **8**: 899
24. Quadri LE, Weinreb PH, Lei M, Nakano MM, Zuber P, Walsh CT (1998) Characterization of Sfp, a *Bacillus subtilis* Phosphopantetheinyl Transferase for Peptidyl Carrier Protein Domains in Peptide Synthetases. Biochemistry **37**: 1585
25. Schneider TL, Shen B, Walsh CT (2003) Oxidase Domains in Epothilone and Bleomycin Biosynthesis: Thiazoline to Thiazole Oxidation During Chain Elongation. Biochemistry **42**: 9722
26. O'Connor SE, Chen HW, Walsh CT (2002) Enzymatic Assembly of Epothilones: The EpoC Subunit and Reconstitution of the EpoA-ACP/B/C Polyketide and Nonribosomal Peptide Interfaces. Biochemistry **41**: 5685
27. Walsh CT, O'Connor SE, Schneider TL (2003) Polyketide-nonribosomal Peptide Epothilone Antitumor Agents: The EpoA, B, C Subunits. J Ind Microbiol Biotechnol **30**: 448
28. Gokhale RS, Khosla C (2000) Role of Linkers in Communication Between Protein Modules. Curr Opin Chem Biol **4**: 22
29. Kumar P, Li Q, Cane DE, Khosla C (2003) Intermodular Communication in Modular Polyketide Synthases: Structural and Mutational Analysis of Linker Mediated Protein-protein Recognition. J Am Chem Soc **125**: 4097
30. Broadhurst RW, Nietlispach D, Wheatcroft MP, Leadlay PF, Weissman KJ (2003) The Structure of Docking Domains in Modular Polyketide Synthases. Chem Biol **10**: 723
31. Weissman KJ (2006) The Structural Basis for Docking in Modular Polyketide Biosynthesis. ChemBioChem **7**: 485
32. Wu N, Cane DE, Khosla C (2002) Quantitative Analysis of the Relative Contributions of Donor Acyl Carrier Proteins, Acceptor Ketosynthases, and Linker Regions to Intermodular Transfer of Intermediates in Hybrid Polyketide Synthases. Biochemistry **41**: 5056

33. Liu F, Garneau S, Walsh CT (2004) Hybrid Nonribosomal Peptide-polyketide Interfaces in Epothilone Biosynthesis: Minimal Requirements at N and C Termini of EpoB for Elongation. Chem Biol **11**: 1533
34. Schneider TL, Walsh CT, O'Connor SE (2002) Utilization of Alternate Substrates by the First Three Modules of the Epothilone Synthetase Assembly Line. J Am Chem Soc **124**: 11272
35. O'Connor SE, Walsh CT, Liu F (2003) Biosynthesis of Epothilone Intermediates with Alternate Starter Units: Engineering Polyketide-Nonribosomal Interfaces. Angew Chem Int Ed **42**: 3917
36. Hicks LM, O'Connor SE, Mazur MT, Walsh CT, Kelleher NL (2004) Mass Spectrometric Interrogation of Thioester-bound Intermediates in the Initial Stages of Epothilone Biosynthesis. Chem Biol **11**: 327
37. Boddy CN, Schneider TL, Hotta K, Walsh CT, Khosla C (2003) Epothilone C Macrolactonization and Hydrolysis are Catalyzed by the Isolated Thioesterase Domain of Epothilone Polyketide Synthase. J Am Chem Soc **125**: 3428
38. Weissman KJ, Bycroft M, Cutter AL, Hanefeld U, Frost EJ, Timoney MC, Harris R, Handa S, Roddis M, Staunton J, Leadlay PF (1998) Evaluating Precursor-Directed Biosynthesis Towards Novel Erythromycins Through in Vitro Studies on a Bimodular Polyketide Synthase. Chem Biol **5**: 743
39. Nagano S, Li HY, Shimizu H, Nishida C, Ogura H, Ortiz de Montellano PR, Poulos TL (2003) Crystal Structures of Epothilone D-bound, Epothilone B-bound, and Substrate-free Forms of Cytochrome P450 EpoK. J Biol Chem **278**: 44886
40. Ogura H, Nishida C, Hoch U, Perera R, Dawson JH, Ortiz de Montellano PR (2004) EpoK, a Cytochrome P450 Involved in Biosynthesis of the Anticancer Agents Epothilones A and B. Substrate-Mediated Rescue of a P450 Enzyme. Biochemistry **43**: 14712
41. Bernhardt R (2004) Cytochrome P-450. In: Encyclopedia of Biological Chemistry, p. 544. Elsevier, Amsterdam
42. Kopp M, Irschik H, Gross F, Perlova O, Sandmann A, Gerth K, Müller R (2004) Critical Variations of Conjugational DNA Transfer into Secondary Metabolite Multiproducing *Sorangium cellulosum* Strains So ce12 and So ce56: Development of a Mariner-Based Transposon Mutagenesis System. J Biotechnol **107**: 29
43. Gross F, Ring MW, Perlova O, Fu J, Schneider S, Gerth K, Kuhlmann S, Stewart AF, Zhang Y, Müller R (2006) Metabolic Engineering of *Pseudomonas putida* for Methylmalonyl-CoA Biosynthesis to Enable Complex Heterologous Secondary Metabolite Formation. Chem Biol **13**: 1253
44. Perlova O, Fu J, Kuhlmann S, Krug D, Stewart F, Zhang Y, Müller R (2006) Reconstitution of Myxothiazol Biosynthetic Gene Cluster by Red/ET Recombination and Heterologous Expression in *Myxococcus xanthus*. Appl Environ Microbiol **72**: 7485
45. Wenzel SC, Gross F, Zhang Y, Fu J, Stewart FA, Müller R (2005) Heterologous Expression of a Myxobacterial Natural Products Assembly Line in Pseudomonads via Red/ET Recombineering. Chem Biol **12**: 349
46. Wenzel SC, Müller R (2005) Recent Developments Towards the Heterologous Expression of Complex Bacterial Natural Product Biosynthetic Pathways. Curr Opin Biotechnol **16**: 594
47. Zirkle R, Ligon JM, Molnar I (2004) Heterologous Production of the Antifungal Polyketide Antibiotic Soraphen A of *Sorangium cellulosum* So ce26 in *Streptomyces lividans*. Microbiology **150**: 2761
48. Julien B, Shah S (2002) Heterologous Expression of Epothilone Biosynthetic Genes in *Myxococcus xanthus*. Antimicrob Agents Chemother **46**: 2772

49. Gross F, Luniak N, Perlova O, Gaitatzis N, Jenke-Kodama H, Gerth K, Gottschalk D, Dittmann E, Müller R (2006) Bacterial Type III Polyketide Synthases: Phylogenetic Analysis and Potential for the Production of Novel Secondary Metabolites by Heterologous Expession in Pseudomonads. Arch Microbiol **185**: 28
50. Arslanian RL, Tang L, Blough S, Ma W, Qiu RG, Katz L, Carney JR (2002) A New Cytotoxic Epothilone from Modified Polyketide Synthases Heterologously Expressed in *Myxococcus xanthus*. J Nat Prod **65**: 1061
51. Tang L, Chung L, Carney JR, Starks CM, Licari P, Katz L (2005) Generation of New Epothilones by Genetic Engineering of a Polyketide Synthase in *Myxococcus xanthus*. J Antibiot **58**: 178
52. Tang L, Ward S, Chung L, Carney JR, Li Y, Reid R, Katz L (2004) Elucidating the Mechanism of cis Double Bond Formation in Epothilone Biosynthesis. J Am Chem Soc **126**: 46
53. Gaitatzis N, Silakowski B, Kunze B, Nordsiek G, Blöcker H, Höfle G, Müller R (2002) The Biosynthesis of the Aromatic Myxobacterial Electron Transport Inhibitor Stigmatellin is Directed by a Novel Type of Modular Polyketide Synthase. J Biol Chem **277**: 13082
54. Perlova O, Gerth K, Hans A, Kaiser O, Müller R (2006) Identification and Analysis of the Chivosazol Biosynthetic Gene Cluster from the Myxobacterial Model Strain *Sorangium cellulosum* So ce56. J Biotechnol **121**: 174
55. Moss SJ, Martin CJ, Wilkinson B (2004) Loss of Co-linearity by Modular Polyketide Synthases: A Mechanism for the Evolution of Chemical Diversity. Nat Prod Rep **21**: 575
56. Mutka SC, Carney JR, Liu Y, Kennedy J (2006) Heterologous Production of Epothilone C and D in *Escherichia coli*. Biochemistry **45**: 1321
57. Lau J, Frykman S, Regentin R, Ou S, Tsuruta H, Licari P (2002) Optimizing the Heterologous Production of Epothilone D in *Myxococcus xanthus*. Biotechnol Bioeng **78**: 280
58. Regentin R, Frykman S, Lau J, Tsuruta H, Licari P (2003) Nutrient Regulation of Epothilone Biosynthesis in Heterologous and Native Production Strains. Appl Microbiol Biotechnol **61**: 451
59. Gerth K, Pradella S, Perlova O, Beyer S, Müller R (2003) Myxobacteria: Proficient Producers of Novel Natural Products with Various Biological Activities – Past and Future Biotechnological Aspects with the Focus on the Genus *Sorangium*. J Biotechnol **106**: 233
60. Müller R, Gerth K (2006) Development of Simple Media Which Allow Investigations into the Global Regulation of Chivosazol Biosynthesis with *Sorangium cellulosum* So ce56. J Biotechnol **121**: 192
61. Rachid S, Gerth K, Kochems I, Müller R (2007) Deciphering Regulatory Mechanisms for Secondary Metabolite Production in the Myxobacterium *Sorangium cellulosum* So ce56. Mol Microbiol (in press)
62. Arslanian RL, Parker CD, Wang PK, McIntire JR, Lau J, Starks C, Licari PJ (2002) Large-scale Isolation and Crystallization of Epothilone D from *Myxococcus xanthus* Cultures. J Nat Prod **65**: 570
63. Frykman S, Tsuruta H, Lau J, Regentin R, Ou S, Reeves C, Carney J, Santi D, Licari P (2002) Modulation of Epothilone Analog Production Through Media Design. J Ind Microbiol Biotechnol **28**: 17
64. Frykman SA, Tsuruta H, Starks CM, Regentin R, Carney JR, Licari PJ (2002) Control of Secondary Metabolite Congener Distributions via Modulation of the Dissolved Oxygen Tension. Biotechnol Prog **18**: 913

4. Total Synthesis of Epothilones A–F

Johann Mulzer and *Kathrin Prantz*

Institut für Organische Chemie der Universität Wien, Wien, Austria

Contents

4.1. Introduction

Although epothilones A–F are available from fermentation, their potential for broad refunctionalization and structural modification is rather limited. Therefore, from early on, the demand of more detailed SAR studies and the search of more specific and higher activity epothilone derivatives have stimulated very intensive synthetic investigations world wide, all the more so as the epothilones possess a far simpler overall chemical structure than, for instance, the taxanes.

A succession of reviews has appeared over the past six years highlighting the progress in the total synthesis of epothilones A and B (**1a** and **1b**) and their deoxy-analogues C and D (**2a** and **2b**) (*1–26*). This overview attempts to give a comprehensive description of all the various approaches that have been developed until the end of 2006. Altogether, 16 complete and formal approaches have been reported for **1a**, **2a** and 20 for **1b**, **2b**: Many of the approaches follow quite similar strategies and feature analogous key reactions and C,C-disconnections. Some of them even use the same key fragments, so that the number of the truly independent syntheses is relatively low.

Over the years, a standard retrosynthesis has been developed (Chart 1) (*2*, *15*, *17*, *18*, *20*). The last step in the synthesis is the more or less stereocontrolled epoxidation of the 12,13-double bond in **2a–2d**. If

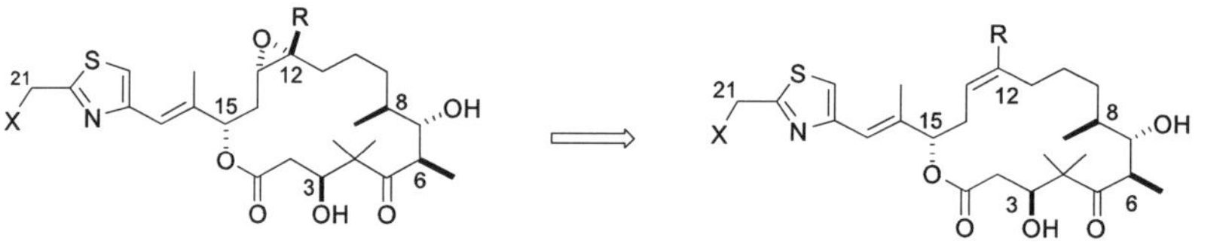

1a R = H, X = H epothilone A
1b R = Me, X = H epothilone B
1c R = H, X = OH epothilone E
1d R = Me, X = OH epothilone F

2a R = H, X = H epothilone C
2b R = Me, X = H epothilone D
2c R = H, X = OH deoxyepothilone E
2d R = Me, X = OH deoxyepothilone F

Fig. 1. Naturally occurring epothilones

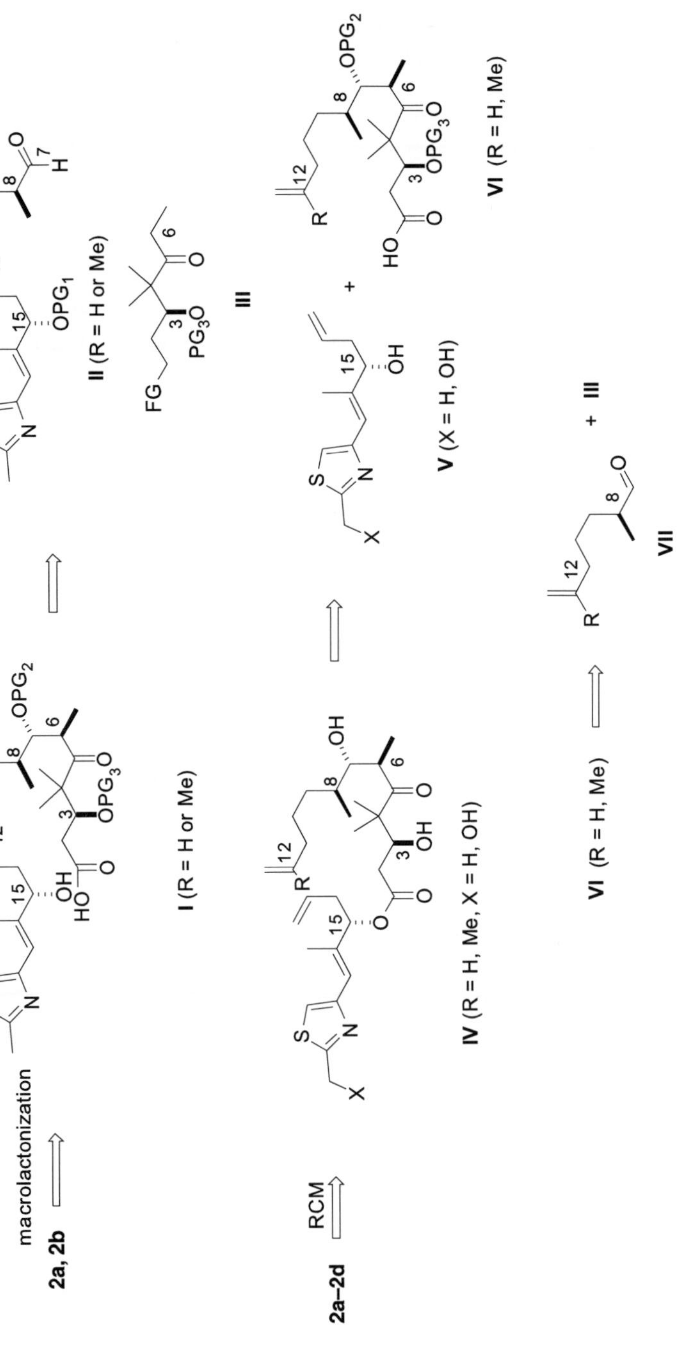

Chart 1. Retrosynthetic considerations

DMDO is used for the epoxidation, a diastereomeric ratio of up to 9:1 may be achieved. The penultimate step in the synthesis of **2a** and **2b** is the macrolactonization of *seco*-acid **I**, normally performed at high dilution (10^{-3} to 10^{-4} *M*) *via* the *Yamaguchi* protocol (*27*). *Seco*-acid **I** is generated by an aldol addition of the C21–C7-aldehyde **II** and the (*Z*)-enolate of ketone **III** (*28–30*). This addition can be performed to get the desired relative configuration for the C6–C8-stereotriad with high stereocontrol. The asymmetric induction crucially hinges on the presence and the configuration of the stereocenter at C3 and the functionalization of the C1–C3-segment in **III**. A second, more general access to **2a–2d** is *via* RCM of diolefin ester **IV**, which is obtained from alcohol **V** and acid **VI**. For the preparation of **VI**, an aldol reaction between **III** and aldehyde **VII** is normally used, similar to the one between **III** and **II**.

4.2. Synthesis Approaches to both the Epothilone A/C- and B/D-Series

The first three syntheses (*Danishefsky*, *Nicolaou*, and *Schinzer*) initially all aimed for **1a**, **2a**. However, the underlying strategies were so designed as to access also **1b**, **2b** without major alterations.

4.2.1. Danishefsky Syntheses (31–42)

The *Danishefsky* group has been particularly imaginative in developing a variety of different syntheses. Their first approach (*31–33*) started with the preparation of the C21–C12-segments **9** and **10** (Chart 2).

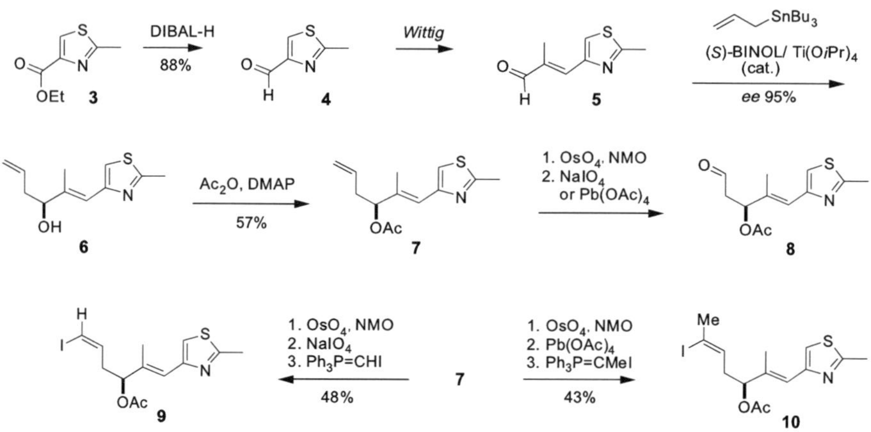

Chart 2. *Danishefsky*'s synthesis of epothilone building blocks

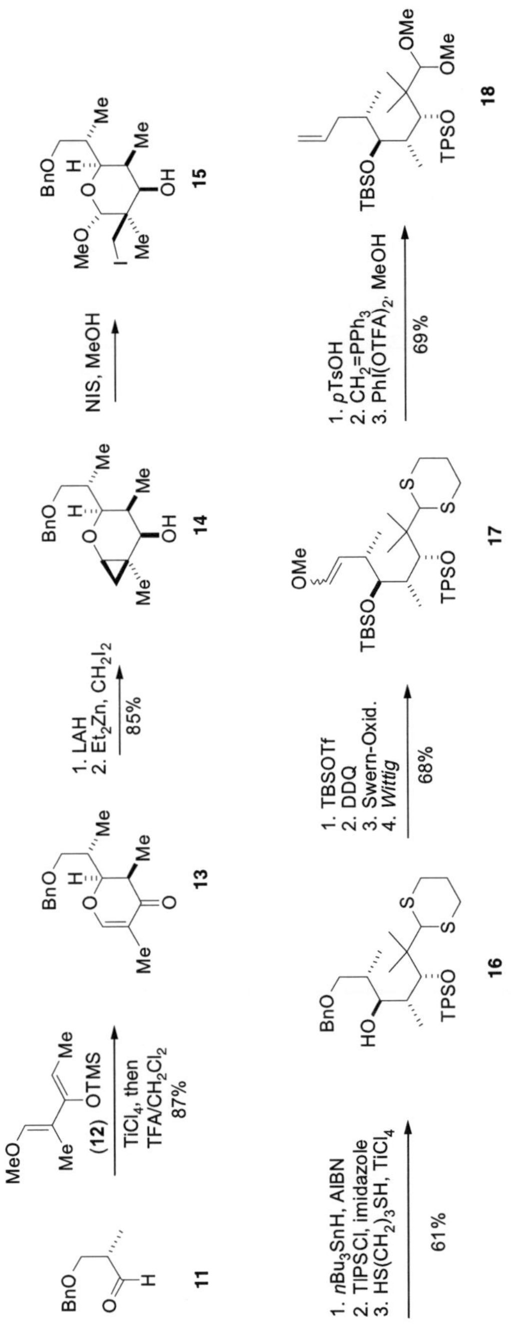

Chart 3. *Danishefsky*'s macroaldolization synthesis of epothilones A and C, Part I

Thiazole ester **3** was reduced to aldehyde **4**, which was converted into **5** by a *Wittig* olefination. Compound **5** was subjected to an enantiocatalyzed allylation to **6** whose acetylation furnished fragment **7**. From **7** aldehyde **8** was prepared from which the (*Z*)-vinyl iodides **9** and **10** were available in moderate yields.

The C3–C11-segment **19** was prepared along a route (Chart 3) deviating from the general strategy described in Chart 1. Thus, the C6–C8-stereotriad was prepared *via* a hetero-*Diels-Alder* reaction between chiral aldehyde **11** and *Danishefsky* diene **12**, which proceeds to dihydropyranone **13** with high *Felkin Anh*-selectivity. The *gem*-dimethyl group at C4 was installed *via* cyclopropanation to **14** and ring opening to iodide **15**, which after deiodination, was converted into the acyclic C3–C9-fragment **16**. Two successive one-carbon chain elongations gave, *via* **17**, the C3–C11-fragment **18**, which was connected with vinyl iodide **9** *via* a B-alkyl-*Suzuki*-coupling to complete the epothilone A/C *seco*-intermediate **20** (Chart 4). Ring closure was achieved between C2 and C3 *via* macroaldolization to give **21** with a 6:1 diastereomeric ratio at C3. Oxidation at C5 and desilylation led to **2a**. The regio- and stereoselective epoxidation of the C12,13-double bond to **1a** was achieved with DMDO (*37*).

A second approach to **2a** (*34*) used macrolactonization as the ring closing step (Chart 5). Hence, acetal **18** was converted into the aldehyde, which was subjected to an aldol addition with *t*-Bu-acetate to give **22** as a separable 1:1-diastereomeric mixture at C3. After oxidation and silylation **23** was obtained whose *Suzuki*-coupling with **9** as before furnished *seco*-acid **24**. *Yamaguchi*-lactonization and deprotection gave

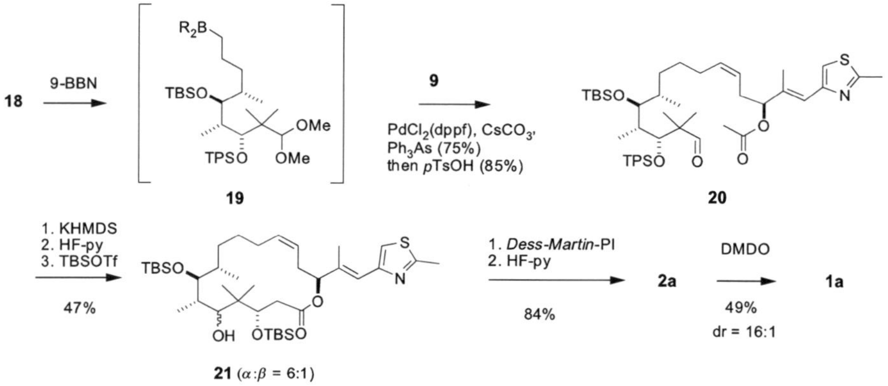

Chart 4. *Danishefsky*'s macroaldolization of epothilones A and C, Part II

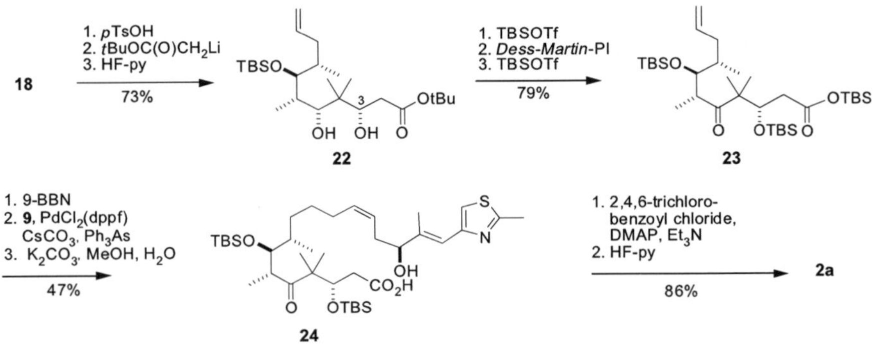

Chart 5. *Danishefsky*'s macrolactonization synthesis of epothilones A and C

2a. Although this second route is four steps longer, the overall yield is about the same (23 *vs*. 25%).

Ring closing metathesis (RCM) was tested as a third mode of macrocyclization (*38*), this time for both **2a** and **2b** (Chart 6). In fact, the epothilones were the first relatively complex substrates for this, at the time, rather novel methodology. Thus, intermediate **16** was converted into C9-aldehyde **25**, which after chain elongation gave olefinic aldehydes **26a** and **26b**. Aldol addition with **8** furnished **27a** and **27b** as 1:1-diastereomeric mixtures at C3. The configuration at C3 was rectified to (3*S*) *via* an oxidation-reduction sequence after which some further modifications led to the diastereomerically pure ketones **28a** and **28b**. Epothilone A precursor **28a** was cyclized with *Grubbs*' first generation RCM catalyst to **29** as a 1:1.7-(*E*)/(*Z*)-mixture that was separated and desilylated to give **2a**. In contrast, the RCM of epothilone B intermediate **28b** had to be performed with *Schrock*'s catalyst to furnish **30** as a 1:1-*E/Z*-mixture. Separation and desilylation gave **2b**.

After these very early reports the *Danishefsky* group concentrated on improvements of their epothilone B syntheses. To this end, OTES-protected vinyl iodide **32** was prepared from **6** *via* **31** along the established route (Chart 7) (*39*). However, the overall yield turned out to be unacceptably low (43%). Therefore, an improved synthesis was develoved using *Evans*' auxiliary **33** for a highly diastereoselective allylation with di-iodide **34** to give **35**, which was converted into **32** over five steps in 63% overall yield (*40*).

For the B-alkyl *Suzuki* coupling the preparation of olefinic ester **44** was required, which was based on an aldol addition between aldehyde **40** and the diisopropyl acetal of aldehyde **39**, which gave **41** with

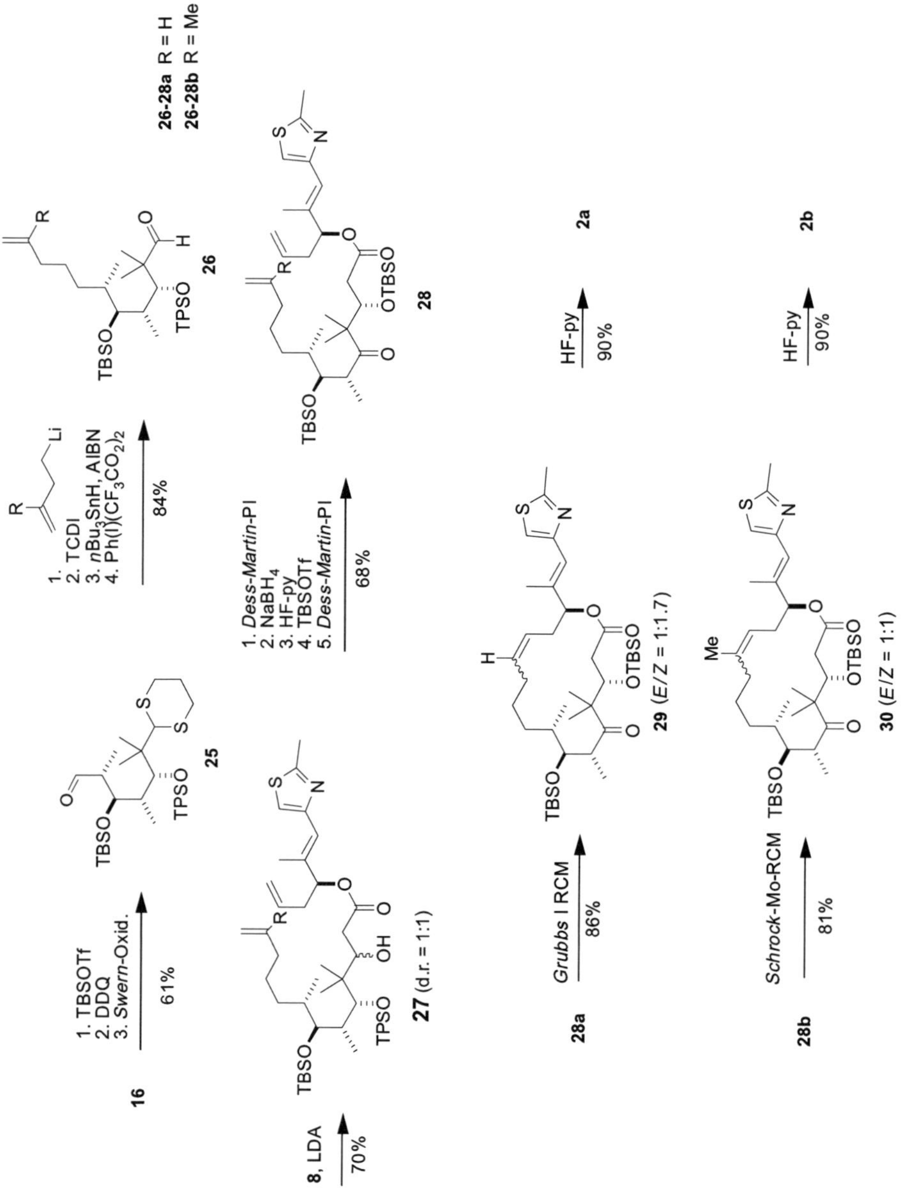

Chart 6. *Danishefsky*'s first RCM synthesis of epothilones A–D

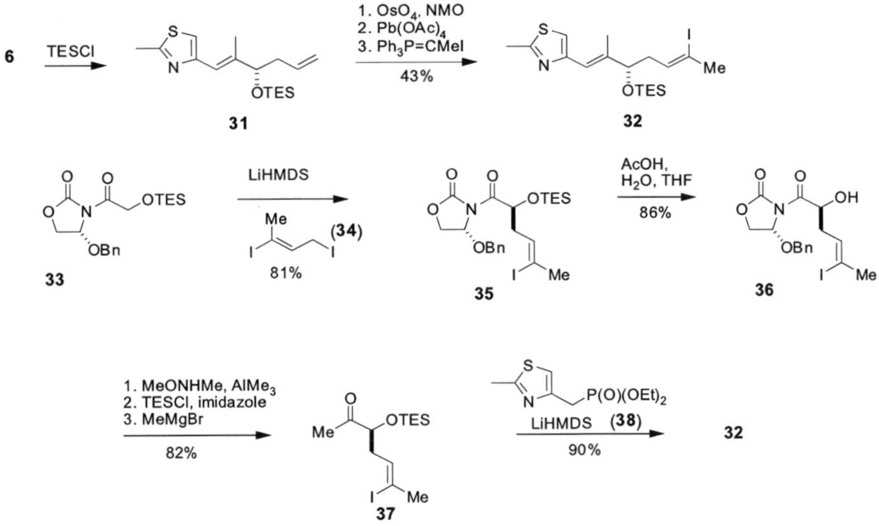

Chart 7. *Danishefsky*'s B-alkyl *Suzuki* approach to epothilone B, Part I

moderate diastereoselectivity (Chart 8). A number of other aldol variants were tried, however without significant improvement. Hence, the main isomer of **41** was separated, Troc-protected to **42** and converted into aldehyde **43**, which was used for another aldol addition, this time with *t*-Bu-acetate, in the presence of *Duthaler*'s complex to give the (3*S*)-epimer **44** after silylation with high stereocontrol. *Suzuki* coupling of **44** with vinyl iodide **32** furnished *seco*-intermediate **45**. Deprotection gave the *seco*-acid **46**, which was macrolactonized and deprotected to give **2a**.

A variation of the original RCM approach (Chart 9) (*37*) was initiated by preparing the terminal olefins **47**, **48** from **32**, **44** and connecting them *via* esterification to the *seco*-intermediate **49**, which on RCM with *Grubbs*' second-generation catalyst gave a 3:1-mixture of the desired macrocycle **50** and the ring-contracted derivative **51**. After separation, **50** was deprotected to give epothilone 490 (**52**) as a highly active non-natural epothilone derivative. Selective hydrogenation of the C10,11-double bond with diimide led to **2a** (*39–41*).

In a second approach to **52** (Chart 10) the *Duthaler* aldol addition of acetate **53** to aldehyde **43** was used as the connecting step. Compound **54** was obtained stereoselectively and macrocyclized by RCM to furnish **55**. In this case, no ring contraction was observed. After removal of the Troc-group **52** was obtained.

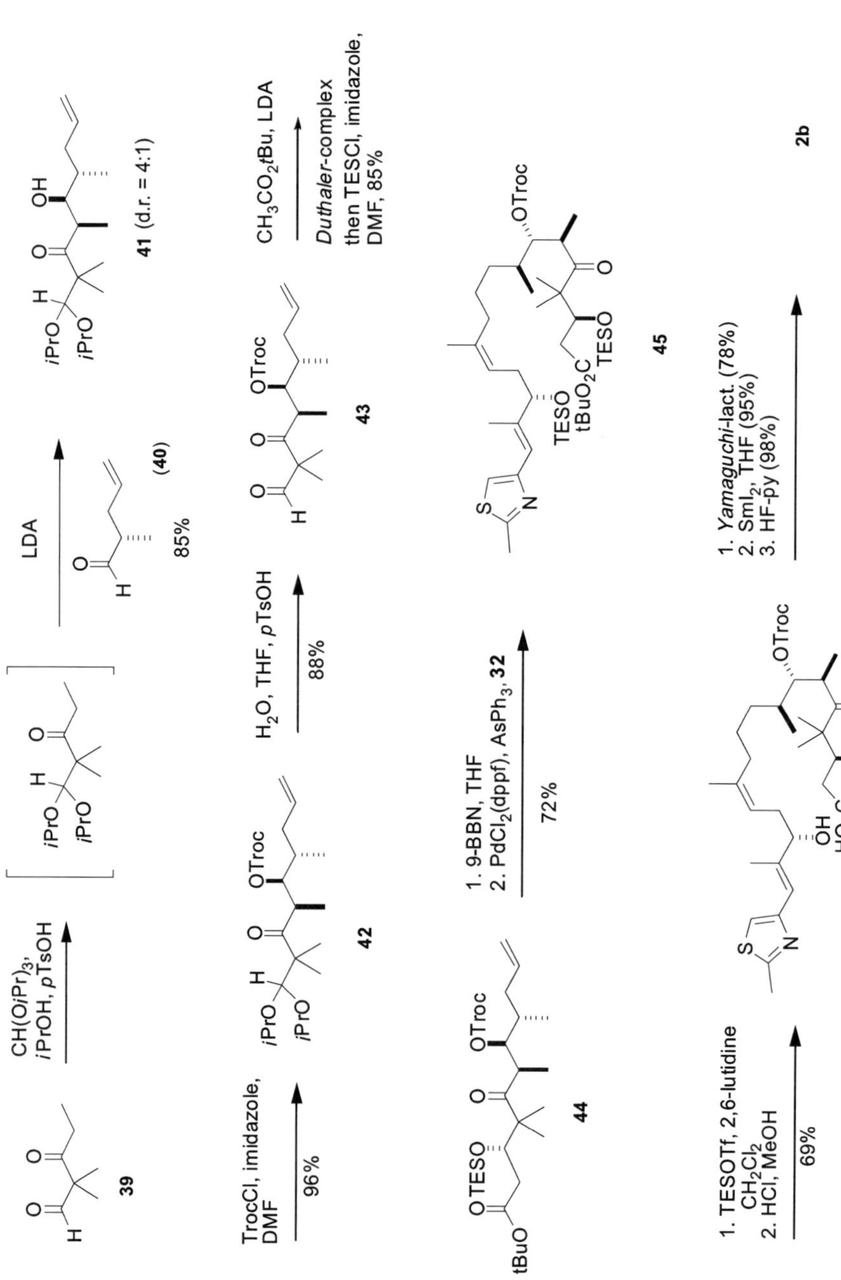

Chart 8. *Danishefsky*'s B-alkyl *Suzuki* approach to epothilone B, Part II

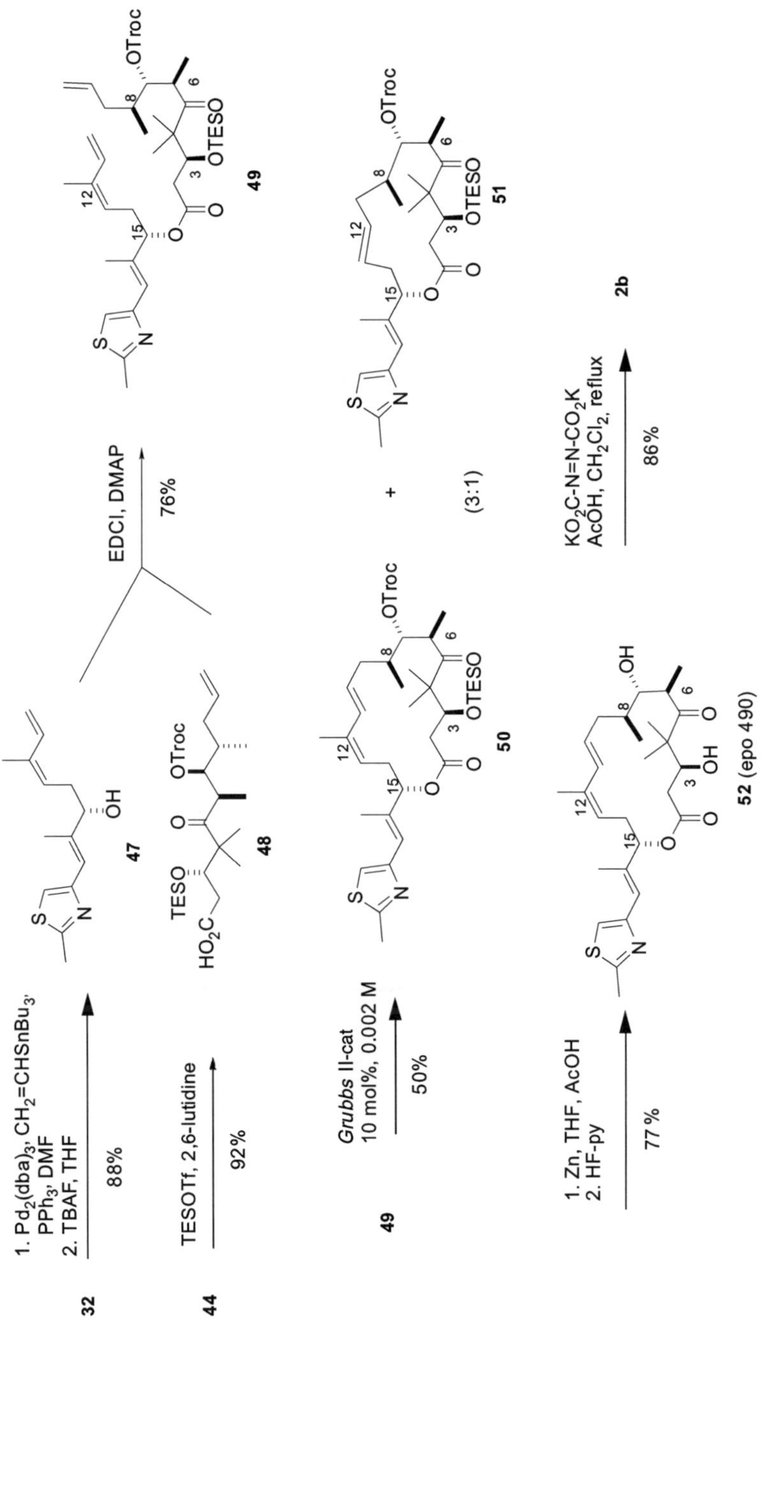

Chart 9. *Danishefsky*'s second RCM synthesis of epothilone D

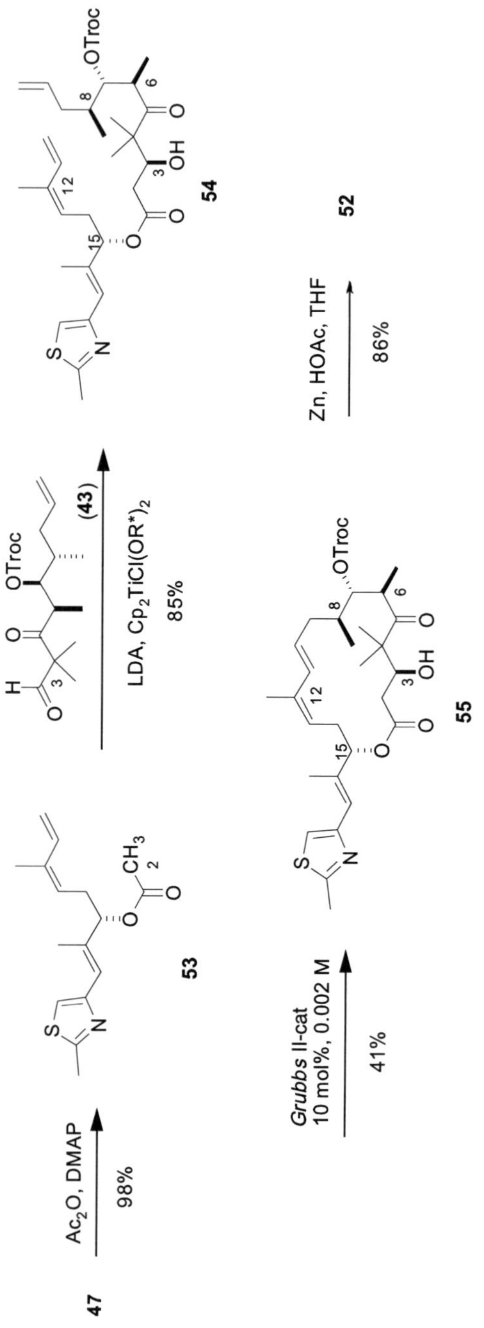

Chart 10. *Danishefsky*'s third RCM synthesis of epothilone D

4.2.2. *Nicolaou Syntheses (43–49)*

The *Nicolaou* group was second to complete total syntheses of both epothilones A and B. Their strategy was focused on an aldol addition between a C6 ketone enolate and a C7 aldehyde. The C12,13-double bond was formed *via* metathesis in the A-series (*43–45*) and *via Wittig* olefination in the B-series (*47*).

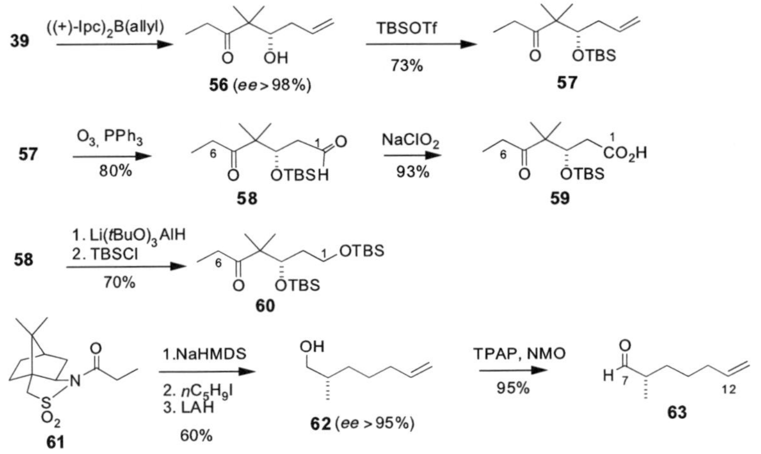

Chart 11. *Nicolaou*'s synthesis of epothilone building blocks

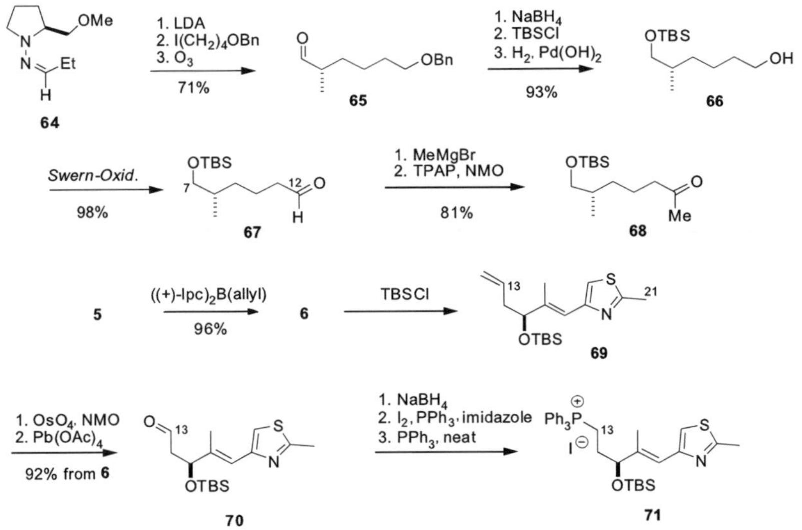

Chart 12. *Nicolaou*'s synthesis of epothilone fragment **71**

The preparation of the building blocks was different from the *Danishefsky* approach. Thus (Chart 11), the C1–C6 fragment **59** was prepared *via* a regio-and enantiocontrolled *Brown* allylation of aldehyde **39** to form **57** after silylation. Ozonolysis led to **58** that was either oxidized to carboxylic acid **59** or reduced to diol **60**. The C7–C12-aldehyde **63** was obtained *via* alcohol **62** by an *Oppolzer* alkylation with sultam **61**.

An alternative route to the C7–C12-fragment (Chart 12) started with an *Enders* alkylation of **64** to form **65** enantioselectively, which was converted into aldehyde **67** (epothilone A series) and ketone **68** (epothilone B series). The C13–C21-phosphonium iodide **71** was prepared *via* a *Brown* allylation of aldehyde **5** to **6**. Silylation to **69** and subsequent oxidation gave aldehyde **70**, from which **71** was obtained in three steps.

Nicolaou's first route to **1a** (*43*) started with a non-stereoselective aldol-type addition of the dianion of **59** to aldehyde **63** (Chart 13). A mixture of diastereomers **72a** and **72b** was obtained, which was esterified with alcohol **6** to give a separable mixture of the di-olefins **73a** and **73b**. RCM of **73a** with *Grubbs*' first-generation catalyst furnished an (*E*)/(*Z*)-mixture of the macrolides **74**, which was separated and desilylated to give **2a**. Epoxidation with DMDO as before led to **1a**.

The RCM approach was also used for a solid-phase supported synthesis of **2a** (Chart 14) (*48*). The ring closure served to disconnect

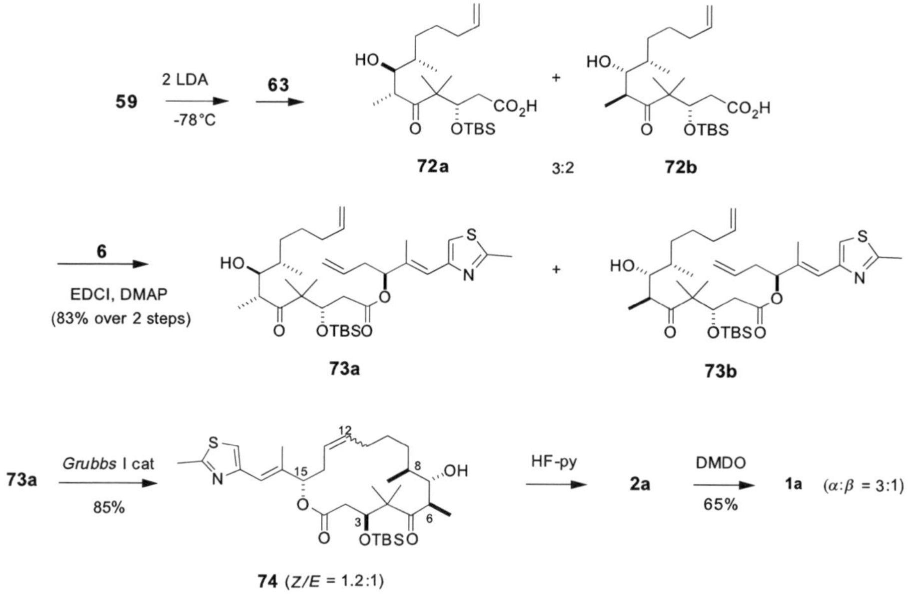

Chart 13. *Nicolaou*'s RCM synthesis of epothilones A and C

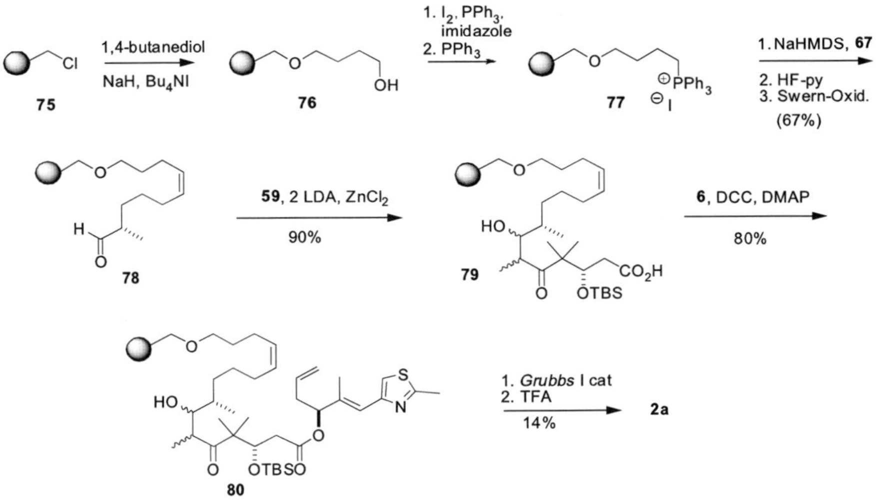

Chart 14. *Nicolaou*'s solid-phase synthesis of epothilone C

the substrate from the solid support by a cyclorelease process. The synthesis started with the *Merrifield* resin attached chloride **75**, which was converted *via* alcohol **76** into phosphonium salt **77**. *Wittig* reaction of **77** with aldehyde **67** followed by desilylation and *Swern* oxidation

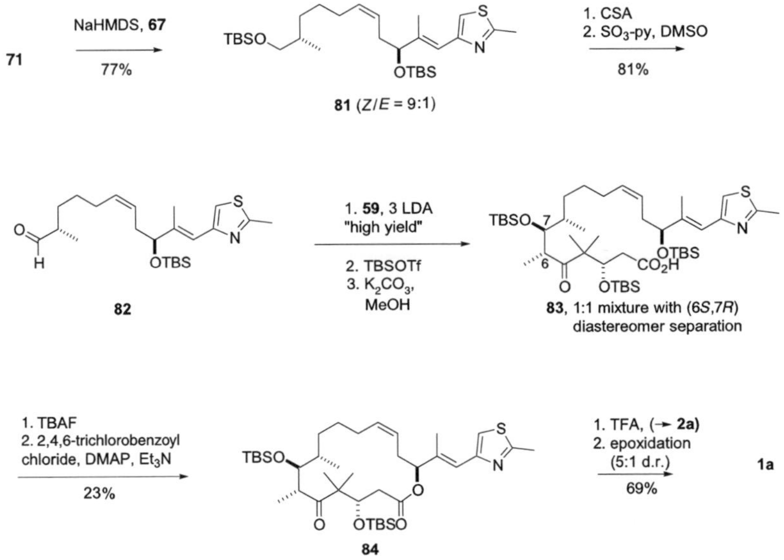

Chart 15. *Nicolaou*'s macrolactonization approach to epothilone A and C

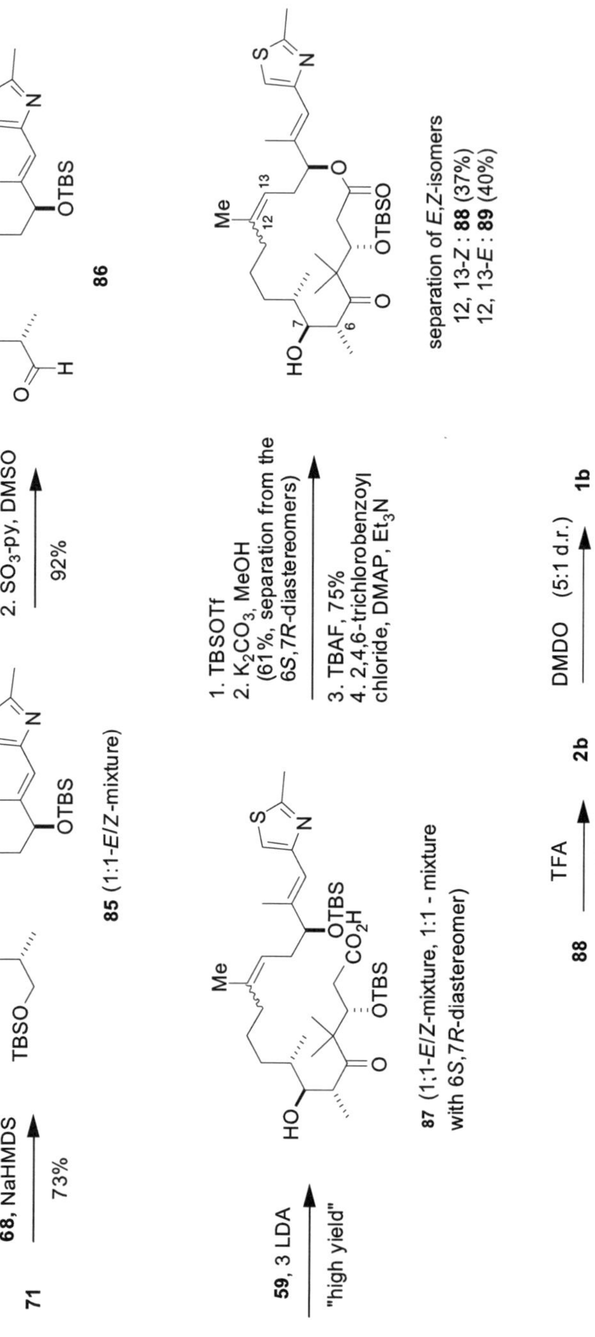

Chart 16. *Nicolaou*'s first synthesis of epothilone B and D

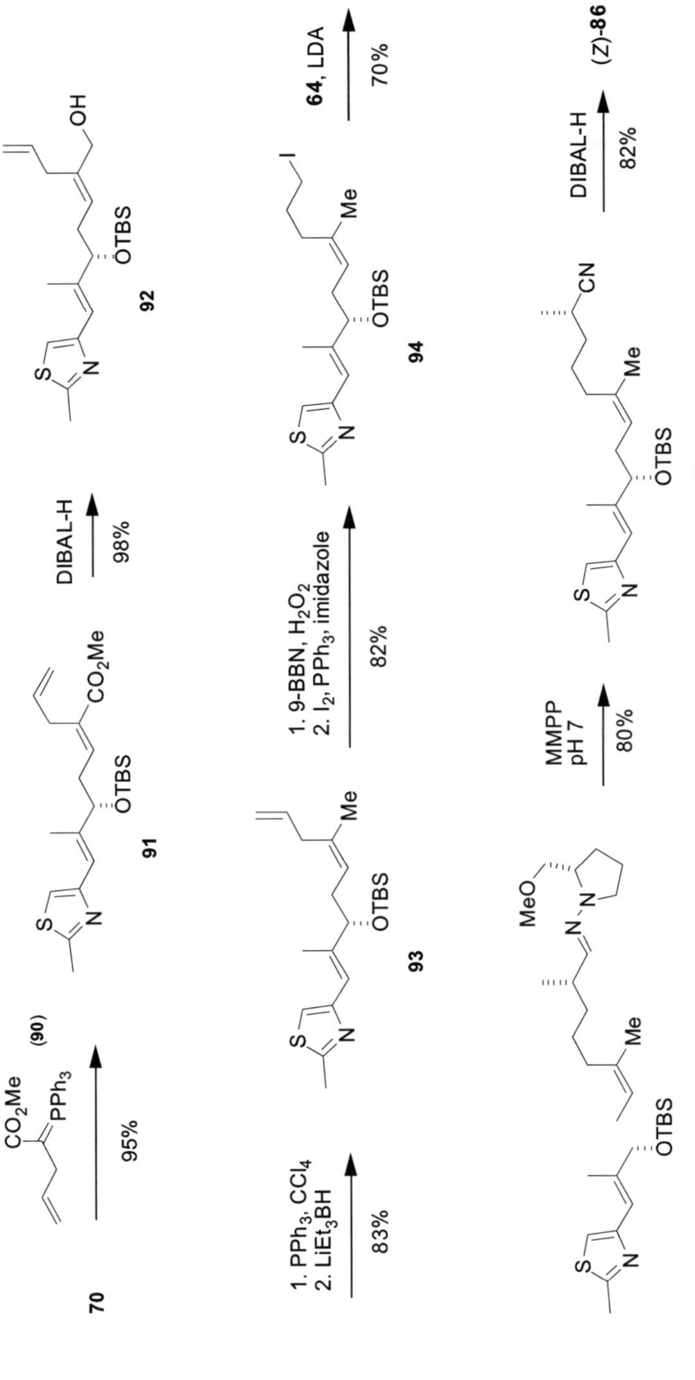

Chart 17. *Nicolaou*'s synthesis of key intermediate (*Z*)-**86**

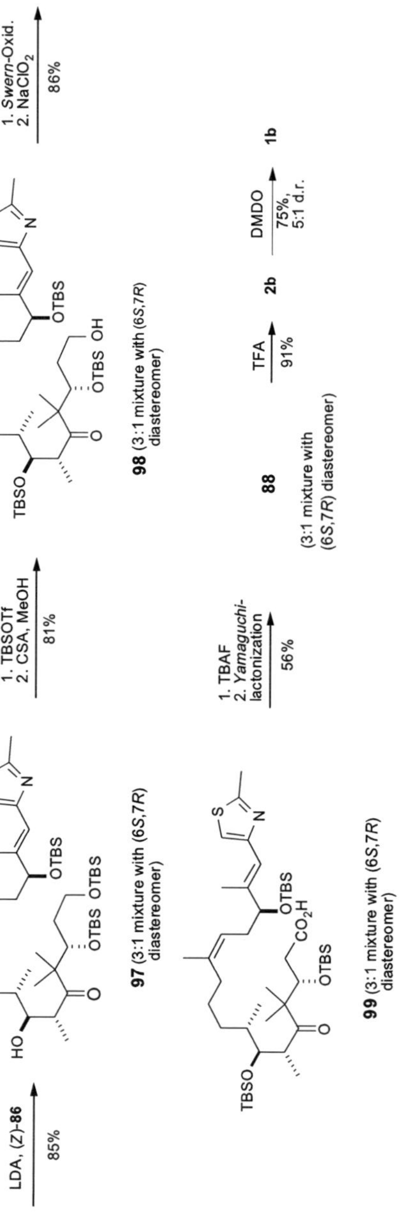

Chart 18. *Nicolaou*'s second synthesis of Epothilone B and D

furnished aldehyde **78**, which was subjected to an aldol addition with keto acid **59** to give carboxylic acid **79** as a mixture of diastereomers. Esterification of **79** with alcohol **6** led to diene **80**, which was converted into **2a** ((*E*)/(*Z*)-mixture) *via* RCM and desilylation.

As an alternative to the RCM, *Yamaguchi* macrolactonization was also used for the synthesis of **1a**, **1c** (Chart 15) (*46*). Thus, *Wittig* reaction of phosphonium salt **71** with aldehyde **67** gave a 9:1-(*Z*)/(*E*)-mixture of olefins **81**. Conversion into aldehyde **82** followed by aldol addition with keto acid **59** led to a 1:1-diastereomeric mixture of aldol adducts **83**, which were separated and subjected to *Yamaguchi* lactonization. Compound **84** was obtained, desilylated to **2a** and then epoxidized to **1a**.

A related protocol was also applied to the synthesis of **1b**, **2b** (Chart 16) (*46*). The only modification was the unselective *Wittig* reaction of **71** with ketone **68**, to give **85** as a 1:1-(*E*)/(*Z*)-mixture that was converted to aldehyde **86**. Aldol addition of **86** with **59** delivered an equimolar mixture of four stereoisomers. The (6*S*,7*R*)-(*E*)/(*Z*)-fraction was removed at this stage and the remaining (6*R*,7*S*)-(*E*)/(*Z*)-olefin mixture was lactonized and then separated into isomers **88** and **89**. The (*Z*)-isomer **88** was then converted into **1a**, **2a**, as usual.

To improve the stereocontrol of this sequence two alterations were undertaken (*46*, *47*). First, aldehyde **86** was prepared as a pure (*Z*)-olefin *via* the sequence shown in Chart 17. The (*E*)-selective *Wittig* olefination of aldehyde **70** with phosphorane **90** gave pure (*E*)-enoate **91** that was reduced to alcohol **92** and deoxygenated to (*Z*)-olefin **93**. Regioselective hydroboration/iodination of the terminal olefin gave iodide **94**, which was used for an enantiocontrolled alkylation with SAMP derivative **64** to furnish **95** selectively. Oxidative removal of the hydrazone gave nitrile **96**, which was reduced to aldehyde (*Z*)-**86** with DIBAL-H.

The aldehyde (*Z*)-**86** was subjected to aldol addition with ketone **60** (Chart 18) to give a 3:1-diastereomeric mixture of **97** in favor of the desired (6*R*,7*S*)-diastereomer. Without separation, the 7-OH group was silylated, the primary alcohol at C1 was deprotected selectively and oxidized to the acid **99**. Selective deprotection of the 15-OTBS group followed by *Yamaguchi* lactonization gave **88**, which was separated into the diastereomers. Deprotection gave **2b** and epoxidation led to **1b** eventually.

4.2.3. Schinzer Synthesis (50–54)

Schinzer's group was chronologically third to complete syntheses of both **1a**, **1b** and **2a**, **2b**, which were conceptually very similar to the *Nicolaou*'s approaches. However, the preparation of the key fragments was different (*50*). For instance the C21–C13-fragment **6** was syn-

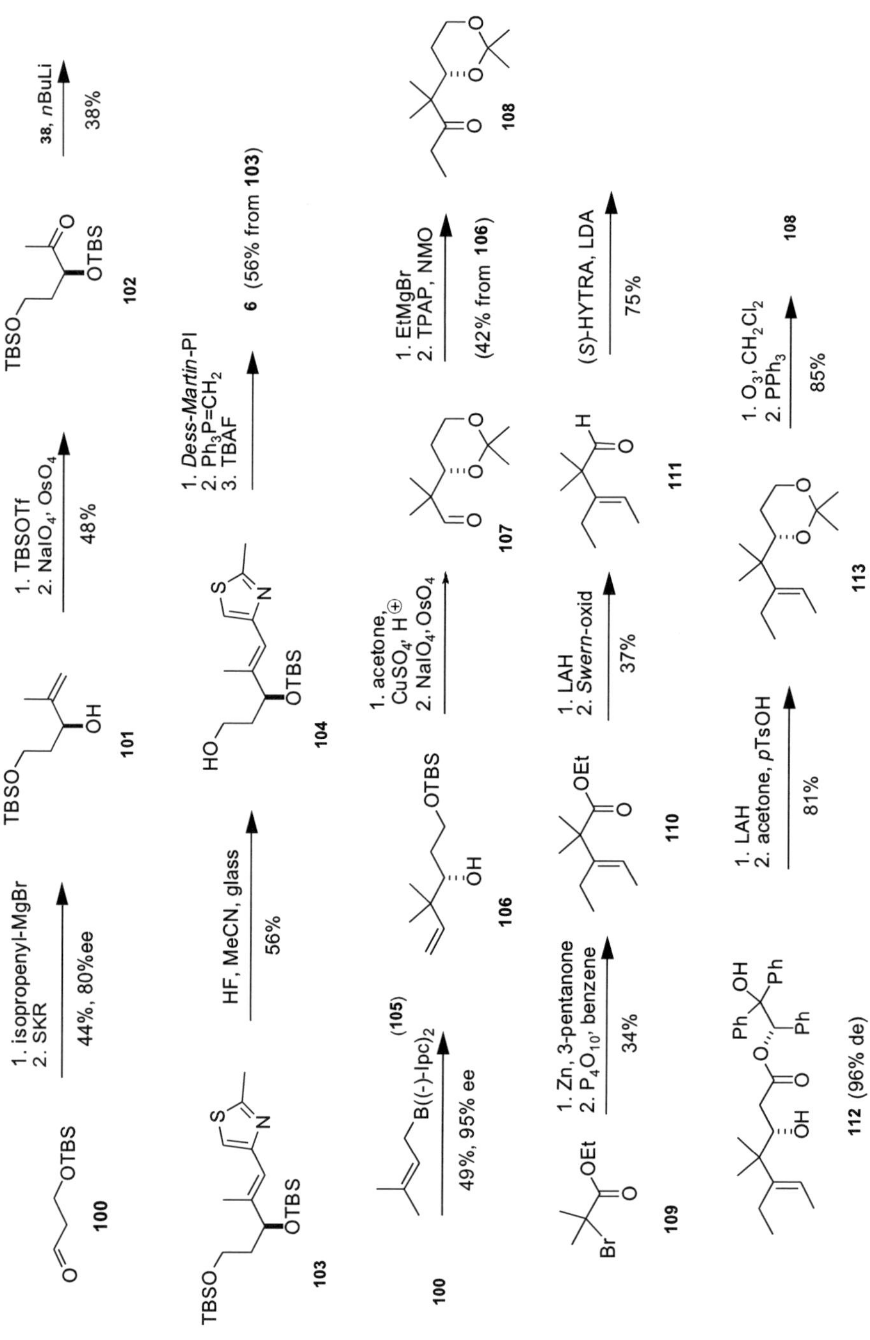

Chart 19. *Schinzer*'s synthesis of epothilone A, Part I

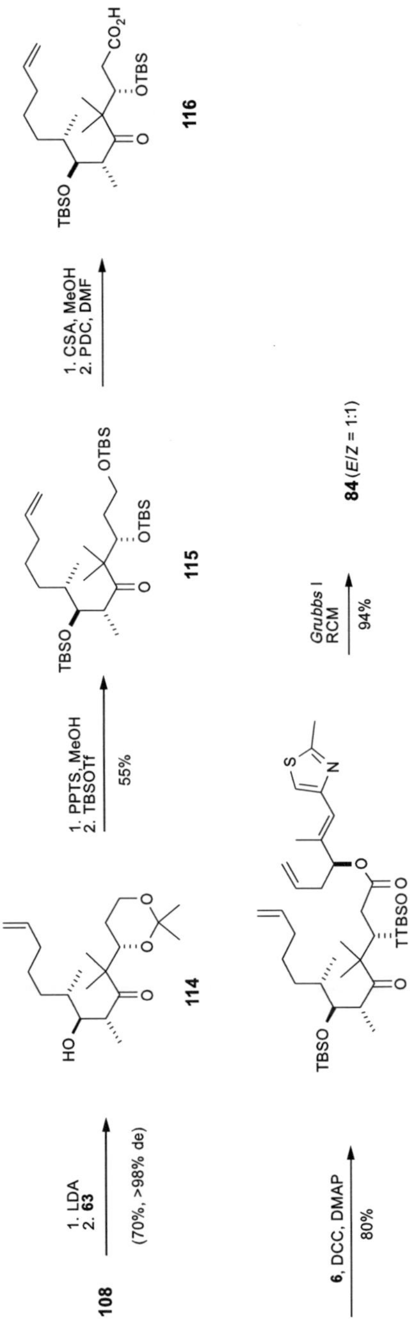

Chart 20. *Schinzer*'s synthesis of epothilone A, Part II

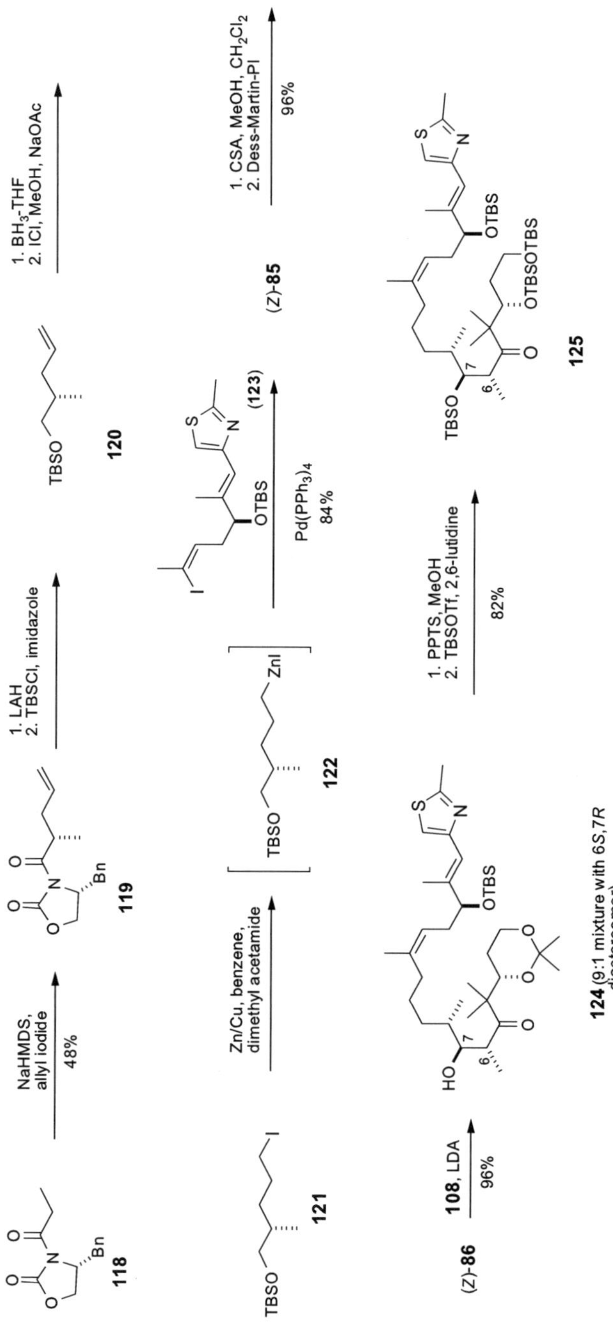

Chart 21. *Schinzer*'s synthesis of epothilone B

thesized from aldehyde **100** (Chart 19), which was converted to the non-racemic alcohol **101** in two steps. Next, the ketone **102** was prepared and subjected to an (*E*)-selective *Horner* olefination with phosphonate **38** to give **103**, which was transformed into C15-alcohol **6** *via* **70** in four steps. The C1–C5-acetonide **108** was available from **100** *via Brown* allylation with borane **105** to provide **106**, which was then converted into **108**. A longer yet more convenient preparation of **108** was started with *Reformatsky* addition of **109** to 3-pentanone to give ester **110**, which was reduced to aldehyde **111** and converted into **112** *via* a *Braun* aldol addition. Four more steps gave **108**.

The acetonide moiety in **108** turned out to be crucial for the stereoselectivity of the aldol addition to aldehyde **63** (Chart 20). In contrast to *Nicolaou*'s experiment a high selectivity in favor of adduct **114** was achieved. The rest of the synthesis (*51*, *52*) closely resembles *Nicolaou*'s RCM approach. Thus, conversion of **114** to acid **116** was followed by esterification with alcohol **6** to give di-olefin **117**, which was cyclized to *Nicolaou*'s intermediate **84** as an 1:1-*E/Z*-mixture.

Schinzer's synthesis of **1b**, **2b** (Chart 21) (*53*, *54*) uses a Zn-modification of *Danishefsky*'s B-alkyl *Suzuki* coupling by converting olefin **120** (obtained *via* allylation of *Evans*' oxazolidinone **118**) into the Zn-derivative **122**. Palladium-assisted coupling of **122** with vinyl iodide **123** gave (*Z*)-**85**, which was transformed into (*Z*)-aldehyde **86**. The aldol addition of (*Z*)-**86** with ketone **108** again was highly stereoselective to form adduct **124** with a d.r. of 9:1. Acetonide hydrolysis followed by global O-silylation delivered *Nicolaou*'s intermediate **125**.

4.2.4. Sinha Syntheses (55–58)

In their synthesis of **2a**, which incorporates modifications of both *Nicolaou*'s RCM and *Danishefsky*'s macrolactonization approach, the *Sinha* group has made use of "in-house" catalytic antibody methodology (Charts 22, 23) (*55–57*). Thus, a racemic mixture of aldol adducts **127a** and **127b** was prepared and subjected to retro-aldol cleavage catalyzed by antibody AB 38C2. Enantiomer **127a** was obtained in 96% *ee*. Conversely, aldehyde **5** was converted into adduct **128** *via* antibody-catalyzed aldol addition with acetone, however in low conversion and only 75% *ee*. Compound **127a** was converted into the C1–C10-segment of **1** *via* catalytic hydrogenation to give a separable mixture of diastereomers **129**, **130**. Pure **129**, after O-silylation and methylation was subjected to an aldol addition to give after silylation and diastereomer separation intermediate **131** in unspecified yield. The phenol ring was oxidized to the carboxylic acid and then converted into C1–C10-aldehyde **132** (Chart 23).

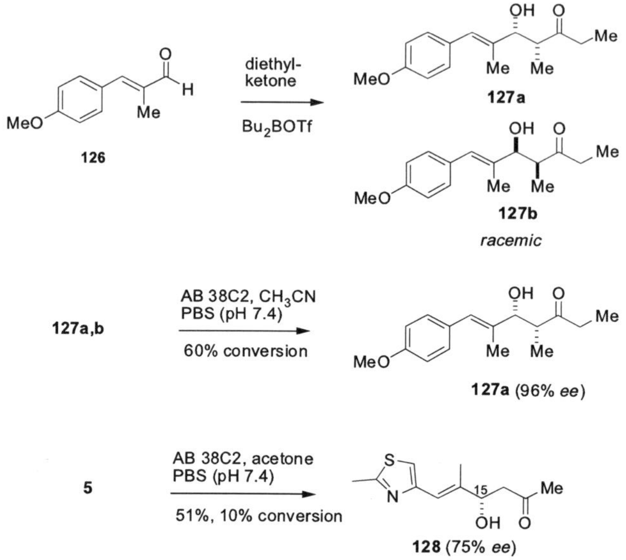

Chart 22. *Sinha*'s synthesis of epothilone C, Part I

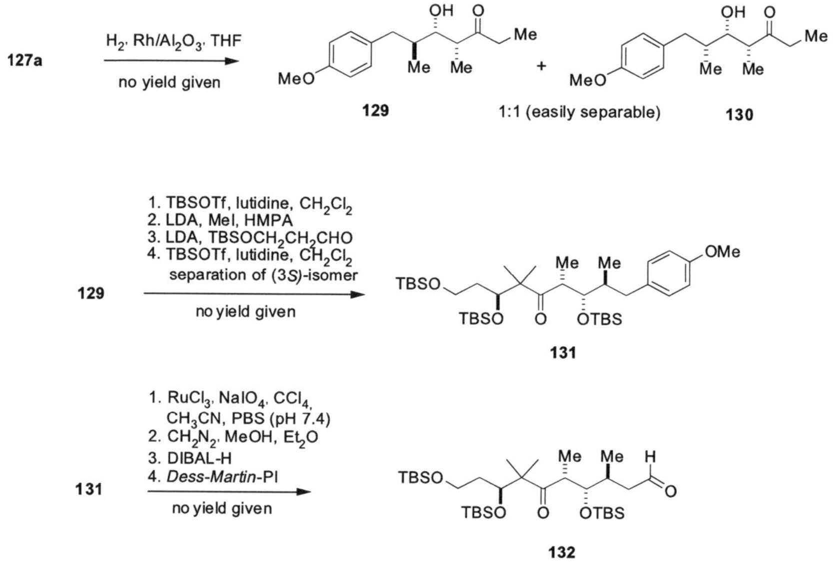

Chart 23. *Sinha*'s synthesis of epothilone C, Part II

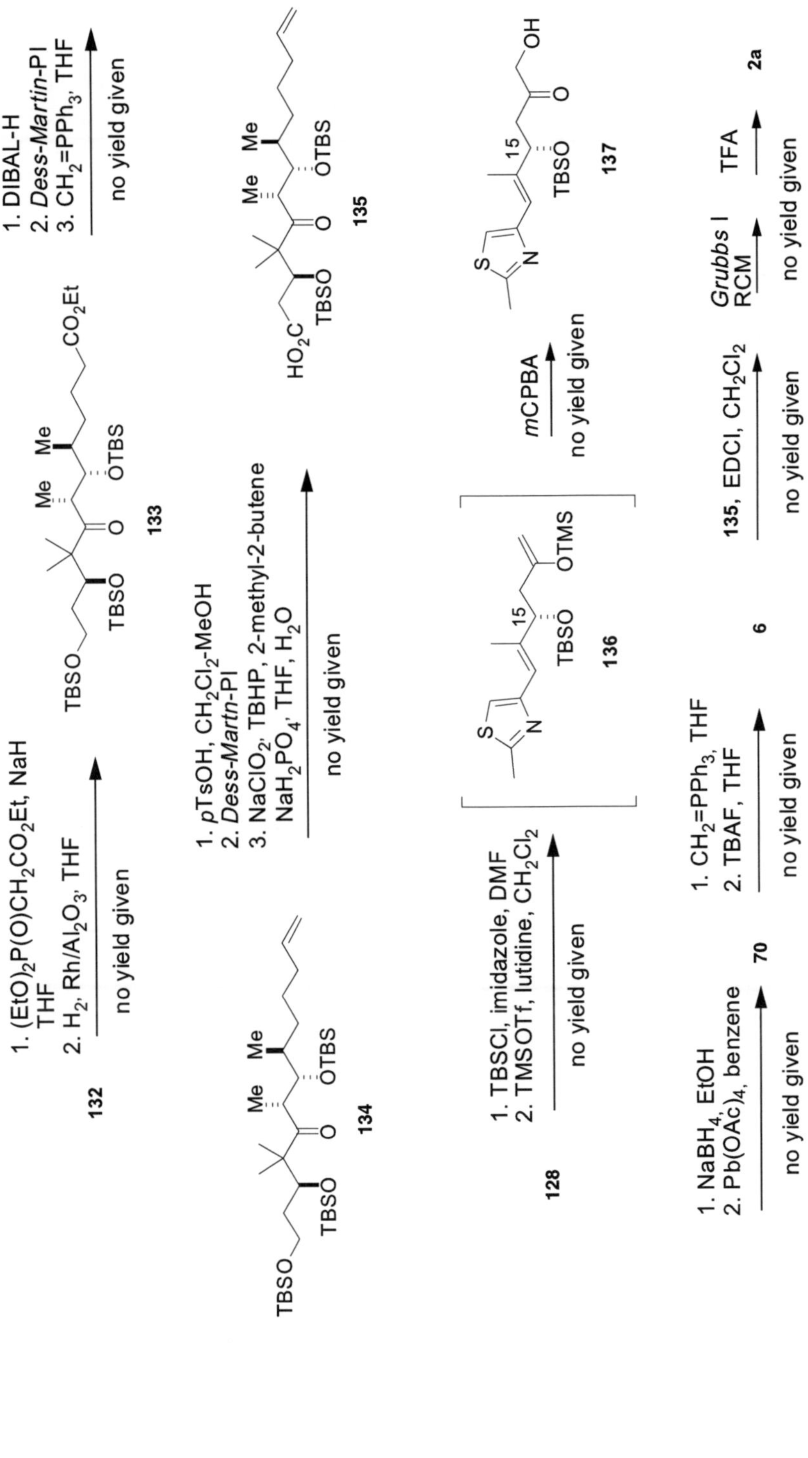

Chart 24. *Sinha's* synthesis of epothilone C, Part III

Via Horner olefination and hydrogenation **132** was transformed into ester **133**, which was then elaborated *via* **134** into the olefinic carboxylic acid **135** (Chart 24). Aldehyde **6** was prepared from **128** *via* enol ether **136** and hydroxyl ketone **137** and olefinated to alcohol **69**. Esterification with acid **135** followed by RCM and desilylation led to **2a**, by analogy to *Nicolaou*'s earlier approach (Chart 13).

In a second approach to **2a** (Chart 25), which parallels *Danishefsky*'s synthesis (Chart 4), the *Sinha* group prepared olefin **138** from **132** and used this material for a *Suzuki* coupling with the (*Z*)-vinyl iodide derived from **70** to give, after desilylation and oxidation *Danishefsky*'s *seco*-acid **24**. For all steps described so far yields are missing.

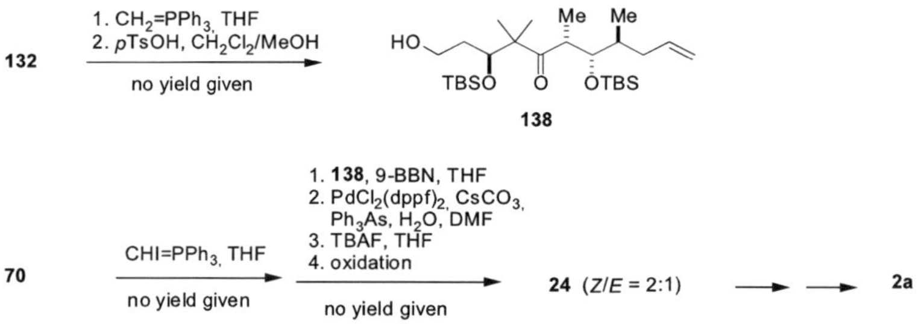

Chart 25. *Sinha*'s first synthesis of epothilone B

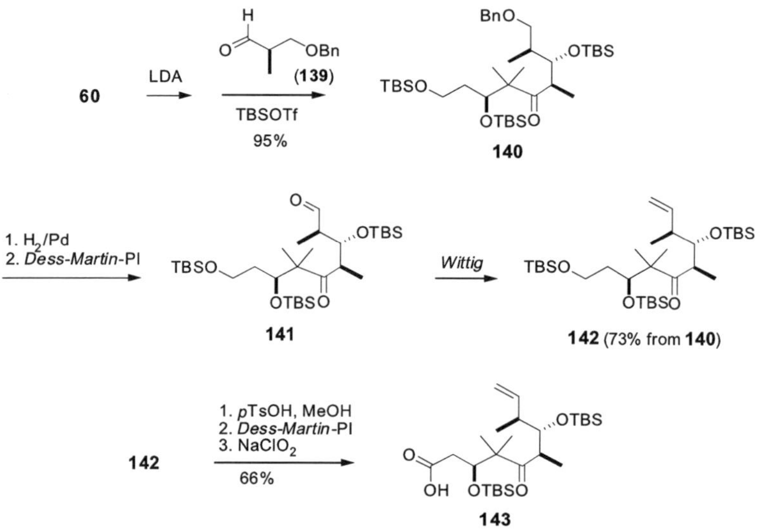

Chart 26. *Sinha*'s second synthesis of epothilone B, Part I

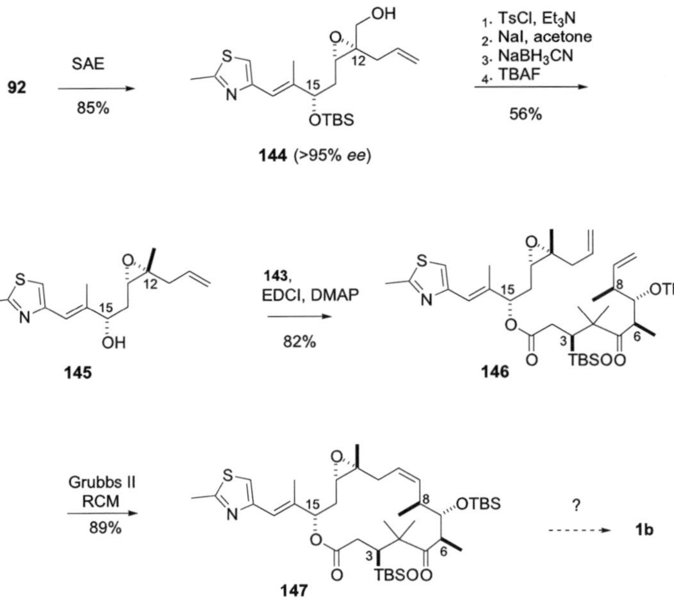

Chart 27. *Sinha*'s second synthesis of epothilone B, Part II

For their synthesis of **2b**, the *Sinha* group avoided antibody methodology and applied a esterification-RCM sequence (Charts 26 and 27) (*58*). The novel feature of this synthesis is the early introduction of the 12,13-epoxide function. However, the authors failed to mention that this approach has been pioneered by the *Mulzer* group (*21*). Following a protocol by *Taylor* (*59*), *Sinha* started with a highly selective aldol-type addition of ketone **60** to aldehyde **139** to give adduct **140**, which was then transformed *via* aldehyde **141** and olefin **142** into olefinic acid **143** (Chart 26). Alcohol **145**, which is required for the esterification with **143**, was obtained from *Nicolaou*'s intermediate **92** (Chart 17) and epoxidized to **144** *via* a highly stereoselective *Sharpless*' asymmetric epoxidation reaction. The superfluous primary OH group was removed reductively to give **145**, after desilylation of the 15-OTBS group. Ester **146** was then formed from **143** and **145** and cyclized to macrolide **147** with *Grubbs* second-generation catalyst. The conversion to **1b** was not described in detail.

*4.2.5. Carreira's Synthesis of **2a** and **2b** (60–62)*

The *Carreira* group has also opted for the "early epoxide" approach originally introduced by the *Mulzer* group (*86*, *87*). Moreover, they

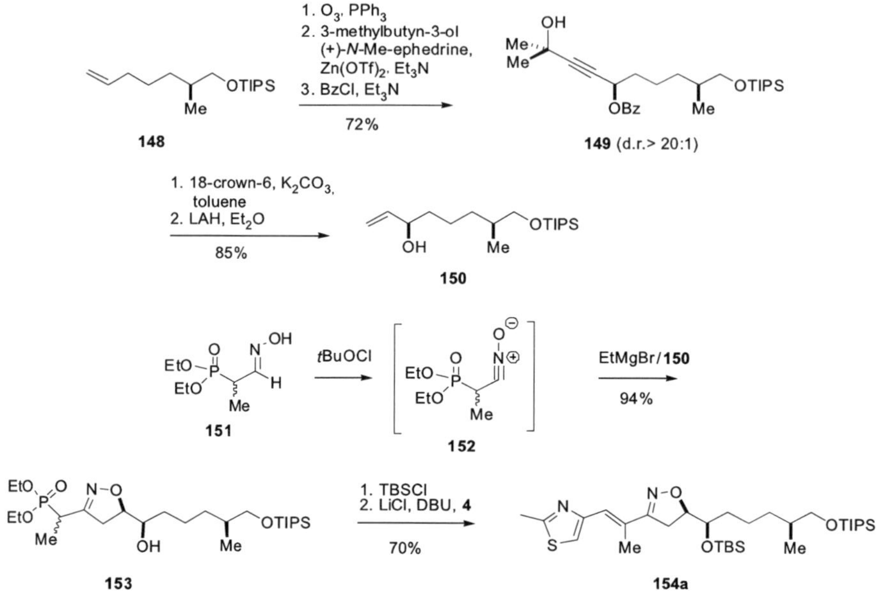

Chart 28. *Carreira*'s synthesis of epothilone A and B, Part I

developed a highly innovative nitrile-oxide-olefin cycloaddition to establish the C12–C15 section in both the epothilone A and B series. Thus (Chart 28), olefin **148** was oxidized to the aldehyde and treated with 3-methylbutyn-3-ol under asymmetric catalysis to give **149** with high selectivity. Additions of this type have been developed earlier in the group. Removal of the terminal teriary alcohol and reduction gave the

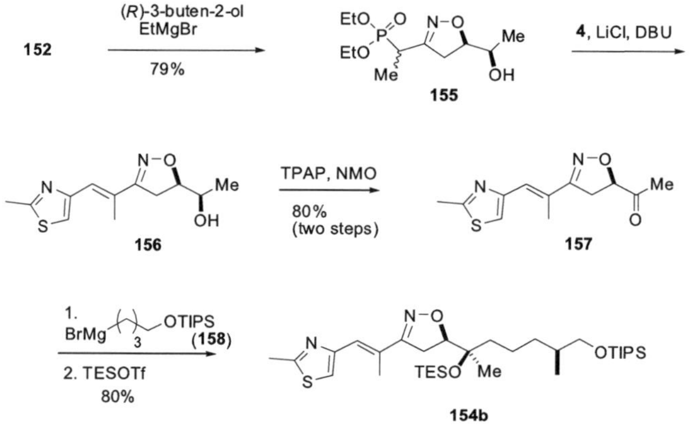

Chart 29. *Carreira*'s synthesis of epothilone A and B, Part II

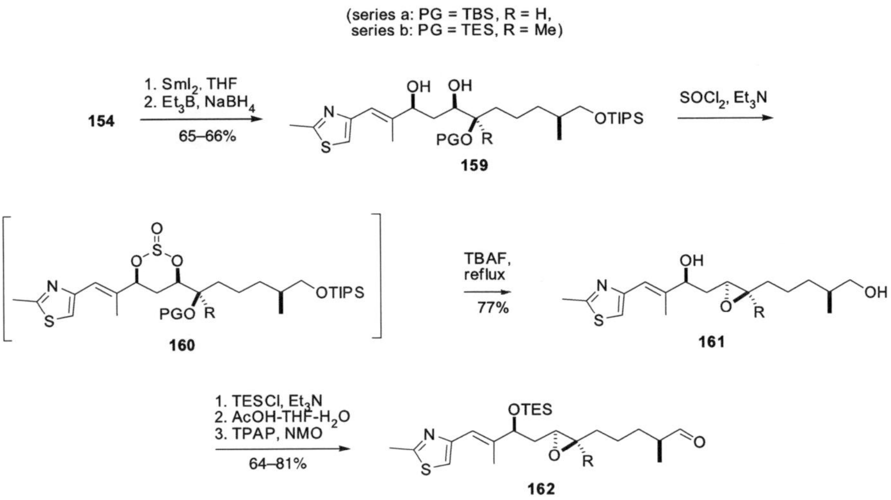

Chart 30. *Carreira*'s synthesis of epothilone A and B, Part III

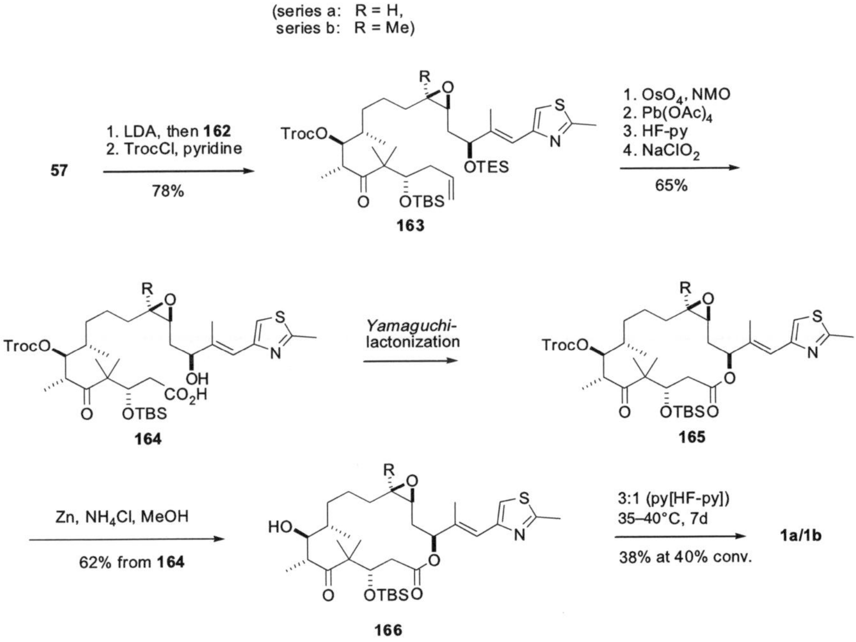

Chart 31. *Carreira*'s synthesis of epothilone A and B, Part IV

vinyl alcohol **150**, which was subjected to a stereocontrolled 1,3-dipolar cycloaddition with nitrile oxide **152**, prepared *in situ* from oxime **151**. The resulting isoxazoline **153** was olefinated with aldehyde **4** to give the C21–C7-epothilone A fragment **154a**.

The corresponding epothilone B intermediate **154b** (Chart 29) was obtained starting from the addition of **152** to (*R*)-3-buten-2-ol. Adduct **155** was converted to ketone **157**. Chelate-*Cram* induced addition of *Grignard* derivative **158** gave **154b** selectively.

In parallel sequences (Charts 30 and 31) intermediates **154a** and **154b** were processed towards the formation of **1a**/**1b**. Specifically, the isoxazoline ring was cleaved reductively to give triol **159**, which was converted into epoxide **161** *via* the cyclic sulfite **160**. Ditesylation followed by selective mono-detesylation and oxidation of the primary OH-function gave aldehyde **162**, which underwent a highly selective aldol addition with ketone **57** to give aldol adduct **163** after Troc-protection of the resulting C7-alcohol function. The endgame was modelled after *Mulzer*'s synthesis (*86*, *87*) to provide **166** *via seco*-acid **164** and macrolactone **165**. The final desilylation required elaborate conditions to give **1a** and **1b** eventually.

4.2.6. Shibasaki Approach (63–68)

The *Shibasaki* group has developed a variety of chiral multifunctional catalysts (*e.g.* **167–169**, Fig. 2) that they apply to natural product synthesis (*63*). Their epothilone A/B synthesis (Charts 32–35) is an adaptation of *Danishefsky*'s B-alkyl *Suzuki* reaction of vinyl iodides **9**, **10** and terminal olefin **187** (*64*). Thus (Chart 32), aldehyde **5** underwent a highly enantioselective cyanosilylation with catalyst **167** (Fig. 2) (*65*) to form **170**, which was converted into aldehyde **171** and then alkyne **173**. Hydromagnesiation-iodination of **173** was used to generate *Danishefsky*'s epothilone A vinyl iodide **9** stereoselectively. The preparation of the epothilone B vinyl iodide **10** (*66*) involves a homologation of aldehyde **171** to *Schinzer*'s aldehyde **70**, which was then elaborated into **10** following the *Danishefsky*/*Schinzer* precedence.

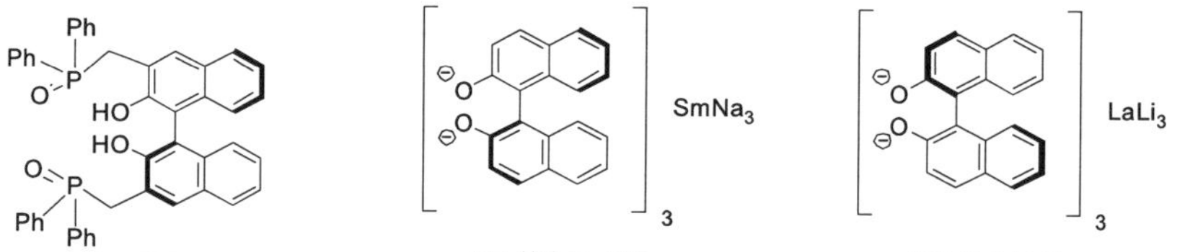

167 **168** ((*S*)-SmSB) **169** ((*R*)-LLB)

Fig. 2. *Shibasaki*'s catalysts

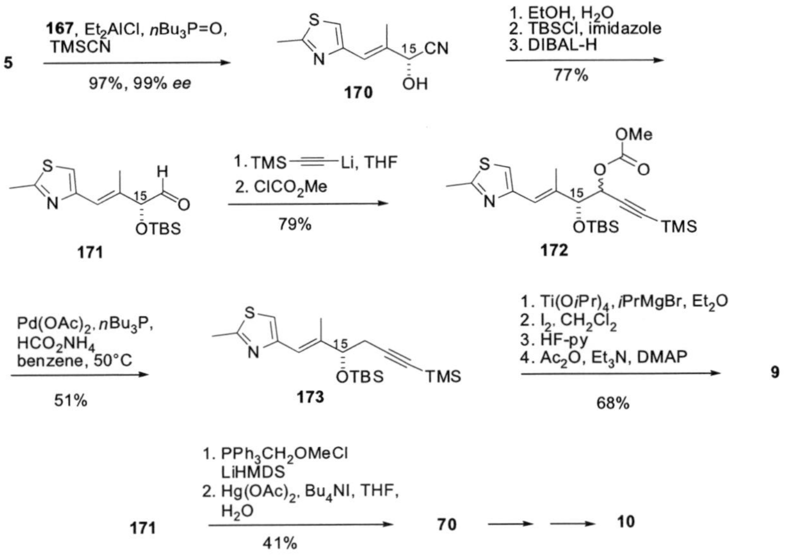

Chart 32. *Shibasaki*'s synthesis of epothilones C and D, Part I

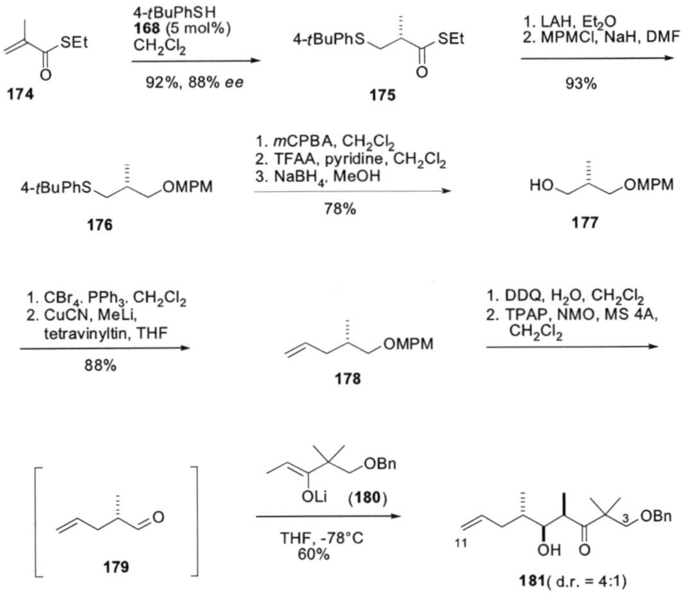

Chart 33. *Shibasaki*'s synthesis of epothilones C and D, Part II

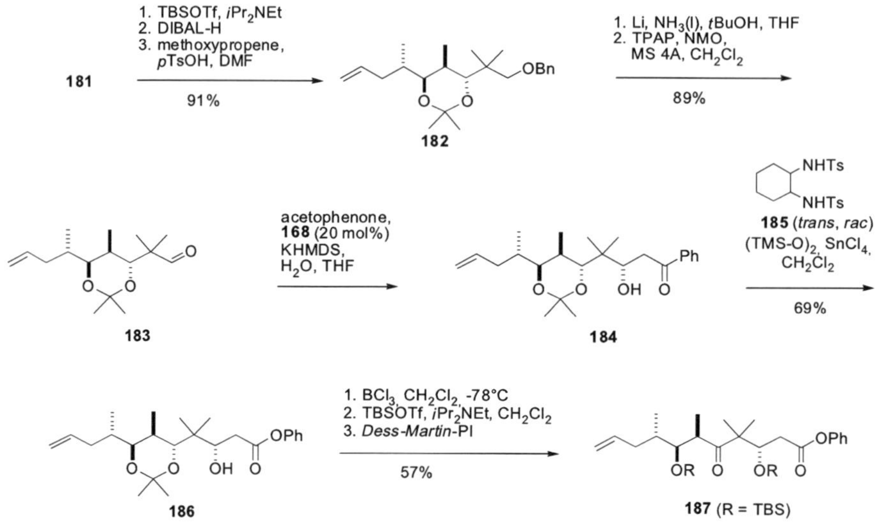

Chart 34. *Shibasaki*'s synthesis of epothilones C and D, Part III

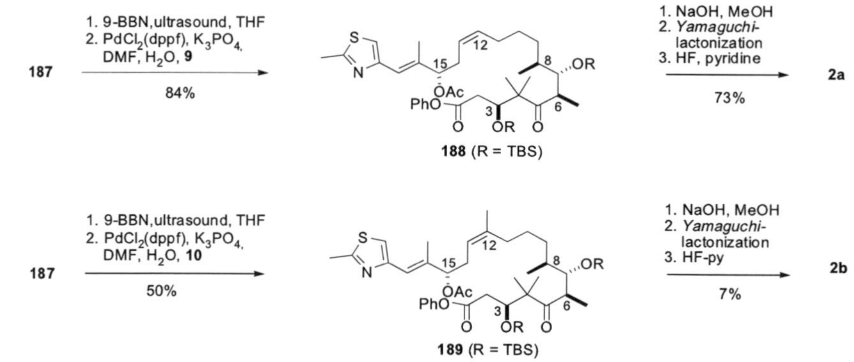

Chart 35. *Shibasaki*'s synthesis of epothilones C and D, Part IV

The synthesis of **187** (Charts 33 and 34) started with a 1,4-addition of 4-*t*BuPh-SH to thiol ester **174** under the enantiocatalysis of **168** to give **175**. Reduction of the thiol ester and O-protection gave thioether **176**, which was converted into **177** *via Pummerer* oxidation-reduction. Chain elongation to olefin **178** was followed by oxidation to the labile aldehyde **179**, which was added to enolate **180** to give the epothilone A/B C3–C11-fragment **181** (Chart 33). Standard methodology was used to convert **181** into acetonide **182** and then aldehyde **183** (Chart 34). Aldol addition with acetophenone under the catalysis of **169** gave phenyl ketone **184**, which was converted to phenyl ester **186** *via* a novel *Baeyer-Villiger* oxidation, catalyzed by diamide **185**. In three additional steps ketone **187** was prepared, which was the coupling partner of **9** and **10** to give the **2a**-precursor **188** and the **2b**-precursor **189** (Chart 35).

4.3. Syntheses of Epothilone A/C (1a, 2a)

4.3.1. Fürstner's Alkyne RCM (67, 68)

The early RCM approaches to **1** by *Danishefsky*, *Nicolaou*, and *Schinzer* were all flawed by the missing (*E*)/(*Z*)-selectivity. In their synthesis of **2a**, *Fürstner et al.* (Chart 36) remedied the situation by applying an alkyne RCM approach to generate macrolide **198**, which was converted into the (*Z*)-olefin exclusively *via Lindlar* hydrogenation. The synthesis started from keto ester **190**, which was converted into acetonide **191** *via Noyori* hydrogenation. The addition of ethyl magnesium bromide led to *Schinzer*'s ketone **108**, which was used for an aldol addition with aldehyde **192**, prepared *via* methylation of the corresponding *Oppolzer* sultam. Adduct **193** was obtained with good selectivity and then converted into acid **194**. Esterification of **194** with alcohol **195**, prepared from *Nicolaou*'s intermediate **6** *via Corey-Fuchs* homologation to introduce the alkyne moiety, gave *seco*-intermediate **196**. The RCM was achieved in good yield with Mo-catalyst **197**.

4.3.2. Liu Synthesis (69, 70)

The *Liu* approach to epothilone A (Chart 37) follows the *Nicolaou* retrosynthesis. Original contributions consist in a novel approach to *Nicolaou*'s aldehyde **82** and the use of the PMB protecting group for the 3-OH function.

The synthesis started with the preparation of the epoxide **199** (fragment C14–C16) from buten-2-ol. Geraniol (**200**) was then converted into diol **201** *via* a known sequence. Routine steps *via* alcohol **203** led to alkyne **204** that was coupled with **199** to give **205**. *Lindlar* hydrogenation generated the C12,13-olefin **206** *Z*-selectively. Debenzylation, oxidation to ketone **207** and (*E*)-selective *Wittig* olefination with the tributyl phosphorane generated from **208** gave acetonide **209**, which was converted into aldehyde **82** by glycol cleavage of the terminal diol function. In a second sequence (Chart 38), keto-acid **214** was prepared from **210** *via* racemic epoxide **211** and *Jacobsen* kinetic resolution (*71*, *72*) to give **212**. Cobalt carbonyl induced carbonylation in methanol led to methyl ester **213**, which was converted into **214**, and connected with

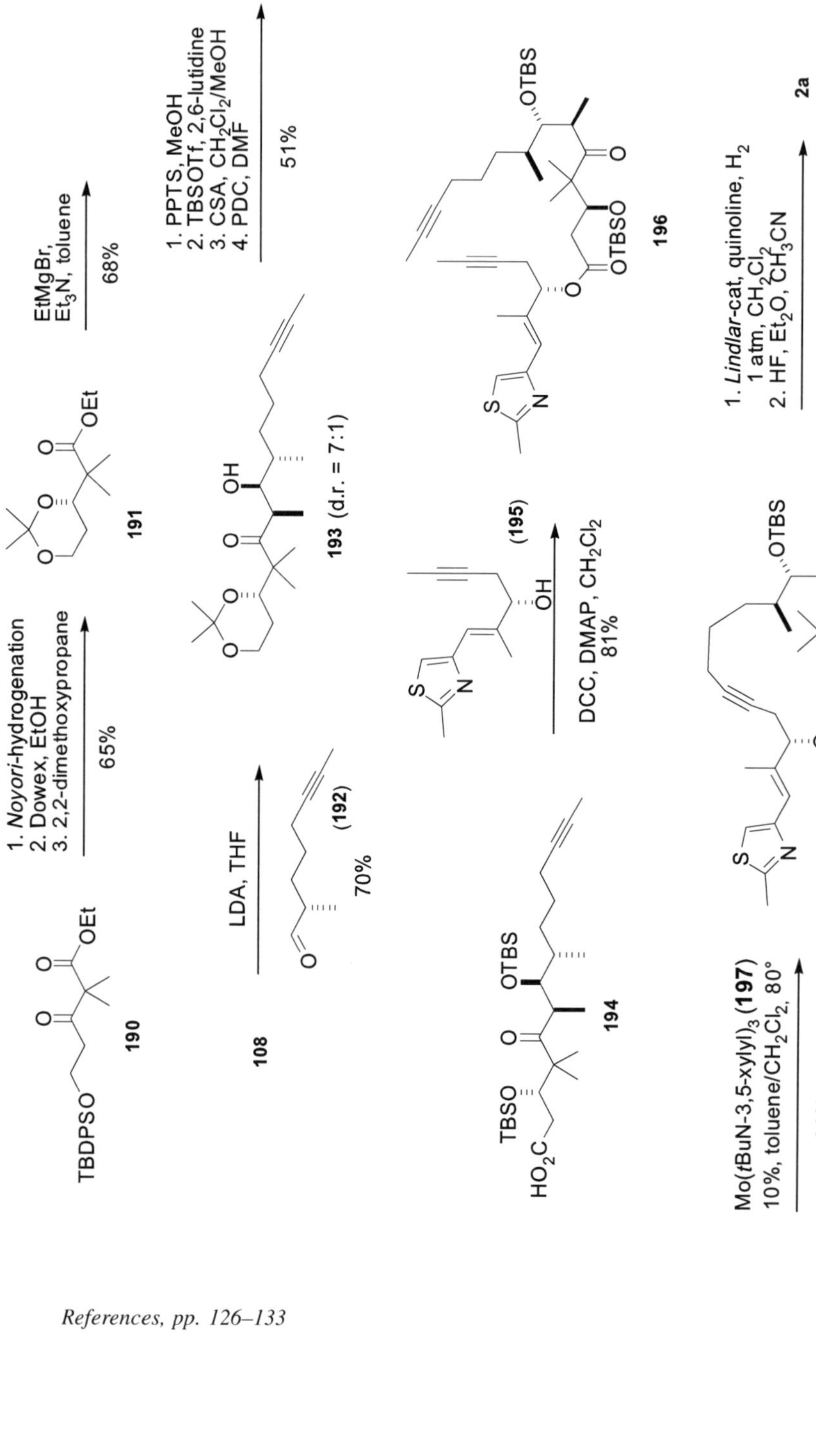

Chart 36. *Fürstner*'s alkyne metathesis approach to epothilone C

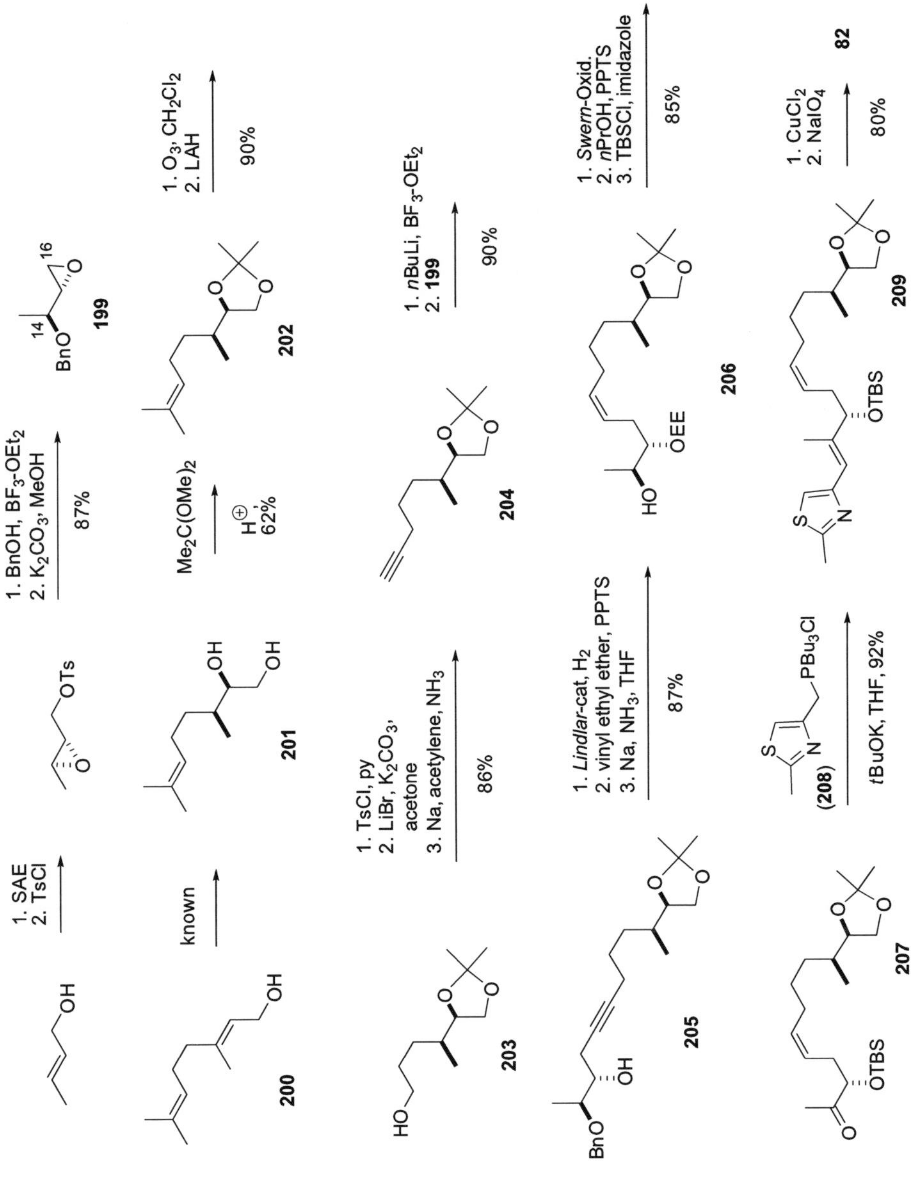

Chart 37. *Liu's* synthesis of epothilone A and C, Part I

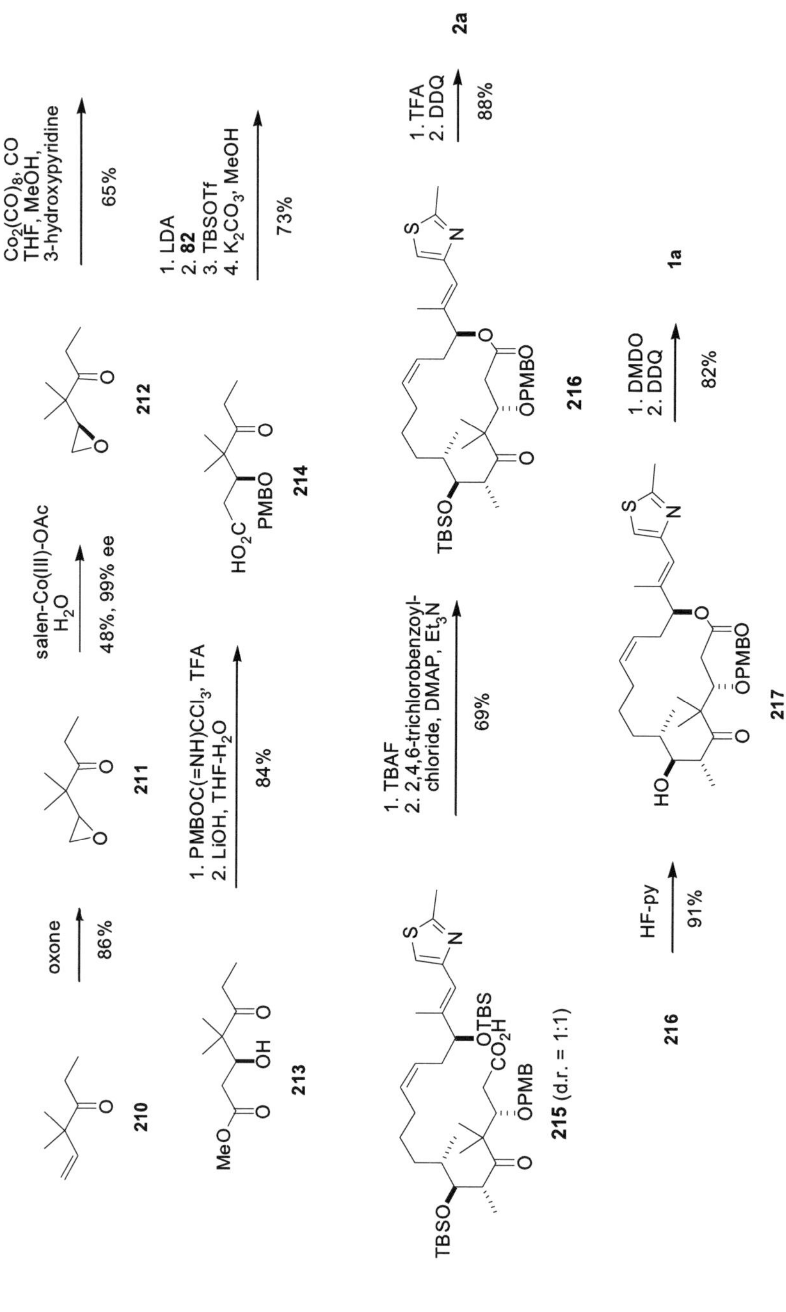

Chart 38. *Liu*'s synthesis of epothilone A and C, Part II

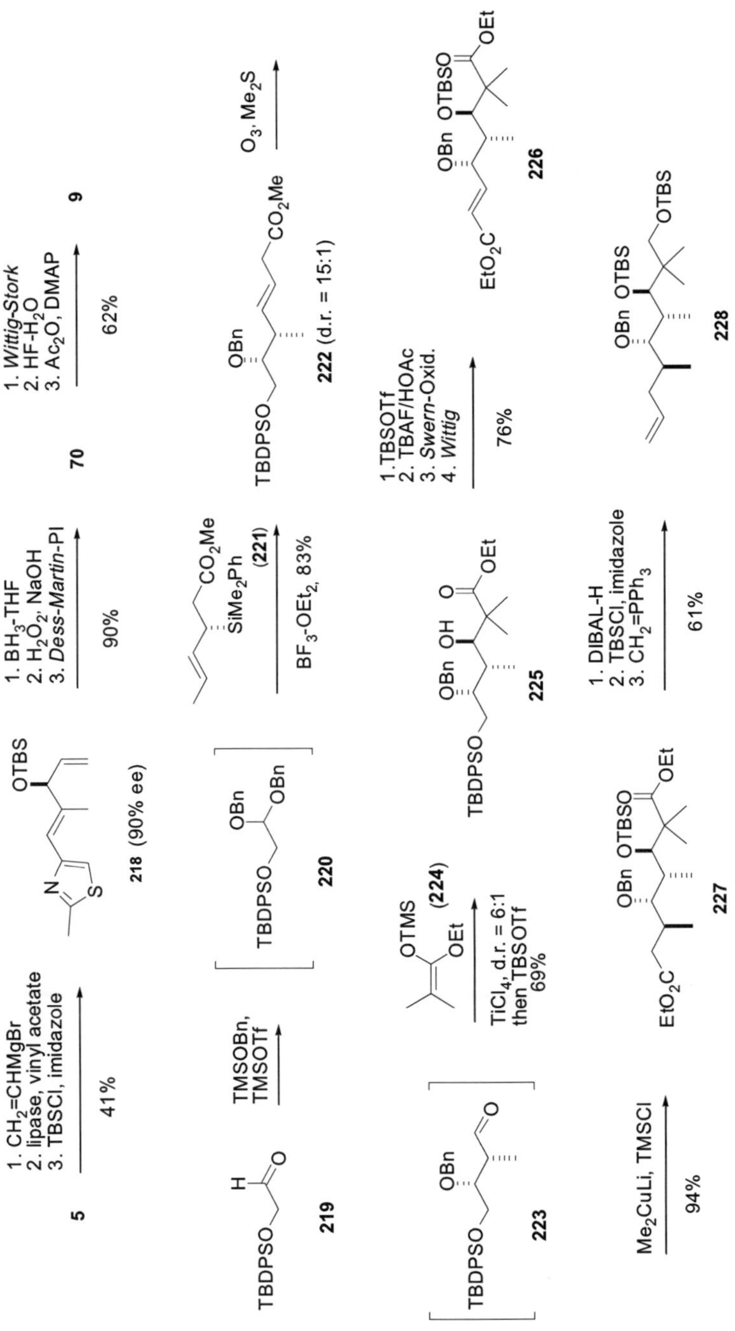

Chart 39. *Panek*'s synthesis of epothilone A and C, Part I

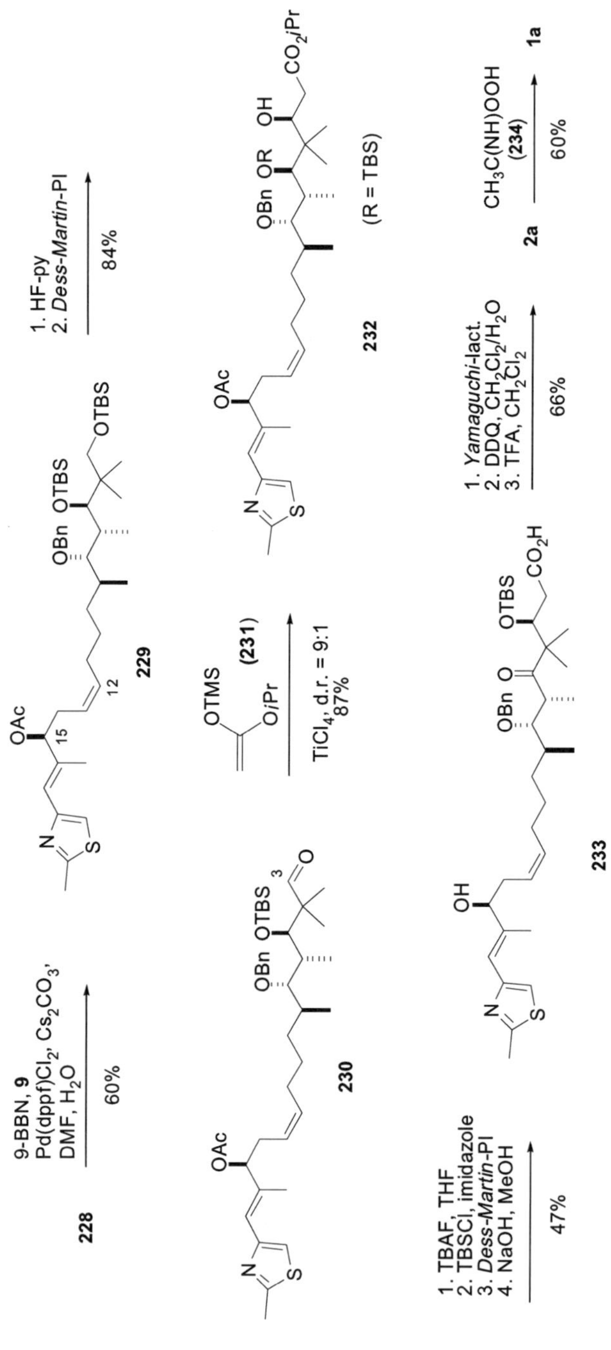

Chart 40. *Panek*'s synthesis of epothilone A and C, Part II

aldehyde **82** *via* an unselective aldol addition. In this way, *seco*-acid **215** was generated, lactonized to macrolide **216** and deprotected to furnish **2a**. The epoxidation to **1a**, however, was performed in a totally stereocontrolled manner *via* PMB – derivative **217**.

4.3.3. Panek Approach (73–75)

Danishefsky's B-alkyl *Suzuki* coupling was used to form the (*Z*)-12,13-double bond. Original methodology, based on asymmetric crotylation with chiral silanes, was applied to generate the stereogenic centers at C6 and C7 (Charts 39 and 40). In a novel approach, vinyl iodide **9** was prepared from aldehyde **5** by adding vinyl magnesium bromide and resolving the resulting racemic allylic alcohol *via* a lipase-catalyzed acetylation. The unreacted alcohol (90% *ee*) was converted into the TBS-ether **218**. Hydroboration and oxidation gave aldehyde **70** that was converted into vinyl iodide **9** as before (Chart 2). For the coupling the C3–C11-olefin fragment **228** was prepared from aldehyde **219** whose di-benzyl-acetal **220** was subjected to a highly stereoselective crotylation with *Panek*'s chiral silane **221**. The resulting olefin **222** was oxidized to aldehyde **223**. *Mukaiyama* addition of ketene acetal **224** led to hydroxyl-ester **225**, from which enoate **226** was prepared. *Hanessian*'s protocol was used to add the methyl group at C6 and the resulting ester **227** was transformed into the desired olefin **228**. Coupling with **9** was achieved by the *Danishefsky* procedure to generate the C3–C21-intermediate **229** (Chart 40). Conversion into aldehyde **230** was followed by another *Mukaiyama* aldol addition, now with ketene acetal **231**, to generate hydroxy ester **232**, which was transformed into *seco*-acid **233** and then into **2a** along the familiar route. Epoxidation to **1a** was performed with imino peracetate **234** (generated *in situ* from acetonitrile and hydrogen peroxide) with good selectivity.

4.3.4. Wong's DERA Approach (76, 77)

The *Wong* synthesis of **1a** also focuses on *Danishefsky*'s B-alkyl *Suzuki* coupling by using vinyl iodide **9** and the C1–C11-olefin **244** (Chart 41). The synthesis of both fragments employs *Wong*'s 2-deoxyribose-5-phosphate aldolase (DERA)-catalyzed formation of furanose and pyranose fragments. Thus, addition of acetaldehyde to aldehyde **235** yielded pyranose **236** with high diastereoselectivity. Oxidation to lactone **237** and stereocontrolled allylation gave **238**, which was transformed to ester **239**. Reduction of the ester group gave acetal **240** with correct configurations at C6–C8. Conversion to aldehyde **241** followed by addition of the dianion from

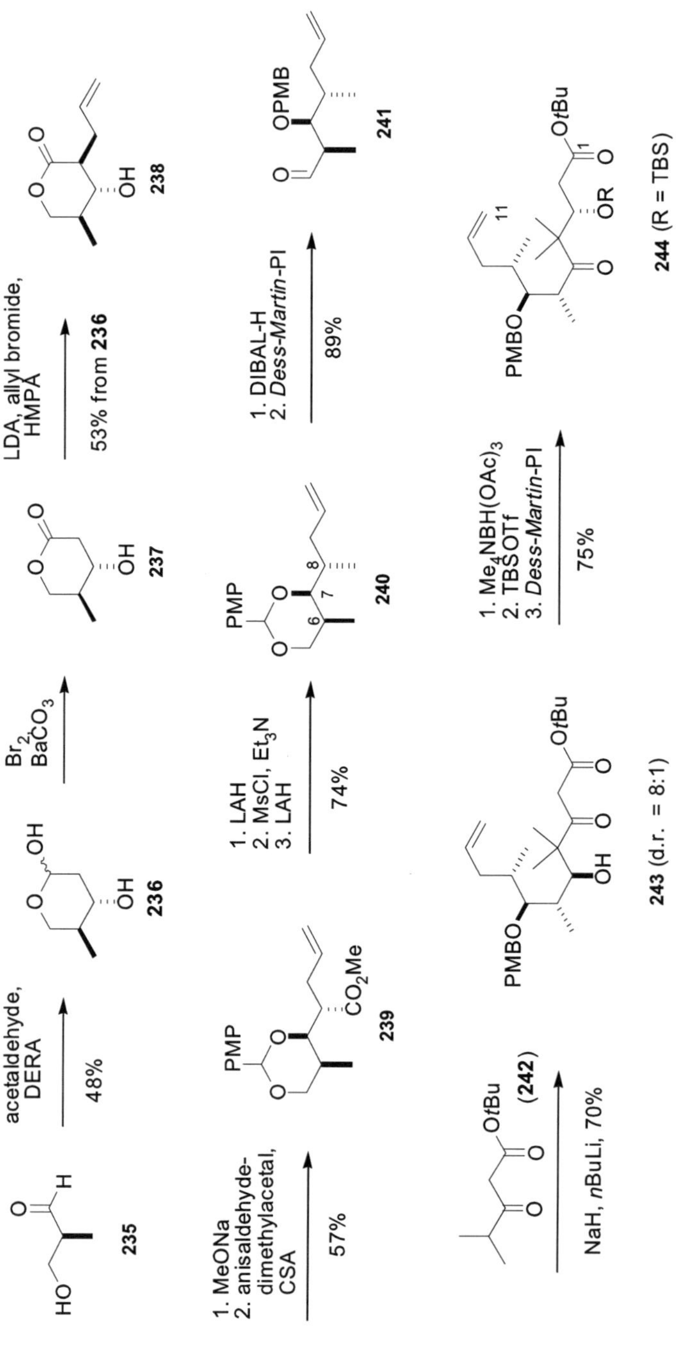

Chart 41. *Wong*'s synthesis of epothilone C, Part I

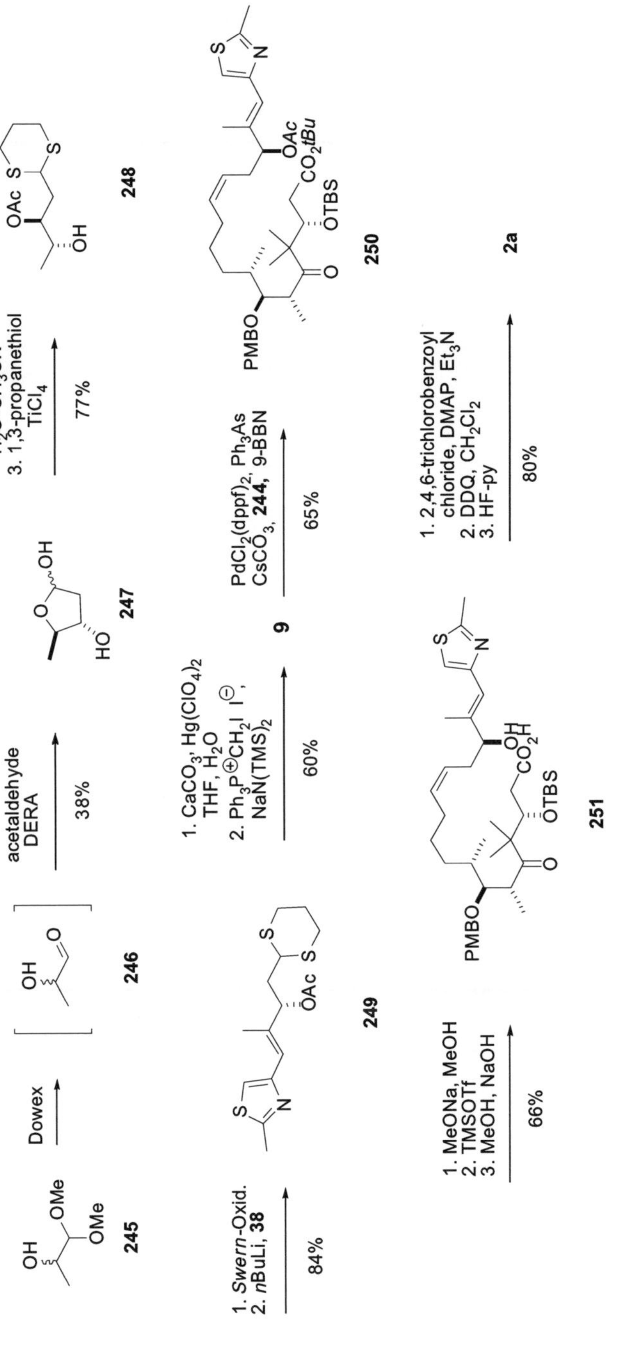

Chart 42. *Wong*'s synthesis of epothilone C, Part II

keto ester **242** led to adduct **243**, which was refunctionalized to **244**. For the synthesis of **9** (Chart 42), DERA-catalyzed addition of acetaldehyde to aldehyde **246**, generated *in situ* from dimethyl acetal **245**, gave furanose **247**. Straightforward transformation to thioketal **248** was followed by *Swern* oxidation and *Horner* olefination with phosphine oxide **38** to give **249**. After deprotection to the aldehyde, *Wittig-Stork* olefination led to **9**, which was coupled with **244** to give **250**, and was converted into *via seco*-acid **251** and transformed into **2a**, as usual.

4.3.5. Ley's Approach (78, 79)

In pioneering studies to avoid tedious workup and purification procedures, the *Ley* group focuses on the use of immobilized reagents and scavengers to improve large-scale preparations of complex natural products. This philosophy is reflected in an approach to **1a** in which (Charts 43 and 44) the key steps are the aldol addition of ketone **60** to aldehyde **63** to form **267** and the (*Z*)-selective *Wittig* reaction of aldehyde

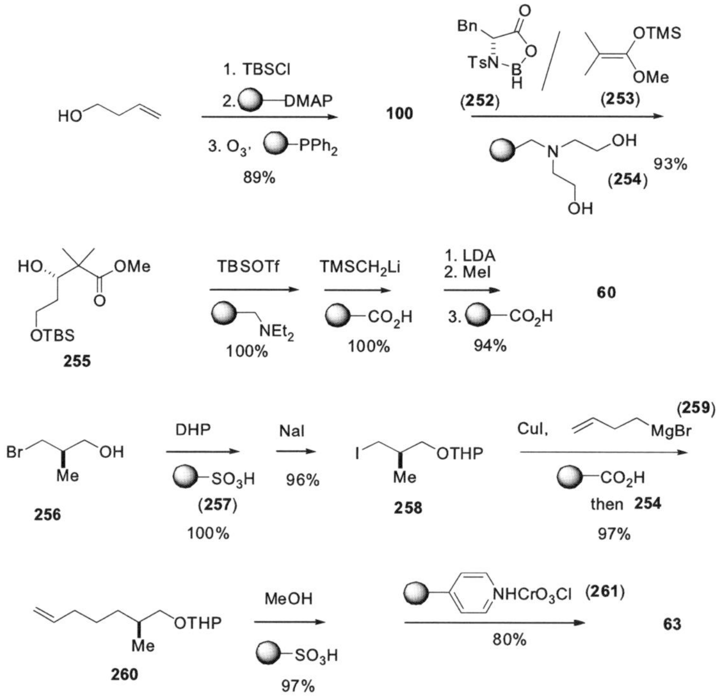

Chart 43. *Ley*'s synthesis of epothilone A and C using immobilized reagent and scavengers, Part I

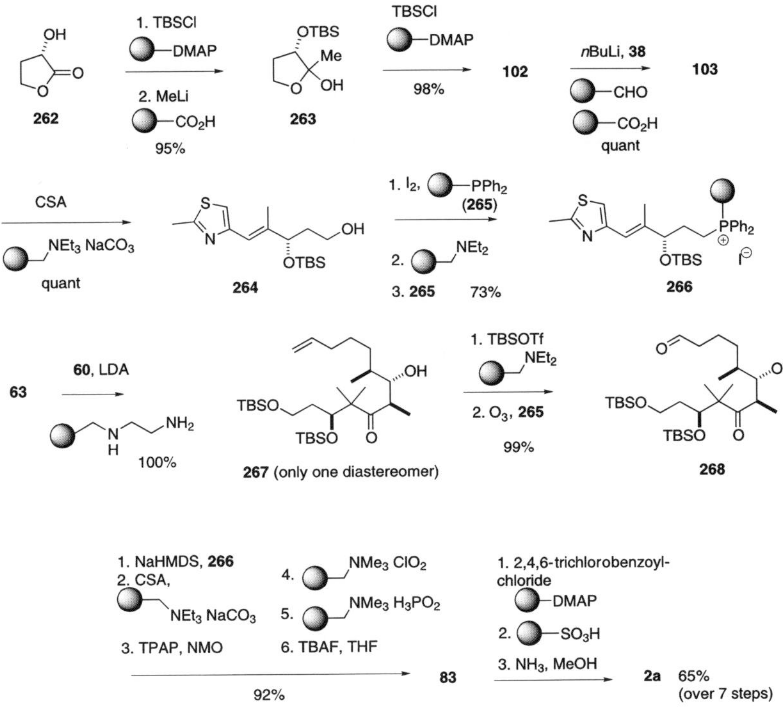

Chart 44. *Ley*'s synthesis of epothilone A and C using immobilized reagent and scavengers, Part II

268 and the ylide generated from the immobilized phosphonium salt **266**. The preparation of ketone **60** and fragment **103** are adapted from the *Mulzer* synthesis of **1b** (*83*) and the *Wittig* reaction between **268** and **266** stems from *Nicolaou* synthesis of **1a** (Chart 15). *Ley*'s novel contribution lies in the fact that the acids, bases and the di-phenyl phosphine previously required for catalyzing reactions, destroying intermediates and isolating reaction products, are fixed to a solid support. In this way aqueous workup, filtration or chromatography are largely avoided and the yields are extremely high, in most steps nearly quantitative.

4.4. Synthesis of Epothilones B/D (1b, 2b)

4.4.1. Mulzer Syntheses (80–87)

The *Mulzer* group has published three different approaches to **1b, 2b**. The first one (Chart 45) (*80–83*), although it was carried through to the

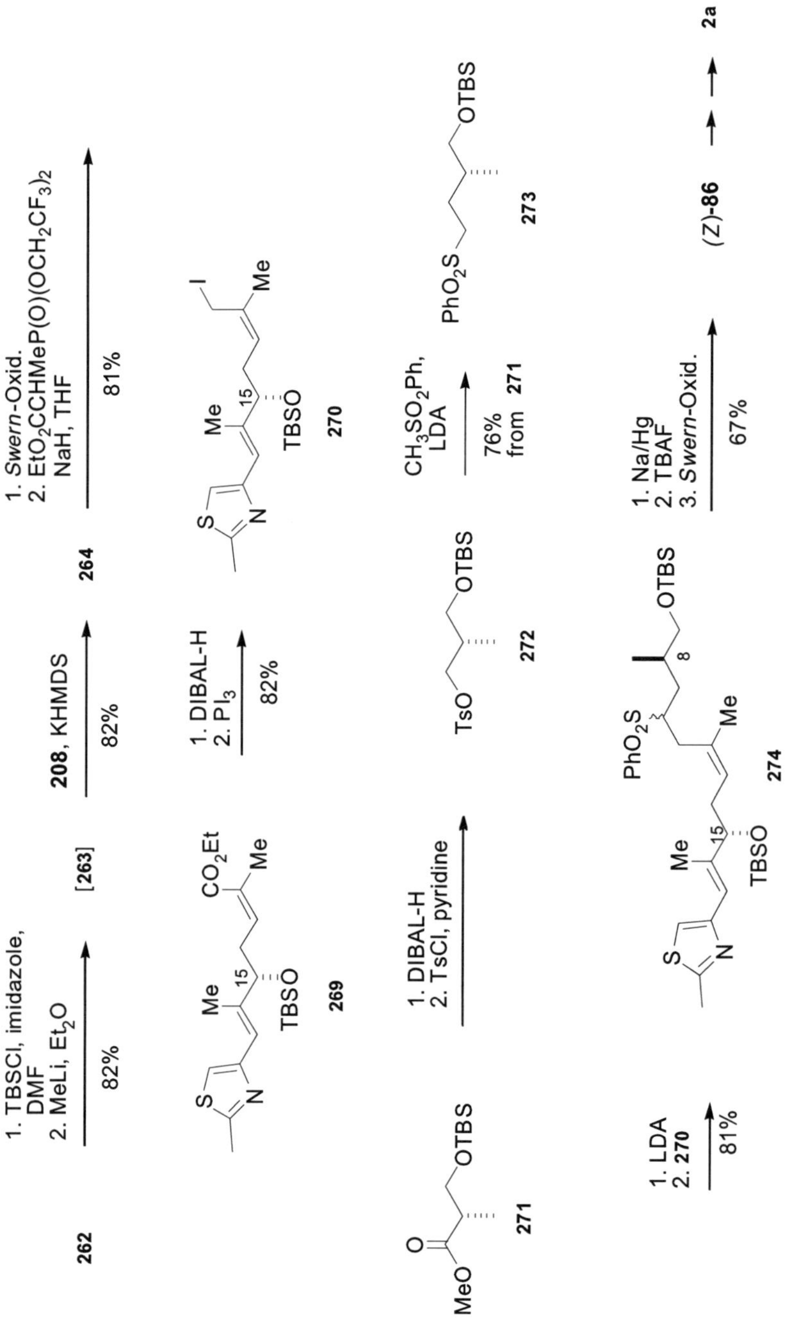

Chart 45. *Mulzer*'s synthesis of epothilone D

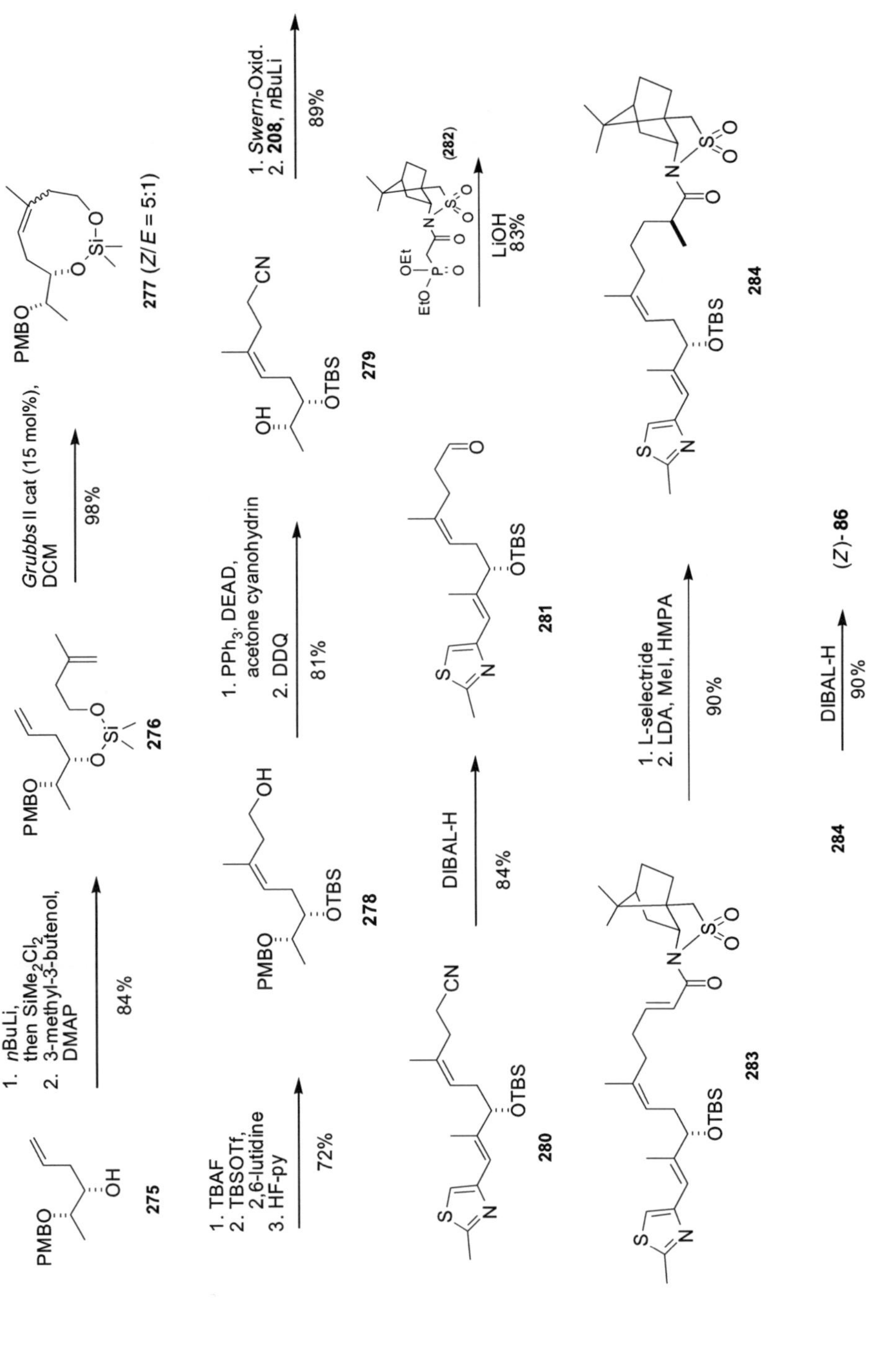

Chart 46. *Mulzer*'s silicon tethered RCM approach to aldehyde (Z)-86

final target, is a formal one and aims for an easier access to compounds **60** and (*Z*)-**86**, which have been used by *Nicolaou* for the aldol addition to give key intermediate **97** (Chart 18). The *Mulzer* contribution lies in a facile introduction of the stereogenic center at C3 in ketone **60**, which has later been used by the *Ley* group in their solid-phase synthesis of **1a** (Chart 45) and a stereocontrolled synthesis of the (*Z*)-12,13-olefin moiety in aldehyde (*Z*)-**86**. Thus, commercially available hydroxyl lactone **262** *via* hemiketal **263** and *Wittig* olefination with **208** gave alcohol **264**. These steps have subsequently also been used by *Ley* (*78*, *79*) (Chart 44). For the introduction of the (*Z*)-olefin a *Still-Gennari* olefination was employed to give enoate **269**, which was converted to iodide **270**. Sulfone **273** was obtained in good yield from ester **271** *via* tosylate **272**. Alkylation of **273** with iodide **270** furnished sulfone **274** that was desulfonylated and converted to (*Z*)-**86**, in which both stereogenic centers at C15 and C8 have thus been derived from the chiral carbon pool.

In an alternative approach to (*Z*)-**86** (*84*, *85*) a silicon tethered RCM reaction of di-olefin **276**, easily available from known alcohol **275** was employed (Chart 46). Cycloolefin **277** was obtained as an easily separable 5:1-(*Z*)/(*E*)-mixture that was converted to alcohol **278**. Chain elongation to the nitrile, *Wittig* olefination with **208** and reduction

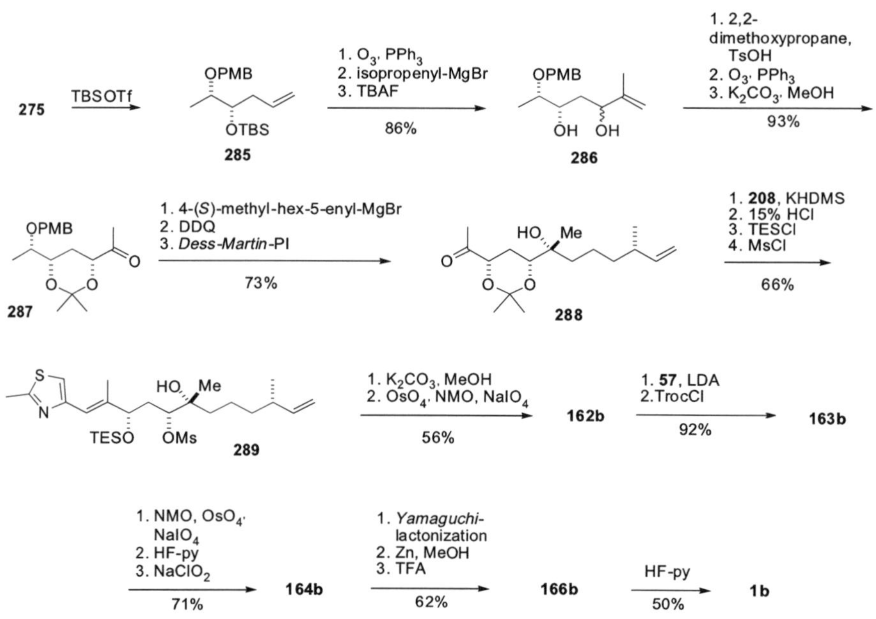

Chart 47. *Mulzer*'s early epoxide approach to epothilone B

with DIBAL-H gave aldehyde **281**, which was olefinated to **283** with *Oppolzer*'s sultam **282**. Enamide **283** was converted into **284** by 1,4-hydride addition/diastereoselective methylation. Reductive removal of the auxiliary gave (*Z*)-**86**.

Whereas these two approaches led to **2b** first, the third approach directly furnished **1b** (early epoxide approach, Chart 47) (*86, 87*). By this concept, which was later adopted in *Sinha*'s and *Carreira*'s syntheses, the low-yielding nonselective epoxidation of **2b** with peroxides was avoided. The synthesis started with the chain elongation of intermediate **285–286**, and its conversion into ketone **287**, which originally was obtained as an epimeric mixture. Base-catalyzed epimerization gave diastereomerically pure **287**. A chelate *Cram*-induced *Grignard* addition furnished tertiary alcohol **288** selectively from which mesylate **289** was easily available. Base-induced cyclization gave the desired epoxide from which aldehyde **162b** was obtained after ozonolysis. The conversion into **1b** has later been used in identical form by the *Carreira* group and has already been described in Section 1.5.

4.4.2. Grieco's Formal Synthesis of ***2b*** *(88)*

The *Grieco* group aimed for a synthesis of *Danishefsky*'s RCM of intermediate **28b** (Chart 6) *via* a different route (Chart 48). To introduce the stereogenic center at C6 they started with a *Sharpless* epoxidation of allylic alcohol **290** to give epoxide **291** with >95% *ee*. The formation of enoate **292** was followed by a regioselective epoxide opening to introduce the 7-Me under inversion of configuration. The stereogenic centers at C8 and C9 were introduced *via* a *Roush* crotylation of the 8-aldehyde to give **294**, which contains the stereotetrad C-6,7,8,9 in correct absolute and relative configuration. Protection, hydroboration and chain elongation led to enone **295**, and after hydrogenation and *Wittig* methylenation olefin **296** was formed. In four additional steps aldehyde **298** was obtained. Following *Danishefsky*'s precedent, **298** was converted into tri-olefin **299** and then into intermediate **28b**.

4.4.3. White's Syntheses (89–92)

The *White* group made novel contributions with respect to the formation of the (*Z*)-12,13-olefin moiety and the *Wittig*-olefination of phosphorane **308** and aldehyde **312** (Charts 49 and 50). The synthesis of **308** (Chart 49) started with a conjugate addition of the cuprate derived from vinyl iodide **300** to *Evans*' oxazolidinone **301** to give the C11–C16-

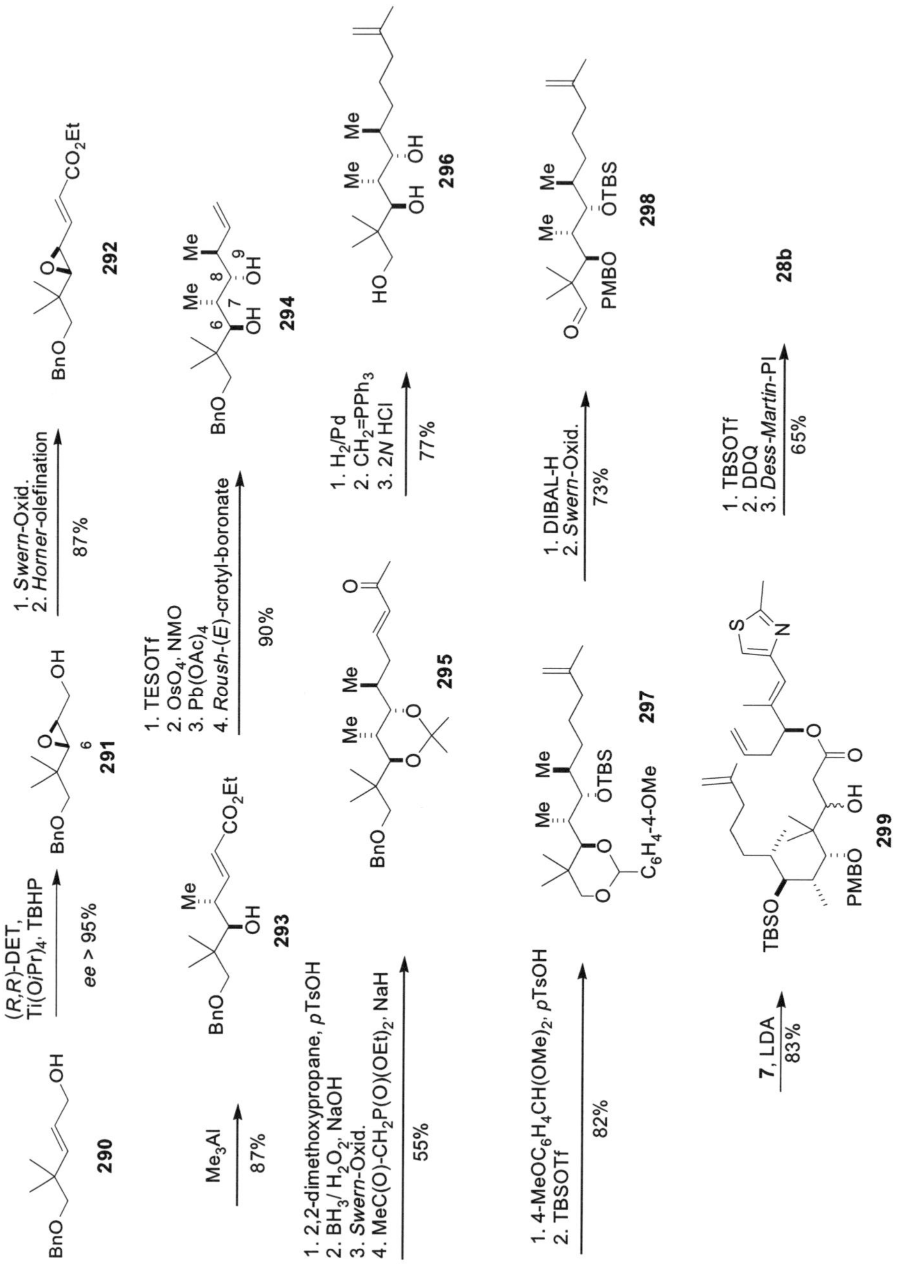

Chart 48. *Grieco*'s formal synthesis of epothilone B

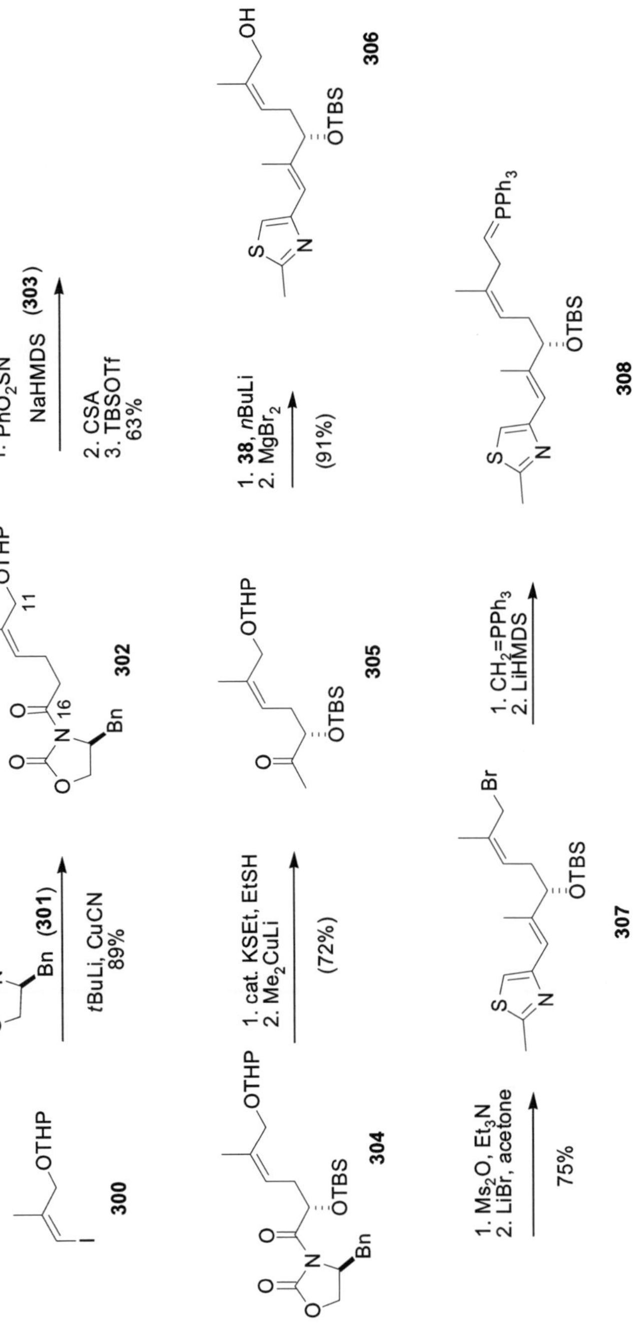

Chart 49. *White*'s synthesis of epothilone D, Part I

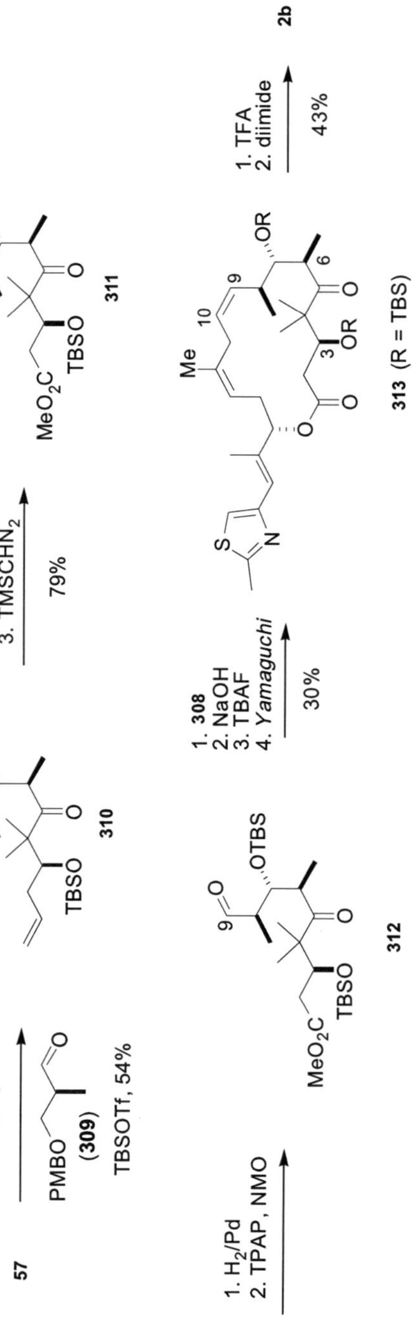

Chart 50. *White*'s synthesis of epothilone D, Part II

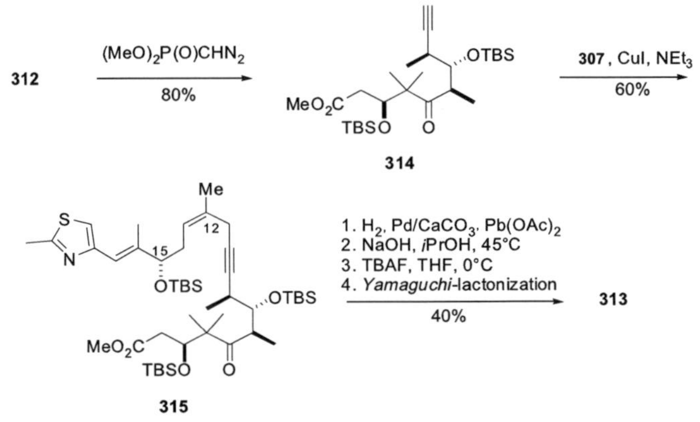

Chart 51. *White*'s synthesis of epothilone D, Part III

fragment **302**. The missing C15–OH-function was introduced with d.r. 98:2 by *Davis*' reagent **303** to give **304**. Removal of the auxiliary gave ketone **305**. *Horner* olefination with phosphonate **38**, deprotection and conversion to the bromide gave intermediate **307**. Chain elongation with *Wittig* ylide led to the C10–C21-phosphorane **308**. Aldehyde **312** was available (Chart 50) from an aldol addition of ketone **57** and aldehyde **309**, first reported by *Mulzer* (*80*), to give **310** with 4:1-selectivity. Oxidation of the double bond furnished carboxylic ester **311**. Formation of the 9-aldehyde led to **312**, whose reaction with **308** gave the (*Z*)-9,10-olefin which was converted to the *seco*-acid. *Yamaguchi* lactonization furnished **313** which was desilylated and selectively hydrogenated with diimide to provide **2b**. In an alternative connection (Chart 51) alkyne **314**, easily available from aldehyde **312** with the *Gilbert-Seyferth* reagent, was alkylated with bromide **307**. The resulting alkyne **315** was hydrogenated to the (*Z*)-9,10-olefin with *Lindlar* catalyst, which was elaborated into **313** as before.

4.4.4. Ermolenko Variation of the White Synthesis (93)

The *Ermolenko* approach (Charts 52 and 53) is a close variation of *White*'s synthesis. It uses analogous fragments **308** and **327** (identical with **312** except for the *t*-Bu-ester moiety) that are both prepared from D-glucose. Thus, in the synthesis of **308** (Chart 52), pyranose **316**, available from D-glucose in four steps, was converted into olefin **317**, the C11–C16-fragment of **2b** *via Chugaev* elimination. Removal of the acetates and oxidation to ketone **318** was followed by *Horner* olefination with **38** to give **319**. Acidic opening of the acetal followed by reduction

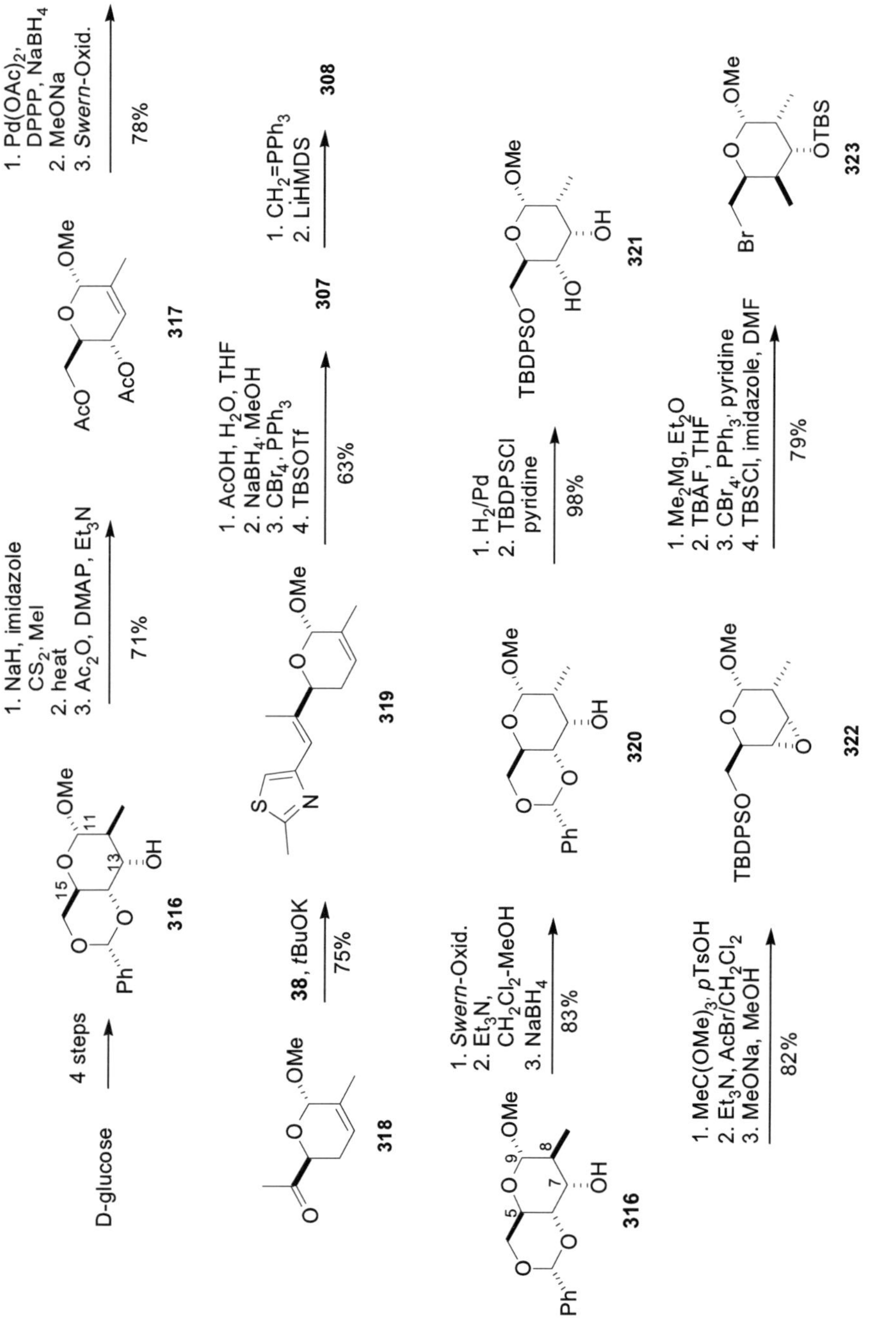

Chart 52. *Ermolenko*'s modification of *White*'s synthesis, Part I

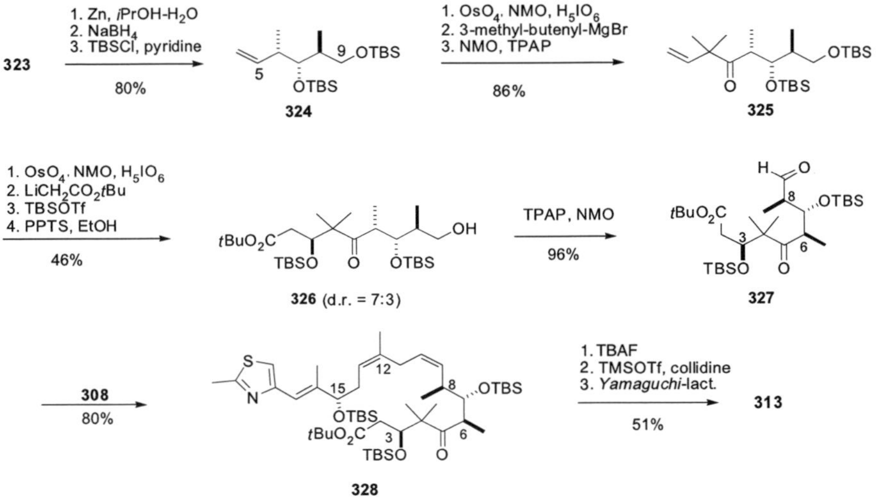

Chart 53. *Ermolenko*'s modification of *White*'s synthesis, Part II

to the alcohol and bromination led to bromide **307**, which was then converted into phosphorane **308** according to the *White* protocol. Pyranose **316** contains also the C5–C9 fragment of **327**. First, the configuration at C8 was inverted *via* equilibration of the 7-ketone. Then, the methyl group at C6 was introduced regioselectively *via* a diaxial ring opening of epoxide **322**, available under retention of configuration from diol **321**. Further elaboration gave bromide **323**, whose reductive opening (Chart 53) afforded acyclic olefin **324**, which was converted to ketone **325**. Oxidation of the olefin to the 5-aldehyde and unselective aldol addition gave **326** as a 7:3-mixture of C3-diastereomers. Separation and oxidation to aldehyde **327** followed by *Wittig* olefination with **308** led to *seco* compound **328**, which was converted to *White*'s intermediate **313**.

4.4.5. Synthesis by E. J. Thomas (94)

Thomas' contribution lies in a novel stereoselective approach to the (*Z*)-12,13-double bond (Charts 54, 55). The sequence started with an alkylation of stannane **330** with iodide **329** to give **331** as a mixture of diastereomers. Deprotection and reprotection *via* **332** followed by an S_R2' addition of tributylstannane gave **333**, which was added to aldehyde **334** under twofold allylic *Thomas* rearrangement to give the homoallylic alcohol **335** as a diastereomeric mixture, however with purely (*Z*)-olefin geometry. From **335**, the superfluous OH function was removed in a *Barton-McCombie* sequence to furnish **336**, which was converted into

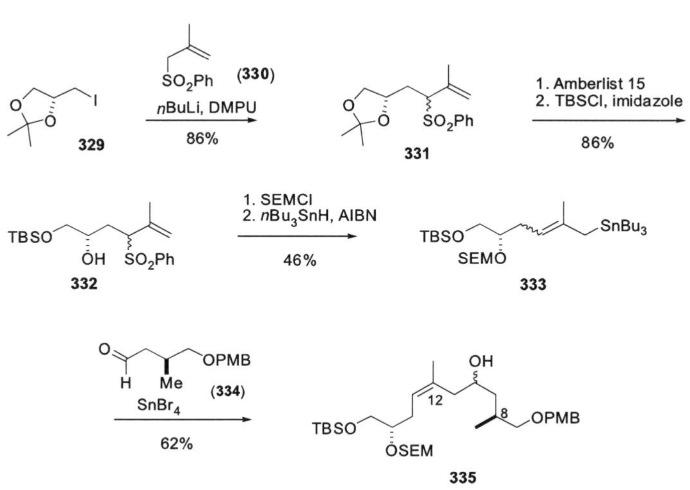

Chart 54. *E. J. Thomas*' synthesis of epothilone D, Part I

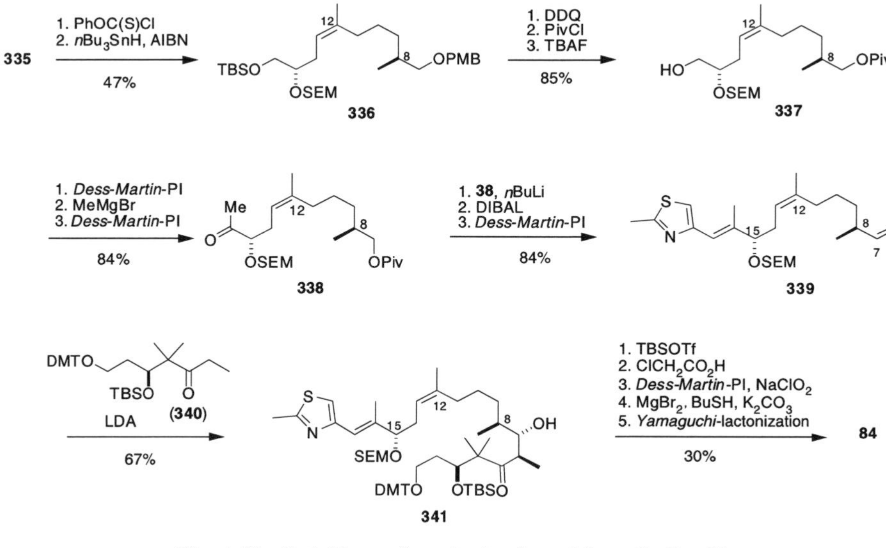

Chart 55. *E. J. Thomas*' synthesis of epothilone D, Part II

alcohol **337** and then ketone **338** by conventional methodology. Usual *Horner* olefination with **38** followed by deprotection and oxidation gave aldehyde **339**, which is identical with *Nicolaou*'s intermediate (*Z*)-**86** except for the 15-OSEM protecting group. Aldol addition with ketone **340**, prepared from pantolactone analogously to *Schering*'s synthesis

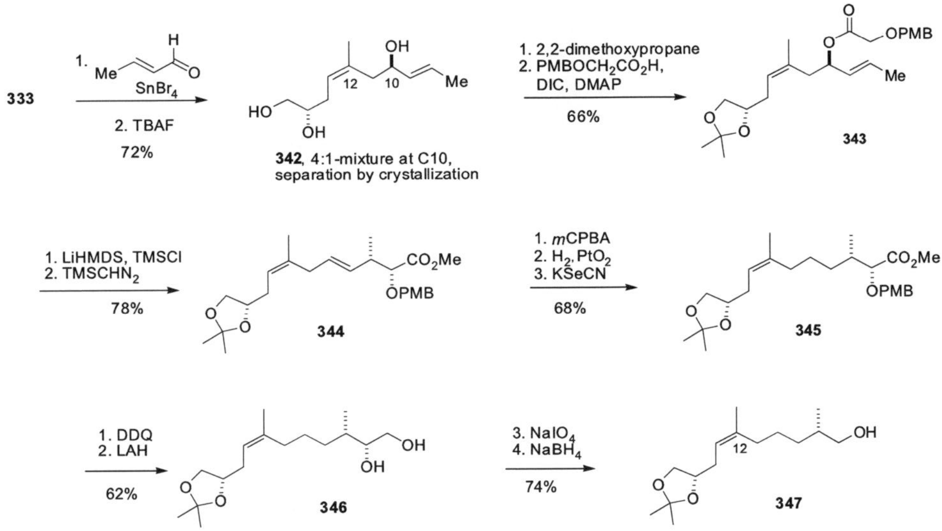

Chart 56. *E. J. Thomas*' synthesis of epothilone D, Part III

(*134*) gave adduct **341** with high diastereoselectivity. The conversion into **84** is essentially identical with *Nicolaou*'s precedence (Chart 15). In a second, yet uncompleted approach to **339**, crotonaldehyde was added to **333** to give a 4:1-mixture of **342**, which was separated by crystallization. *Ireland-Claisen* rearrangement was then used to convert ester **343** to **344** stereoselectively. The undesired C9,10-olefin was removed by hydrogenation after protection of the C12,13-double bond as an epoxide. Ester **345** was reduced to diol **346** that was then converted into alcohol **347** *via* glycol cleavage and reduction (Chart 56).

4.4.6. Avery's Synthesis (95–99)

Avery's contribution lies in a novel approach to the (*Z*)-12,13-olefin moiety by carbo-cupration. First, the C14–C16-epoxide **349** was prepared by *Sharpless* kinetic resolution (*S*KR) of allylic alcohol **348**. In the key step, bromide **351** was converted into the cuprate and treated with propyne and hexynyl lithium to give cuprate **352** as an intermediate, which was trapped with epoxide **349** to furnish C7–C16-fragment **353**. The synthesis then aimed for *Thomas*' aldehyde **339** by conversion of **353** into ketone **354**, which was olefinated with **38** to give **355**. Regioselective asymmetric hydroboration of the C7,8-olefin followed by *Swern* oxidation led to **339** (Chart 57). In a parallel sequence, *Nicolaou*'s keto acid **59** was prepared *via* aldol addition of *Oppolzer*'s sultam **350** to

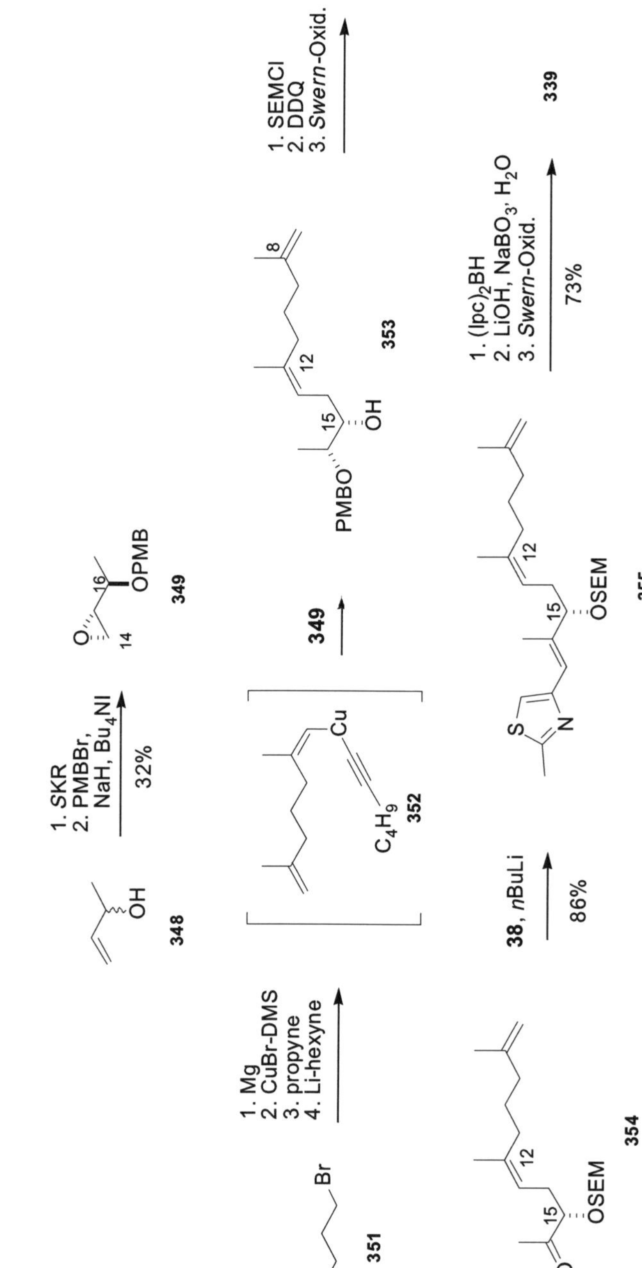

Chart 57. *Avery*'s synthesis of epothilone D, Part I

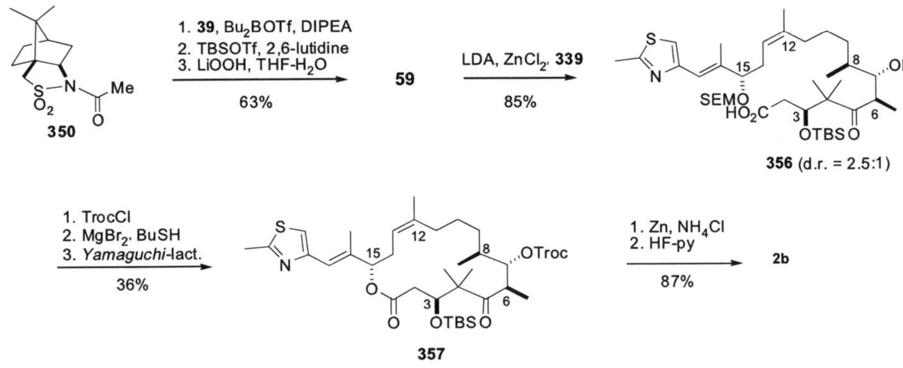

Chart 58. *Avery*'s synthesis of epothilone D, Part II

aldehyde **39**. Aldol addition of **59** to aldehyde **339** gave *seco*-acid **356** with low stereocontrol. Protection of the C9–OH function followed by deprotection of the 15-position and *Yamaguchi* lactonization furnished **357**, which was deprotected to give **2b**.

4.4.7. R. E. Taylor's Synthesis (100–102)

The innovation in *Taylor*'s approach again lies in the preparation of the (*Z*)-12,13-double bond. Thus, aiming for *Nicolaou*'s aldehyde (*Z*)-**86**,

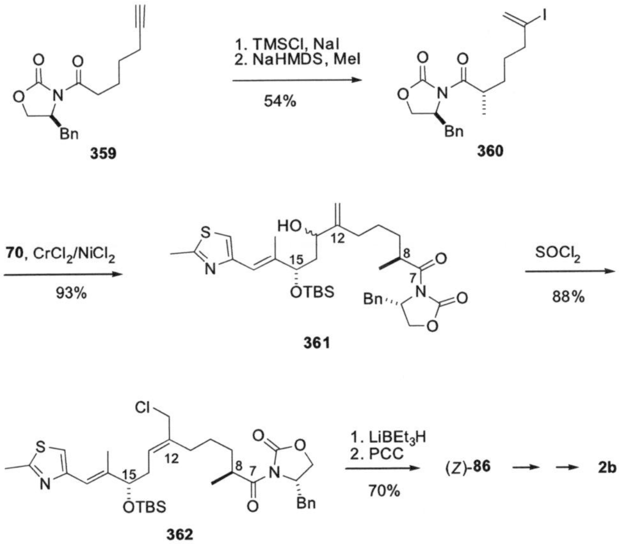

Chart 59. *R. E. Taylor*'s synthesis of aldehyde (*Z*)-**86**

Evans' oxazolidinone **359** was converted into vinyl iodide **360**. *Nozaki-Kishi* addition of **360** to aldehyde **70** (Chart 12) gave allylic alcohol **361** as a mixture of diastereomers. Allylic rearrangement with thionyl chloride led to allylic chloride **362** stereoselectively. Reduction with superhydride removed both the chloride and the auxiliary. The resulting C7-alcohol was oxidized with PCC to furnish (*Z*)-**86**, which was then converted into **2b** according to *Nicolaou*'s protocol.

4.5. Syntheses of Fragments

*4.5.1. Kalesse's Synthesis of Nicolaou's Intermediates **60** and **82** (103–107)*

The synthesis of ketone **60** started with diol **363** that was converted into enoate **364**. Reduction and SAE gave epoxide **365**, which was reduced to triol **366**. Debenzylation and oxidation gave aldehyde **367**, from which ketone **60** was prepared. For the synthesis of aldehyde **82**, a highly innovative RCM approach was developed. In view of the low *E/Z*-selectivity of the early RCM approaches the *Kalesse* group chose a tethered version to enforce the (*Z*)-geometry *via* formation of a 10-membered macrolide **370**. Thus, diolefin **369** was prepared from lactic aldehyde **368** *via* the allyl adduct **275** (which was later also used in *Mulzer*'s synthesis, Chart 47), and subjected to a *Grubbs* I catalyzed RCM that furnished **370** with 12:1 (*Z*)-selectivity. A similar concept was later pursued in *Mulzer*'s

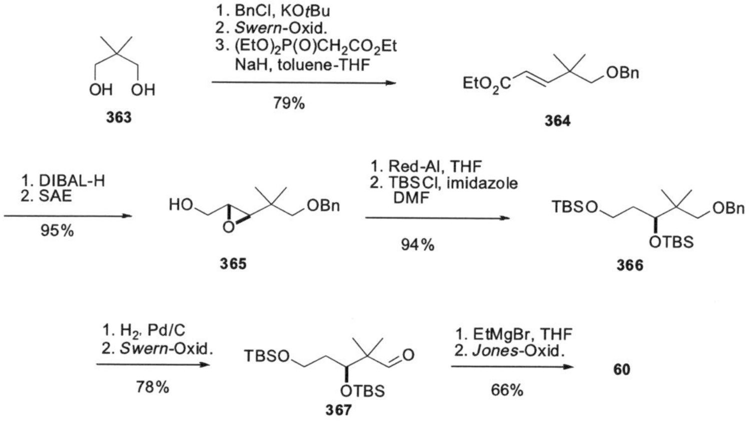

Chart 60. *Kalesse*'s synthesis of epothilone C fragment **60**

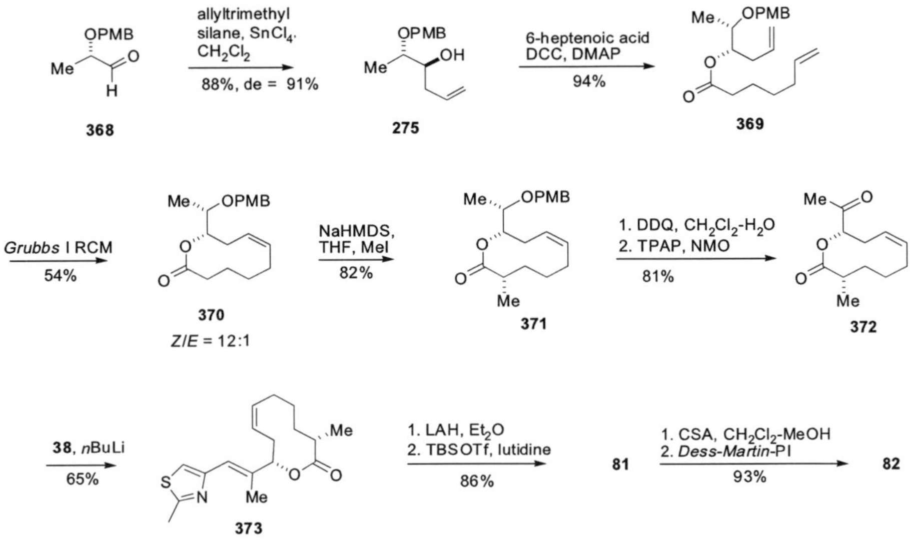

Chart 61. *Kalesse*'s synthesis of epothilone C fragment **82**

silicon tethered synthesis of aldehyde (*Z*)-**86** (Chart 46). Conformation controlled methylation of **370** gave **371** stereoselectively. The conversion of **371** to alcohol **81** involved olefination of ketone **372** with **38** to give **373**, followed by reductive cleavage of the lactone and di-silylation. The conversion of **81** into aldehyde **82** was already performed by *Nicolaou* (Chart 15).

*4.5.2. Chandrasekhar's Synthesis of Keto Acid **378** and C7–C16 Fragment **381** (108)*

In an approach aiming for *Nicolaou*'s key aldol addition (cf. Chart 15), *Chandrasekhar et al.* reported alternative syntheses for the C1–C5-fragment **378** and the C7–C16 fragment **381**. Thus, cyclohexane-1,3-dione (**374**) was converted into ketone **375**, which was dimethylated and reduced with *Corey*'s CBS reagent to give enantiomerically enriched acetate **376**. Allylic oxidation followed by dihydroxylation of the double bond furnished **377**. Glycol cleavage and oxidation led to keto acid **378**. The preparation of fragment **381** is very similar to *Liu*'s approach (Chart 37) by alkylating alkyne **379** with epoxide **199** to give **380**, which was hydrogenated with *Lindlar*'s catalyst and silylated to deliver (*Z*)-olefin **381**.

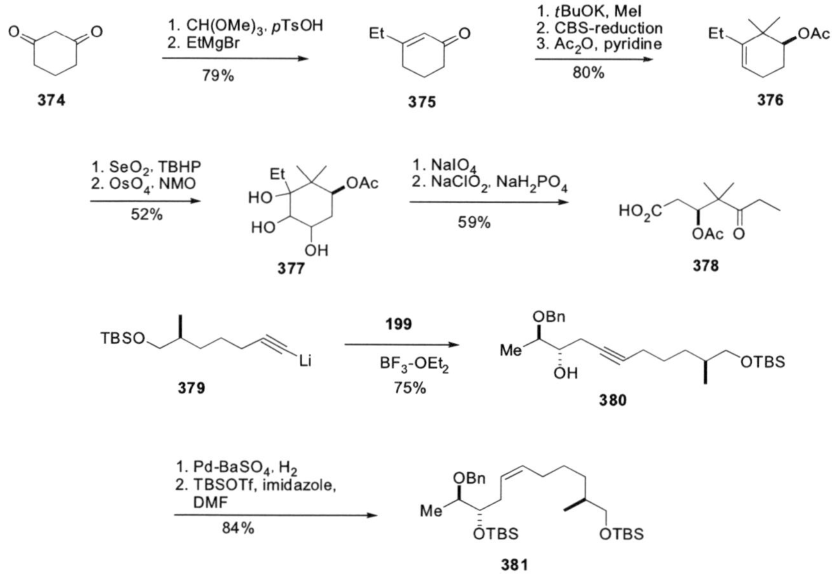

Chart 62. *Chrandrasekhar*'s synthesis of epothilone C fragment **381**

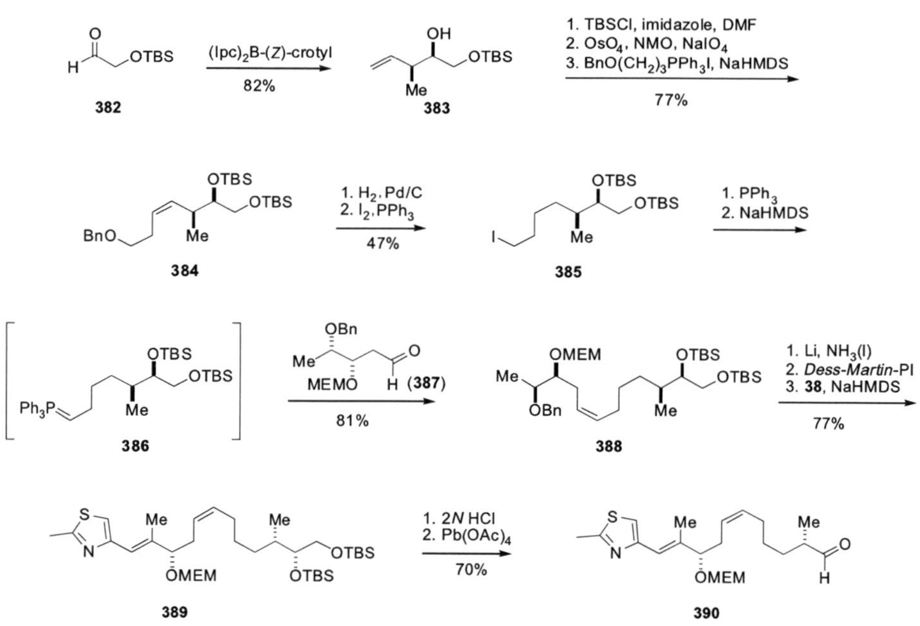

Chart 63. *Ramachandran*'s synthesis of epothilone C fragment **390**

References, pp. 126–133

*4.5.3. Ramachandran's Synthesis of the MEM-protected Nicolaou-Aldehyde (**390**) (109, 110)*

To generate the 12,13-olefin (*Z*)-selectively, *Ramachandran* made use of an inverted version of *Nicolaou*'s *Wittig* olefination (cf. Chart 15). Thus, asymmetric *Brown* crotylation of aldehyde **382** gave adduct **383**, which was converted into (*Z*)-olefin **384** *via Wittig* olefination of the C7-aldehyde. Hydrogenation and iodination gave **385**, which was transformed into phosphorane **386**, and *Wittig* olefination with aldehyde **387** led to (*Z*)-olefin **388**. Debenzylation, oxidation to the ketone and *Horner* olefination gave **389**, which converted to aldehyde **390** *via* desilylation and glycol cleavage.

*4.5.4. De Brabander's Synthesis of Aldehyde **63** and Acid **59** (111)*

Aldehyde **63** was prepared in a stereocontrolled manner *via* methylation of *Oppolzer*'s sultam **392**, prepared from **391**. Similarly, the aldol addition of *ent*-**350** and aldehyde **39** gave **393**, which was O-silylated and hydrolyzed to **59** (Chart 64).

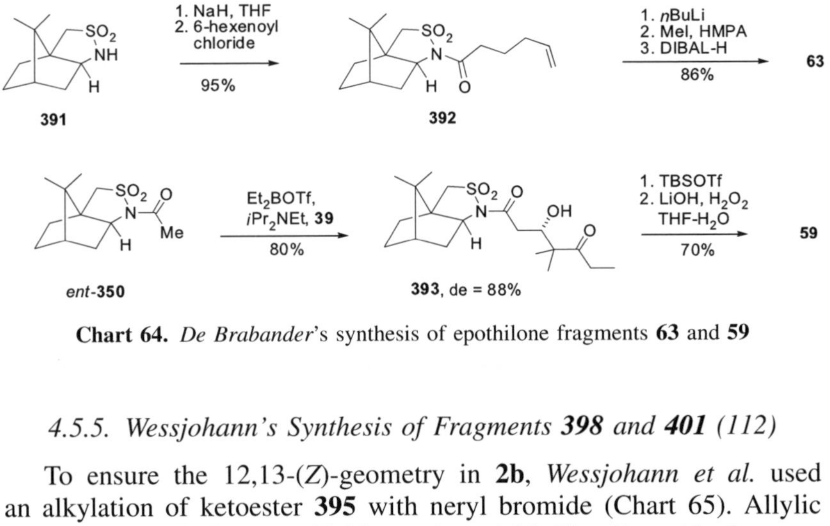

Chart 64. *De Brabander*'s synthesis of epothilone fragments **63** and **59**

*4.5.5. Wessjohann's Synthesis of Fragments **398** and **401** (112)*

To ensure the 12,13-(*Z*)-geometry in **2b**, *Wessjohann et al.* used an alkylation of ketoester **395** with neryl bromide (Chart 65). Allylic oxidation with SeO_2 gave **396** in modest yield. The *Noyori*-hydrogenation of the C8,9-olefin proceeded with only 21% ee to furnish acetate **397**, which was chirally resolved with lipase to deliver keto alcohol **398**. For the synthesis of the C1–C5 keto acid **401** a highly enantioselective *Reformatsky* reaction of an *Evans* bromoacetyl oxazolidinone **399** with ketoaldehyde **39** was used.

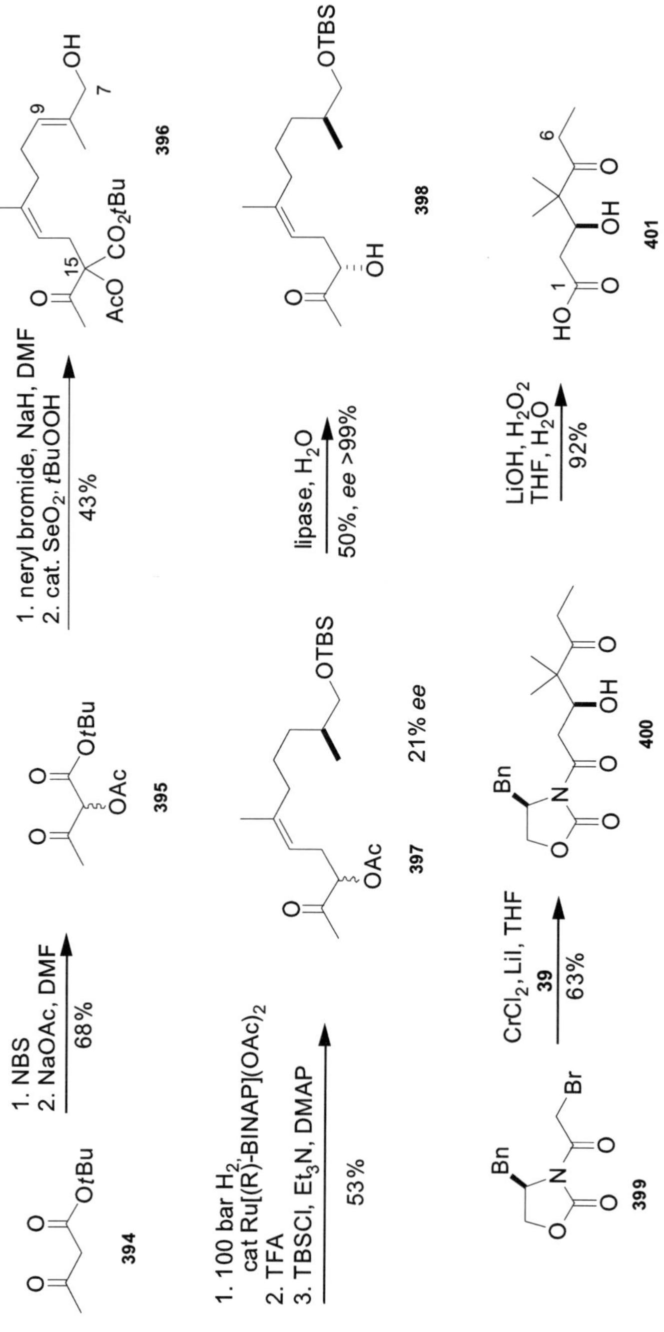

Chart 65. *Wessjohann*'s synthesis of epothilone fragments **398** and **401**

*4.5.6. Kulinkovich's Synthesis of Aldehyde **408** (113)*

In application of the *Kulinkovich* reaction malic ester **402** (Chart 66) was converted into di-cyclopropanolyl derivative **403**, which was transformed into allylic bromide **404** and coupled with *Grignard* reagent **405** to give **406** in high yield. The *exo* double bond was isomerized and the cyclopropanol moiety was oxidized to give ester **407**, which was converted into aldehyde **408**.

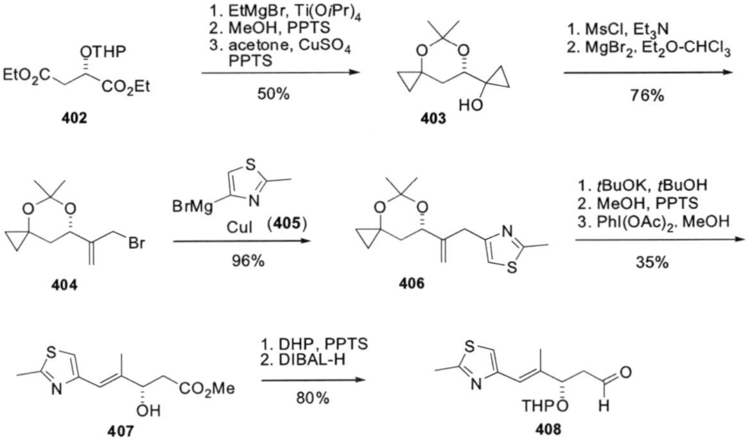

Chart 66. *Kulinkovich*'s synthesis of epothilone fragment **408**

*4.5.7. Georg's Synthesis of Aldehyde **70** (114)*

Aldehyde **4** was converted into enone **410** by *HWE* olefination with phosphonate **409**. Enantioselective CBS-reduction gave **411** that was converted into **70** (Chart 67), previously prepared by *Nicolaou* (Chart 12).

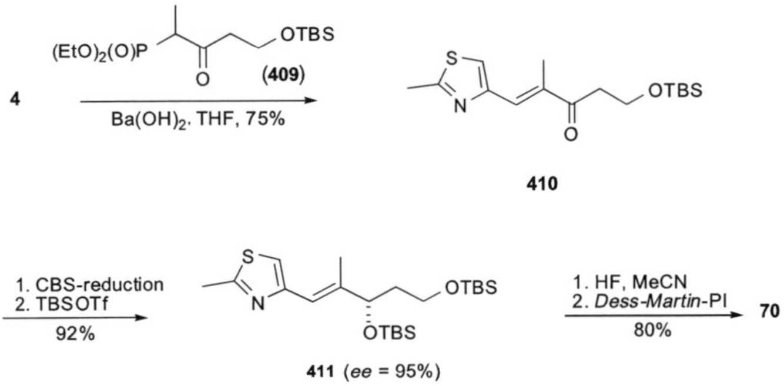

Chart 67. *Georg*'s synthesis of epothilone fragment **70**

*4.5.8. Lipase-Catalyzed Synthesis of the C1–C6 Fragment **414** (115)*

Racemic acetoxy ester **413** was obtained from aldol addition of **412** to **39** and subjected to a lipase-catalyzed deacetylation to give enantiomerically pure hydroxyl ester **414**. A similar enzyme-catalyzed route was used for the C7–C12 segment *(116)*.

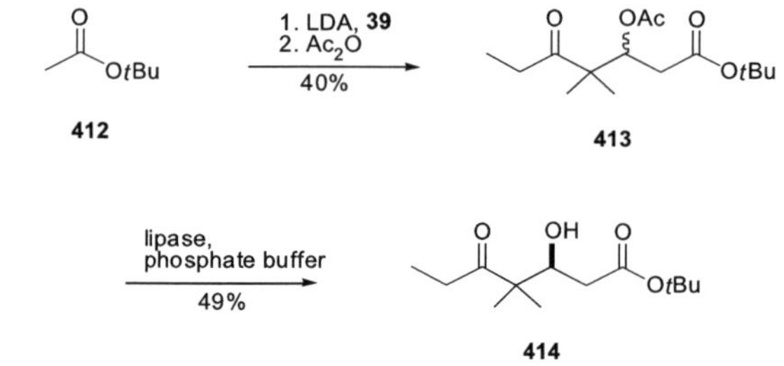

Chart 68. Lipase catalyzed synthesis of fragment **414**

4.6. Semisynthetic Degradation/Reconstruction of 2b *(117, 118)*

In a very interesting and novel approach, **2b** was degraded to the acyclic keto ester **415**, which still contains the C1–C12 backbone of **2b**. *Tebbe* methylenation and saponification led to acid **416**, which was esterified with alcohol **6** to give *seco*-ester **28b**. *Grubbs* RCM regenerated macrocycle **30** as a 1:1-(*E*)/(*Z*)-mixture. It was shown that instead of **6**, other olefinic alcohols could also be incorporated.

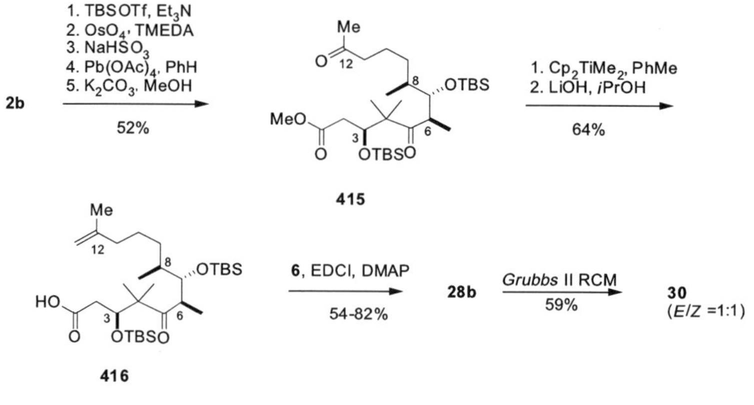

Chart 69. Synthesis of epothilone fragment **70** by degradation of epothilone D

In addition, several other fragment syntheses have been reported *(119–125)*, which are not discussed in detail.

4.7. Syntheses of Epothilones E and F (1c, 1d) and Their 12,13-Deoxy Derivatives (2c, 2d) (*126*)

4.7.1. *Nicolaou's Synthesis of* **1d** *(Chart 70)*

The synthesis started with the *Wittig* olefination of aldehyde **418** with phosphorane **417** to give ester **419** (*E*)-selectively. Conversion to **420** was followed by oxidation to aldehyde **421**, which was subjected to an aldol addition with ketone **60** to give adduct **422** with high diastereoselectivity. Detritylation and *Sharpless* epoxidation furnished **423** that was elaborated along familiar lines to macrolide **424**. Reduction of the primary alcohol to the methyl group and *Stille*-cross coupling with stannane **426** furnished **1d**. A similar *Stille*-coupling of vinyl iodide **425** with pyrazolyl-stannane **427** led to a highly active epothilone B derivative **428** (*127*).

4.7.2. *Sinha's Synthesis of* **1c**, **1d** *and* **2c**, **2d** *(Chart 71, 128)*

The synthesis is essentially the same as for **1a**, **1b** and **2a**, **2b** (cf. Charts 22–44). Thus, racemic hydroxyl ketone **427** was resolved chirally to give the pure enantiomer **428** that was converted into **430** *via* **429** in eight steps. Esterification with carboxylic acids **135** and **431** delivered di-olefin esters **432**, **433**, which were cyclized by RCM to give **434** and **435**. Global desilylation led to **2c** and **2d**, whose epoxidation finally gave **1c** and **1d** along with the corresponding α-epoxides (α/β ratio 5:1).

4.8. *Nicolaou*'s Synthesis of Epothilone Analogues (*1, 129–133*)

For SAR studies (see Chapter 6.1.4) a broad variety of epothilone derivatives was desirable. From early on, the *Nicolaou* group, together with the *Danishefsky* group, applied their synthetic approach for structural variations in essentially all sections of the epothilone skeleton. From the methodological viewpoint basically nothing new was developed. Thus, it may suffice to give one representative example from the *Nicolaou* manifold (*133*) (Chart 72). From a broad library of analogues *cis*-(15*S*)-CP-py epo **447** was the most active one. The synthesis started with a *Wittig* reaction of phosphonium salt **436** and *O*-benzyl-glycol aldehyde to give (*Z*)-olefin **437** which was converted into alkynol **438** in four steps. Reduction with nickel boride furnished

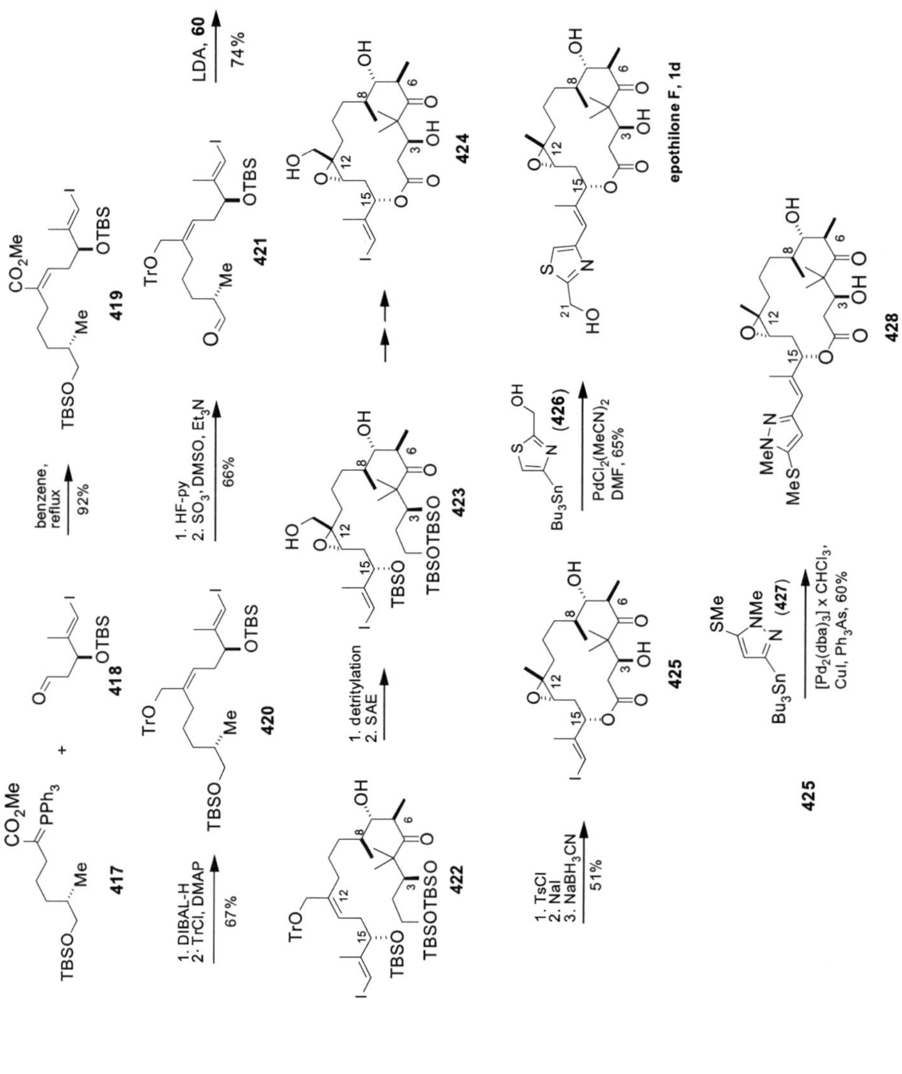

Chart 70. *Nicolaou*'s synthesis of epothilone F

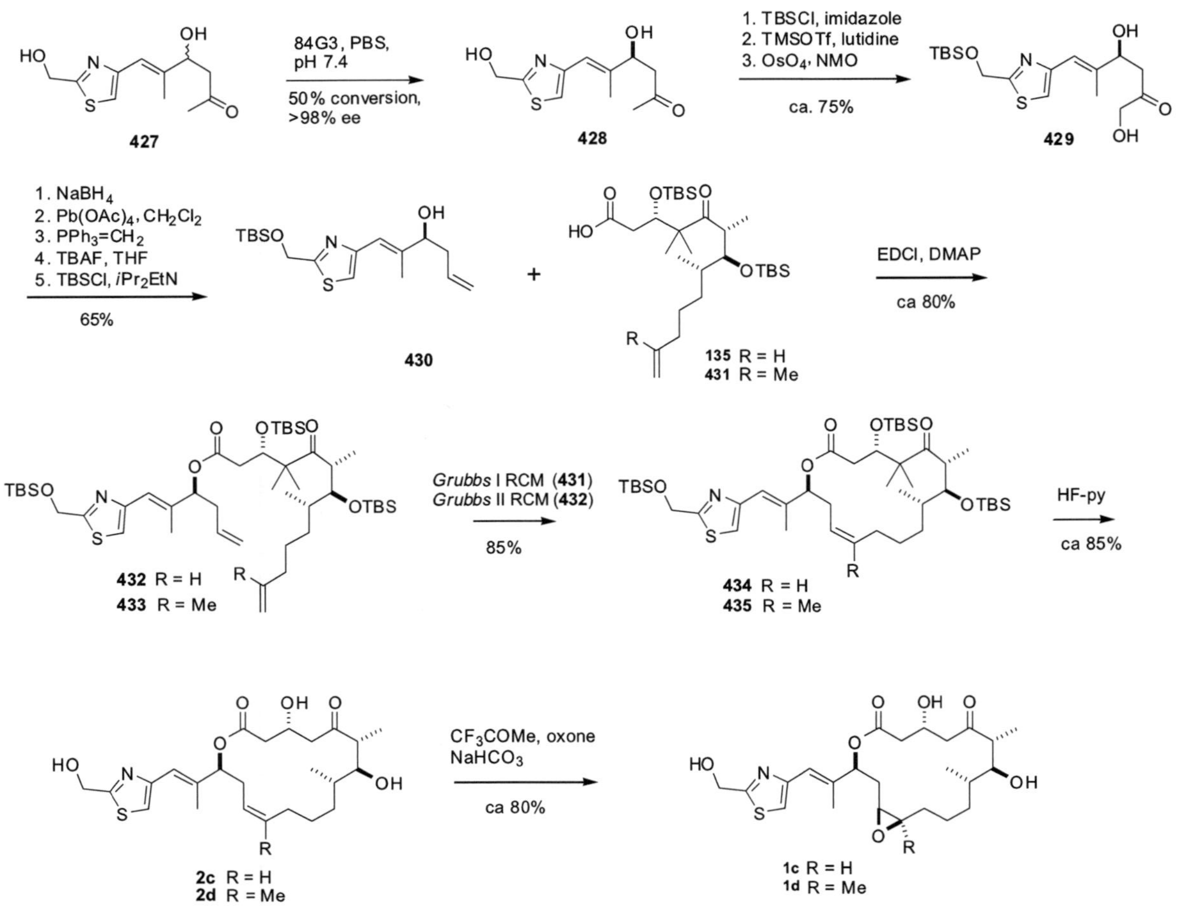

Chart 71. *Sinha*'s synthesis of epothilone E and F and their 12,13-deoxy derivatives

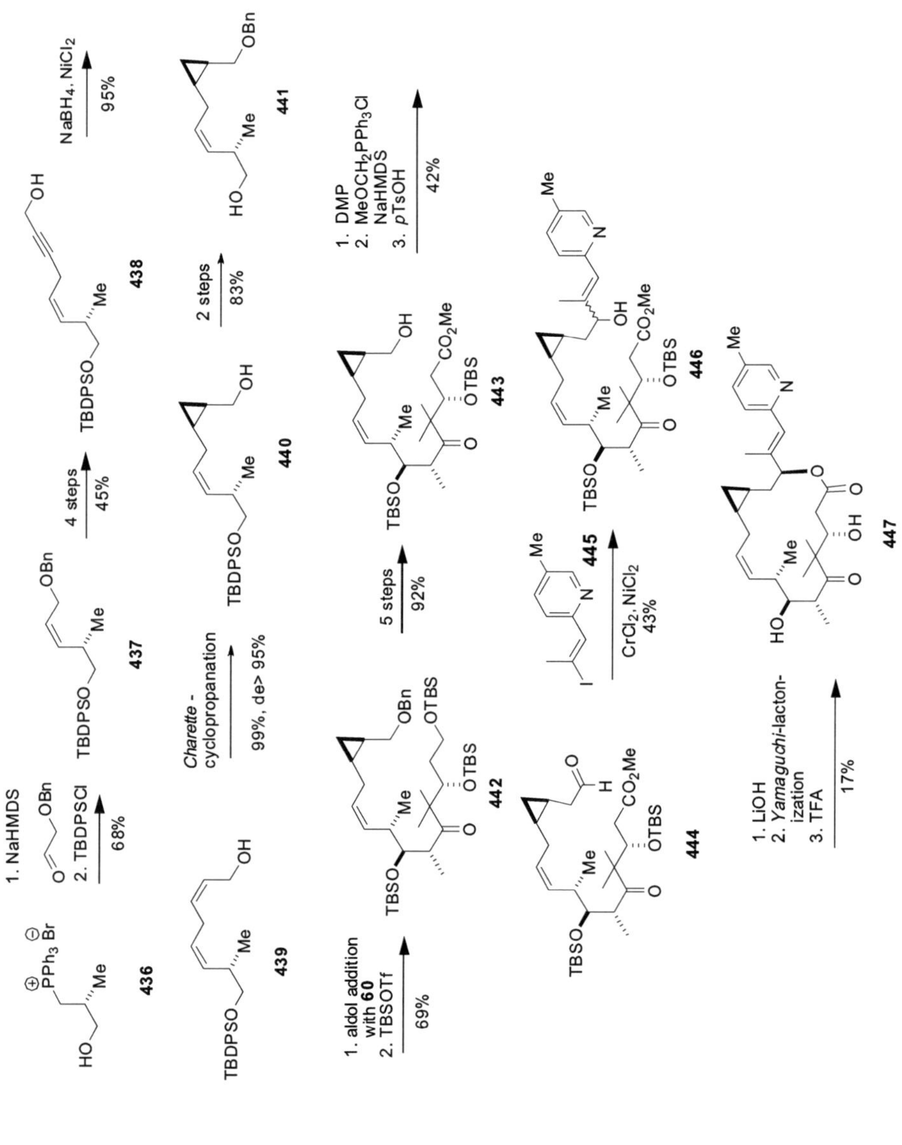

Chart 72. *Nicolaou*'s synthesis of *cis*-(15*S*)-CP-py-epo

allylic alcohol **439**, which was subjected to an asymmetric *Charette* cyclopropanation. Cyclopropane alcohol **440** was generated with high *ee*, which was then elaborated to aldol aduct **442** in the usual manner. For the introduction of the C-16-sidechain, aldehyde **444** was generated from alcohol **443**, and subjected to a stereo-uncontrolled *Nozaki-Hiyama-Kishi* addition with vinyl iodide **445**. A mixture of C-15-diastereomers was generated that was separated and converted into **447** along the routine sequence.

4.9. Conclusion and Industrial Application (ZK-Epo (Sagopilone))

The total synthesis of the epothilones has undoubtedly been one of the largest efforts ever in the history of organic chemistry. The urgency

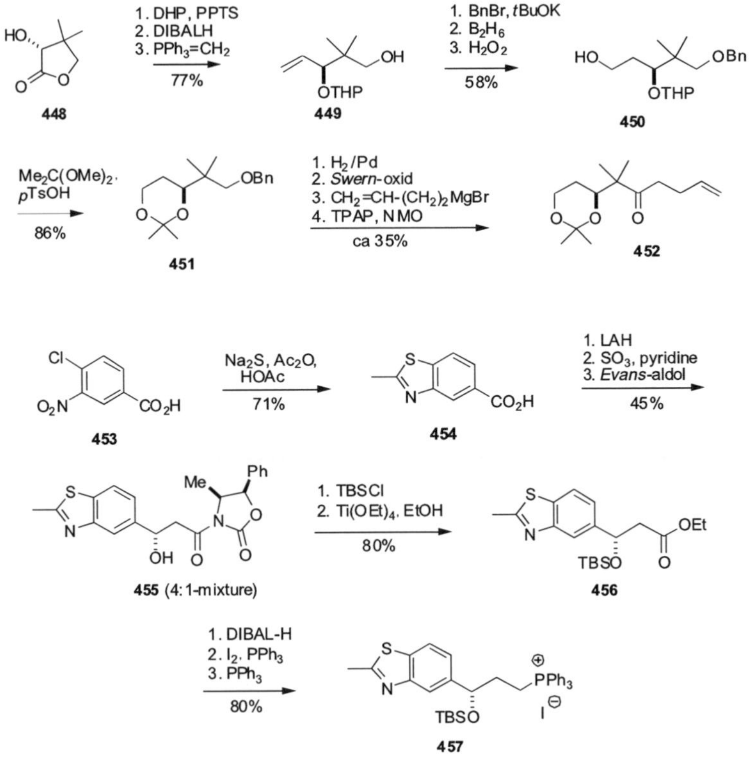

Chart 73. Schering synthesis of ZK-Epo (Sagopilone), Part I

to come up with the first synthesis and to prepare as many derivatives for pharmaceutical testing was evident from the very beginning. In fact the first three syntheses (*Danishefsky*, *Nicolaou*, *Schinzer*), published within a few weeks of each other, were seminal to such a degree that they established disconnections and key fragments, which were used by later authors and rendered most of the subsequent approaches formal ones. At first sight relatively little new methodology has emanated from this huge worldwide project. However, several advances may be gleaned: application of RCM reactions to rather complex substrates, all kinds of stereoselective approaches to generate a (*Z*)-tri-substituted double bond, long-range effects on stereoselective aldol additions that connect oxygenated fragments, the robustness and generality of *Yamaguchi* macrolactonization, and last but not least, stereo- and regioselective epoxidation with DMDO and the stability of epoxides towards a variety of aggressive reagents.

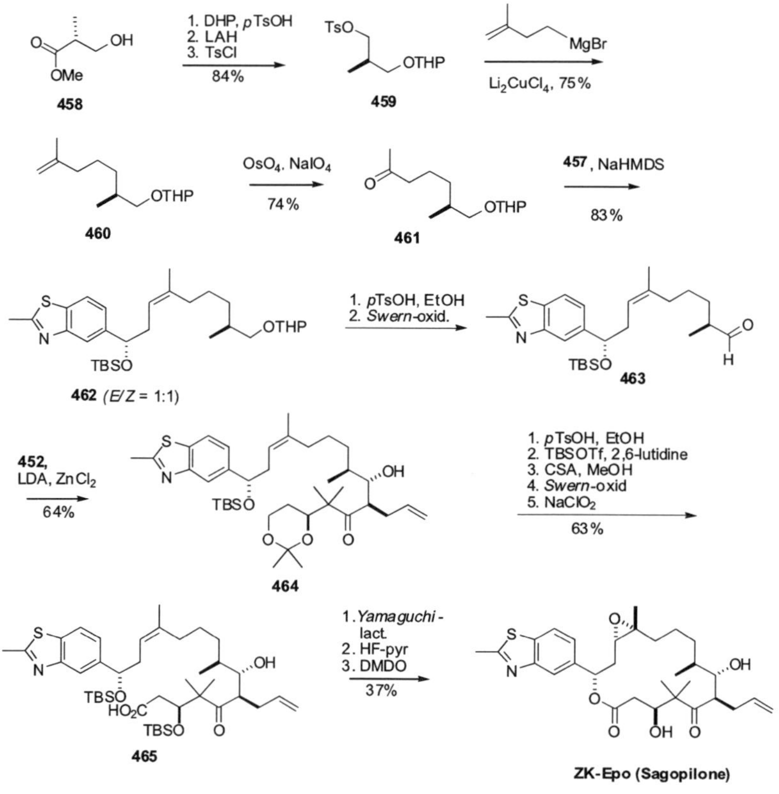

Chart 74. Schering synthesis of ZK-Epo (Sagopilone), Part II

It is an interesting point to see how much of the academic contributions has been transferred to the kilogram scale industrial synthesis of ZK-Epo (Sagopilone) (*134*), a highly potent unnatural analog of **1b**, which is in phase II clinical trial (see Chapter 5.3) (Charts 73 and 74). The synthesis essentially is a hybrid of *Schinzer* and *Nicolaou* components based on three major fragments **452**, **457**, and **461**. The *Wittig* olefination of ketone **461** and phosphonium salt **457** is modelled after *Nicolaou*'s precedence (Chart 16), and like that, is entirely unselective, furnishing a 1:1-(*E*)/(*Z*)-mixture of **462**, which is separated and the undesired (*E*)-olefin recycled by photochemical (*E*)/(*Z*)-equilibration. The aldol addition of ketone **452** and aldehyde **463** is adapted from *Schinzer*'s approach (Charts 20, 21) to give adduct **464** with "good selectivity", yet only 64% yield. The endgame also follows the routine protocol, namely, *Yamaguchi* lactonization, desilylation, and DMDO epoxidation (selectivity 7:1). The overall sequence is surprisingly long, no less than 44 steps all told. For instance the laborious preparation of **452** from pantolactone **448** requires nine steps, three different protecting groups and repeated chromatographic separations. Similarly, the synthesis of ketone **462** makes use of an expensive *Grignard* reagent and toxic osmium tetroxide. Nevertheless, the synthesis is an impressive achievement and may be understood as a general and encouraging signal to the chemical community that multi-step synthesis can be performed on an industrial scale if so desired.

Abbreviations

Ac, Acetyl; AIBN, 2,2′-Acabisisobutyronitrile; 9-BBN, 9-Borobicyclo[3.3.1]nonane; Bn, Benzyl; Bz, Benzoyl; *CBS*, *Corey-Bakshi-Shibata*; CSA, Camphorsulfonic Acid; DBU, 1,8-Diazabicyclo[5.4.0]undec-7-ene; DCC, Dicyclohexylcarbodiimide; DDQ, 2,3-Dichloro-5,6-dicyano-1,4-benzoquinone; DERA, 2-Deoxyribose-5-Phosphate Aldolase; *Dess-Martin*-PI, *Dess-Martin* periodinane; DHP, Dihydropyran; DIBAL-H, Diisobutylaluminum hydride; DIC, *N,N′*-Diisopropylcarbodiimide; DIPEA, Diisopropylethylamine; DMAP, 4-(Dimethylamino)pyridine; DMDO, Dimethyldioxirane; DMF, *N,N′*-Dimethylformamide; DMPU, *N,N′*-Dimethylpropylenurea; DMS, Dimethylsulfide; DMSO, Dimethylsulfoxide; EDCI, 1-Ethyl-3-(3-dimethylaminopropyl)carbodiimide; HMPA, Hexamethylsilylphosphoramide; Ipc, Isopinocamphenyl; KHMDS, Potassium Hexamethylsilylazide; LAH, Lithium Aluminum Hydride; LDA, Lithium Diisopropylamide; LiHMDS, Lithium Hexamethylsilylazide; *m*CPBA, *m*-Chlorperoxybenzoic acid; MEM, 2-Methoxyethoxymethyl; MMPP, Magnesium Monoperphthalate; MPM, *para*-Methoxyphenylmethyl; MS, Molecular Sieves; Ms, Methansulfonyl; NaHMDS, Sodium Hexamethylsilylazide; NBS, *N*-Bromosuccinimide; NIS, *N*-Iodosuccinimide; NMO, *N*-Methylmorpholine-*N*-oxide; PBS, Phosphate-buffered Saline; PCC, Pyridinium Chlorochromate; PDC, Pyridinium Dichromate; PG, Protective Group; Ph, Phenyl; Piv, Pivaloyl; PMB, *para*-Methoxybenzyl; PPTS, Pyridinium *p*-toluenesulfonate; py, Pyridine; R, Organic Residue; RCM, Ring Closing Metathesis; RedAL, $NaAlH_2(O–CH_2–CH_2–OMe)_2$; *SAE*, *Sharpless* Asymmetric Epoxidation; SEM, 2-(trimethylsilyl)-

ethoxymethyl; *SKR*, *Sharpless* Kinetic Resolution; TBAF, Tetra-*n*-Butylammonium Fluoride; TBDPS, *tert*-Butyldiphenylsilyl; TBHP, *tert*-Butyl Hydroperoxide; TBS, *tert*-Butyldimethylsilyl; TPS, Triphenylsilyl; TCDI, Thiocarbonyl Diimidazole; TES, Triethylsilyl; Tf, Triflate; TFA, Trifluoroacetic Acid; TFAA, Trifluoroacetic Anhydride; THP, Tetrahydropyran; TIPS, Triisopropylsilyl; TMS, Trimethylsilyl; TPAP, Tetra-*n*-Propylammonium Perruthenate; Tr, Trityl; Troc, 2,2,2-Trichloroethyl Carbonate; Ts, Tosyl; *p*TsOH, *p*-Toluenesulfonyl Acide.

References

1. Altmann KH, Pfeiffer B, Arseniyadis S, Pratt BA, Nicolaou KC (2007) The Chemistry and Biology of Epothilones – the Wheel Keeps Turning. Chem Med Chem **2**: 396
2. Watkins EB, Chittiboyina AG, Avery MA (2006) Recent Developments in the Syntheses of the Epothilones and Related Analogues. Eur J Org Chem **18**: 4071
3. Altmann KH, Cachoux F, Caravatti G, Isarno T, Wartmann M (2006) Recent Developments in the Chemistry and Biology of Epothilones. ARKIVOC **8**: 131
4. Luduvico I, Le Hyaric M, De Almeida MV, Da Silva AD (2006) Synthetic Methodologies for the Preparation of Epothilones and Analogs. Mini-Rev Org Chem **3**: 49
5. Watkins EB, Chittiboyina AG, Jung J-C, Avery MA (2005) The Epothilones and Related Analogues – A Review of Their Syntheses and Anti-Cancer Activities. Curr Pharm Design **11**: 1615
6. Altmann KH (2004) The Merger of Natural Product Synthesis and Medicinal Chemistry: On the Chemistry and Chemical Biology of Epothilones. Org Biomol Chem **2**: 2137
7. Altmann KH, Flörsheimer A, O'Reilly T, Wartmann M (2004) The Natural Products Epothilones A and B as Lead Structures for Anticancer Drug Discovery: Chemistry, Biology, and SAR Studies. Prog Med Chem **42**: 171
8. Zhang R, Liu Z (2004) Recent Progress in the Synthesis of Epothilones. Curr Org Chem **8**: 267
9. Mulzer J, Martin HJ (2004) Lessons Learned from Macrolide Synthesis. Chem Record **3**: 258
10. Altmann KH (2003) Epothilone B and its Analogs – a New Family of Anticancer Agents. Mini-Rev Med Chem **3**: 149
11. Nicolaou KC, Snyder SA (2003) Classics in Total Synthesis II, p. 161. Wiley VCH, Weinheim
12. Wartmann M, Altmann KH (2002) The Biology and Medicinal Chemistry of Epothilones. Curr Med Chem Anti-Cancer Agents **2**: 123
13. Stachel SJ, Biswas K, Danishefsky SJ (2001) The Epothilones, Eleutherobins, and Related Types of Molecules. Curr Pharm Design **7**: 1277
14. Nicolaou KC, Ritzen A, Namoto K (2001) Recent Developments in the Chemistry, Biology and Medicine of the Epothilones. Chem Commun: 1523
15. Harris CR, Kuduk SD, Danishefsky SJ (2001) Chemical Synthesis and Biological Studies of the Epothilones – Microtubule Stabilizing Agents with Enhanced Activity Against Multidrug-Resistant Cell Lines and Tumors. Chemistry for the 21st Century: 8
16. Florsheimer A, Altmann KH (2001) Epothilones and Their Analogues – a New Class of Promising Microtubule Inhibitors. Expert Opinion on Therapeutic Patents **11**: 951
17. Wessjohann LA, Scheid G (2000) Synthetic Access to Epothilones – Natural Products with Extraordinary Anticancer Activity. In: Schmalz HG (ed.) Organic Synthesis Highlights IV, p. 251. Wiley VCH, Weinheim

18. Mulzer J (2000) Epothilone B and its Derivatives as Novel Antitumor Drugs: Total and Partial Synthesis and Biological Evaluation. Monatsh Chem **131**: 205

19. a) Harris CR, Balog A, Savin K, Danishefsky SJ, Chou TC, Zhang X-G (1999) Epothilones: Microtubule Stabilizing Agents with Enhanced Activity Against Multidrug-Resistant Cell Lines and Tumors. Act Chim Therap **25**: 187; b) Harris CR, Danishefsky SJ (1999) Complex Target-Oriented Synthesis in the Drug Discovery Process: A Case History in the dEpoB Series. J Org Chem **64**: 8434

20. Jiang J (1999) Advances in Research on Novel Natural Anticancer Compounds: Epothilones. Tianran Chanwu Yanjiu Yu Kaifa **11**: 77

21. Nicolaou KC, Roschangar F, Vourloumis D (1998) Chemical Biology of Epothilones. Angew Chem Int Ed **37**: 2014

22. Chou TC, Zhang XC, Balog A, Su DS, Meng D, Savin K, Bertino JR, Danishefsky SJ (1998) Desoxyepothilone B: An Efficacious Microtubule-Targeted Antitumor Agent with a Promising in Vivo Profile Relative to Epothilone B. Proc Natl Acad Sci USA **95**: 9642

23. Appendino G, Casiraghi G (1998) The Synthesis of Epothilones: Highlights from a Year's Race. Chemtracts **11**: 678

24. Buck SB, Victory SF, Himes RH, Georg GI (1998) Epothilones, a New Class of Microtubule-Stabilizing Agents with a Taxol-Like Mechanism of Action. Chemtracts **11**: 671

25. Nakamura S, Hashimoto S (1997) Total Synthesis of Antitumor Antibiotic Epothilone Having Same Action Mechanism with Taxol. Kagaku (Kyoto) **52**: 70

26. Marshall A, Castellino A (1997) Total Synthesis of Epothilone. Nature Biotechnol **15**: 205

27. Inanaga J, Hirata K, Saeki H, Katsuki T, Yamaguchi M (1979) A Rapid Esterification by Means of Mixed Anhydride and Its Application to Large-Ring Lactonization. Bull Chem Soc Jpn **52**: 1989

28. Schinzer D (2004) The Aldol Reaction in Natural Product Synthesis. The Epothilone Story. Modern Aldol Reactions **1**: 311

29. Harris CR, Kuduk SD, Balog A, Savin K, Glunz PW, Danishefsky SJ (1999) New Chemical Synthesis of the Promising Cancer Chemotherapeutic Agent 12,13-Desoxyepothilone B: Discovery of a Surprising Long-Range Effect on the Diastereoselectivity of an Aldol Condensation. J Am Chem Soc **121**: 7050

30. Lee CB, Wu Z, Zhang F, Chappell MD, Stachel SJ, Chou T-C, Guan Y, Danishefsky SJ (2001) Insights into Long-Range Structural Effects on the Stereochemistry of Aldol Condensations: A Practical Total Synthesis of Desoxyepothilone F. J Am Chem Soc **123**: 5249

31. Bertinato P, Sorensen EJ, Meng D, Danishefsky SJ (1996) Synthesis of Epothilone A: Stereocontrolled Assembly of the Acyl Region and Models for Macrocyclization. J Org Chem **61**: 8000

32. Meng D, Sorensen EJ, Bertinato P, Danishefsky SJ (1996) Studies Toward a Synthesis of Epothilone A: Use of Hydropyran Templates for the Management of Acyclic Stereochemical Relationships. J Org Chem **61**: 7998

33. Balog A, Meng D, Kamenecka T, Bertinato P, Su D-S, Sorensen EJ, Danishefsky SJ (1997) Total Synthesis of (−)-Epothilone A. Angew Chem Int Ed **35**: 2801

34. Meng D, Bertinato P, Balog A, Su D-S, Kamenecka T, Sorensen EJ, Danishefsky SJ (1997) Total Syntheses of Epothilones A and B. J Am Chem Soc **119**: 10073

35. Balog A, Harris CR, Savin K, Zhang X-G, Chou T-C, Danishefsky SJ (1998) A Novel Aldol Condensation with 2-Methyl-4-Pentenal and its Application to an Improved Total Synthesis of Epothilone B. Angew Chem Int Ed **37**: 2675

36. a) Harris CR, Kuduk SD, Savin K, Balog A, Danishefsky SJ (1999) Dianion Equivalents Corresponding to the Polypropionate Domain of Epothilone B. Tetrahedron Lett **40**: 2263; b) Su D-S, Meng D, Bertinato P, Balog A, Sorensen EJ, Danishefsky SJ, Zheng H, Chou T-C, He L, Horwitz SB (1997) Total Synthesis of (−)-Epothilone B: An Extension of the Suzuki Coupling Method and Insights into Structure-Activity Relationships of the Epothilones. Angew Chem Int Ed **36**: 757
37. Stachel SJ, Danishefsky SJ (2001) Chemo- and Stereoselective Epoxidation of 12,13-Desoxyepothilone B Using 2,2′-Dimethyldioxirane. Tetrahedron Lett **42**: 6785
38. Meng D, Su D-S, Balog A, Bertinato P, Sorensen EJ, Danishefsky SJ, Zheng H, Chou, TC, He L, Horwitz SB (1997) Remote Effects in Macrolide Formation Through Ring-Forming Olefin Metathesis: An Application to the Synthesis of Fully Active Epothilone Congeners. J Am Chem Soc **119**: 2733
39. Rivkin A, Cho YS, Gabarda AE, Yoshimura F, Danishefsky SJ (2004) Application of Ring-Closing Metathesis Reactions in the Synthesis of Epothilones. J Nat Prod **67**: 139
40. a) Chappell MD, Stachel SJ, Lee CB, Danishefsky SJ (2000) En Route to a Plant Scale Synthesis of the Promising Antitumor Agent 12,13-Desoxyepothilone B. Org Lett **2**: 1633; b) Rivkin A, Njardarson JT, Biswas K, Chou T-C, Danishefsky SJ (2002) Total Syntheses of [17]- and [18]-Dehydrodesoxyepothilones B *via* a Concise Ring-Closing Metathesis-Based Strategy: Correlation of Ring Size with Biological Activity in the Epothilone Series. J Org Chem **67**: 7737
41. Biswas K, Lin H, Njardarson JT, Chappell MD, Chou T-C, Chao G, Yongbiao T, William P, He L, Horwitz SB, Danishefsky SJ (2002) Highly Concise Routes to Epothilones: The Total Synthesis and Evaluation of Epothilone 490. J Am Chem Soc **124**: 9825
42. Rivkin A, Yoshimura F, Gabarda AE, Chou T-C, Dong H, Tong WP, Danishefsky SJ (2003) Complex Target-Oriented Total Synthesis in the Drug Discovery Process: The Discovery of a Highly Promising Family of Second Generation Epothilones. J Am Chem Soc **125**: 2899
43. Nicolaou KC, He Y, Vourloumis D, Vallberg H, Yang Z (1996) An Approach to Epothilones Based on Olefin Metathesis. Angew Chem Int Ed **35**: 2399
44. Nicolaou KC, He Y, Vourloumis D, Vallberg H, Roschangar F, Sarabia F, Ninkovic S, Yang Z, Trujillo JI (1997) The Olefin Metathesis Approach to Epothilone A and Its Analogs. J Am Chem Soc **119**: 7960
45. Nicolaou KC, King NP, He Y (1998) Ring-Closing Metathesis in the Synthesis of Epothilones and Polyether Natural Products. Top Organomet Chem **1** (Alkene Metathesis in Organic Synthesis): 73
46. Nicolaou KC, Sarabia F, Ninkovic S, Yang Z (1997) Total Synthesis of Epothilone A: The Macrolactonization Approach. Angew Chem Int Ed **36**: 525
47. Nicolaou KC, Ninkovic S, Sarabia F, Vourloumis D, He Y, Vallberg H, Finlay MRV, Yang Z (1997) Total Syntheses of Epothilones A and B *via* a Macrolactonization-Based Strategy. J Am Chem Soc **119**: 7974
48. Nicolaou KC, Winssinger N, Pastor J, Ninkovic S, Sarabia F, He Y, Vourloumis D, Yang Z, Li T, Giannakakou P, Hamel E (1997) Synthesis of Epothilones A and B in Solid and Solution Phase. Nature **387**: 268; Erratum: Nature **390**: 100
49. Nicolaou KC, Hepworth D, Finlay MRV, King NP, Werschkun B, Bigot A (1999) Synthesis of 16-Desmethyl Epothilone B: Improved Methodology for the Rapid, Highly Selective and Convergent Construction of Epothilone B and Analogs. Chem Commun: 519
50. Schinzer D, Limberg A, Boehm OM (1996) Studies Towards the Total Synthesis of Epothilones: Asymmetric Synthesis of the Key Fragments. Chemistry Eur J **2**: 1477

51. Schinzer D, Limberg A, Bauer A, Boehm OM, Cordes M (1997) Total Synthesis of (−)-Epothilone A. Angew Chem Int Ed **36**: 523

52. Schinzer D, Bauer A, Boehm OM, Limberg A, Cordes M (1999) Total Synthesis of (−)-Epothilone A. Chemistry Eur J **5**: 2483

53. Schinzer D, Bauer A, Schieber J (1998) Synthesis of Epothilones. Stereoselective Routes to Epothilone B. Synlett: 861

54. Schinzer D, Bauer A, Schieber J (1999) Syntheses of (−)-Epothilone B. Chemistry Eur J **5**: 2492

55. Sinha SC, Barbas CF III, Lerner LA (1998) The Antibody Catalysis Route to the Total Synthesis of Epothilones. Proc Natl Acad Sci USA **95**: 14603

56. Sinha SC, Sun J, Miller G, Barbas CF III, Lerner RA (1999) Sets of Aldolase Antibodies with Antipodal Reactivities. Formal Synthesis of Epothilone E by Large-Scale Antibody-Catalyzed Resolution of Thiazole Aldol. Org Lett **1**: 1623

57. Sinha SC, Sun J, Miller GP, Wartmann M, Lerner RA (2001) Catalytic Antibody Route to the Naturally Occurring Epothilones: Total Synthesis of Epothilones A–F. Chem Eur J **7**: 1691

58. Sun J, Sinha SC (2002) Stereoselective Total Synthesis of Epothilones by the Metathesis Approach Involving C9–C10 Bond Formation. Angew Chem Int Ed **41**: 1381

59. Taylor RE, Haley JD (1997) Towards the Synthesis of Epothilone A: Enantioselective Preparation of the Thiazole Sidechain and Macrocyclic Ring Closure. Tetrahedron Lett **38**: 2061

60. Bode JW, Carreira EM (2001) Stereoselective Syntheses of Epothilones A and B *via* Nitrile Oxide Cycloadditions and Related Studies. J Org Chem **66**: 6410

61. Bode JW, Carreira EM (2001) Stereoselective Syntheses of Epothilones A and B *via* Directed Nitrile Oxide Cycloaddition. J Am Chem Soc **123**: 3611

62. Review of the Nitrile Oxide Approach: Blackman B, Georg GI (2003) Stereoselective Syntheses of Epothilones A and B *via* Nitrile Oxide Cycloadditions. Chemtracts **16**: 97

63. Shibasaki M, Kanai M (2001) Multifunctional Asymmetric Catalysis. Chem Pharm Bull **49**: 511

64. Sawada D, Shibasaki M (2000) Enantioselective Total Synthesis of Epothilone A Using Multifunctional Asymmetric Catalyses. Angew Chem Int Ed **39**: 209

65. Hamashima Y, Sawada D, Nogami H, Kanai M, Shibasaki M (2001) Highly Enantioselective Cyanosilylation of Aldehydes Catalyzed by a Lewis Acid–Lewis Base Bifunctional Catalyst. Tetrahedron **57**: 805

66. Sawada D, Kanai M, Shibasaki M (2000) Enantioselective Total Synthesis of Epothilones A and B Using Multifunctional Asymmetric Catalysis. J Am Chem Soc **122**: 10521

67. Fürstner A, Mathes C, Grela K (2001) Concise Total Syntheses of Epothilone A and C Based on Alkyne Metathesis. Chem Commun: 1057

68. Fürstner A, Mathes C, Lehmann CW (2001) Alkyne Metathesis: Development of a Novel Molybdenum-Based Catalyst System and its Application to the Total Synthesis of Epothilone A and C. Chem Eur J **7**: 5299

69. Liu Z-Y, Yu C-Z, Wang R-F, Li G (1998) Epoxide Opening with Acetylide for Synthesis of Epothilone A C7-21 Segment. Tetrahedron Lett **39**: 5261

70. Liu Z-Y, Chen Z-C, Yu C-Z, Wang R-F, Zhang R-Z, Huang C-S, Yan Z, Cao D-R, Sun J-B, Li G (2002) Total Synthesis of Epothilone A Through Stereospecific Epoxidation of the *p*-Methoxybenzyl Ether of Epothilone C. Chem Eur J **8**: 3747

71. Tokunaga M, Larrow JF, Kakiuchi F, Jacobsen EN (1997) Asymmetric Catalysis with Water: Efficient Kinetic Resolution of Terminal Epoxides by Means of Catalytic Hydrolysis. Science **277**: 936

72. Hinterding K, Jacobsen EN (1999) Regioselective Carbomethoxylation of Chiral Epoxides: A New Route to Enantiomerically Pure β-Hydroxy Esters. J Org Chem **64**: 2164
73. Zhu B, Panek JS (2000) Enzymatic Resolution of Thiazole-Containing Vinyl Carbinols. Synthesis of the C12–C21 Fragment of the Epothilones. Tetrahedron Lett **41**: 1863
74. Zhu B, Panek JS (2000) Total Synthesis of Epothilone A. Org Lett **2**: 2575
75. Zhu B, Panek JS (2001) Methodology Based on Chiral Silanes in the Synthesis of Polypropionate-Derived Natural Products – Total Synthesis of Epothilone A. Eur J Org Chem: 1701
76. Machajewski TD, Wong C-H (1999) Chemoenzymic Synthesis of Key Epothilone Fragments. Synthesis: 1469
77. Liu J, Wong C-H (2002) Aldolase-Catalyzed Asymmetric Synthesis of Novel Pyranose Synthons as a New Entry to Heterocycles and Epothilones. Angew Chem Int Ed **41**: 1404
78. Storer RJ, Takemoto T, Jackson PS, Ley SV (2003) A Total Synthesis of Epothilones Using Solid-Supported Reagents and Scavengers. Angew Chem Int Ed **42**: 2521
79. Storer RJ, Takemoto T, Jackson PS, Brown DS, Baxendale IR, Ley SV (2004) Multi-Step Application of Immobilized Reagents and Scavengers: A Total Synthesis of Epothilone C. Chem Eur J **10**: 2529
80. Mulzer J, Mantoulidis A (1996) Synthesis of the C(1)–C(9) Segment of the Cytotoxic Macrolides Epothilone A and B. Tetrahedron Lett **37**: 9179
81. Mulzer J, Mantoulidis A, Öhler E (1997) Synthesis of the C(11)–C(20) Segment of the Cytotoxic Macrolide Epothilone B. Tetrahedron Lett **38**: 7725
82. Mulzer J, Mantoulidis A, Öhler E (1998) Easy Access to the Epothilone Family – Synthesis of Epothilone B. Tetrahedron Lett **39**: 8633
83. Mulzer J, Mantoulidis A, Öhler E (2000) Total Syntheses of Epothilones B and D. J Org Chem **65**: 7456
84. Gaich T, Mulzer J (2005) Synthesis of Epothilones *via* a Silicon-Tethered RCM Reaction. Org Lett **7**: 1311
85. Gaich T, Karig G, Martin HJ, Mulzer J (2006) New Solutions to the C-12,13 Stereoproblem of Epothilones B and D; Synthesis of a 12,13-Diol-Acetonide Epothilone B Analog. Eur J Org Chem: 3372
86. Martin HJ, Drescher M, Mulzer J (2000) How Stable are Epoxides? A Novel Synthesis of Epothilone B. Angew Chem Int Ed **39**: 581
87. Martin HJ, Pojarliev P, Kahlig H, Mulzer J (2001) The 12,13-Diol Cyclization Approach for a Truly Stereocontrolled Total Synthesis of Epothilone B and the Synthesis of a Conformationally Restrained Analog. Chem Eur J **7**: 2261
88. May SA, Grieco PA (1998) Total Synthesis of (−)-Epothilone B. Chem Commun: 1597
89. White JD, Carter RG, Sundermann KF (1999) A Highly Stereoselective Synthesis of Epothilone B. J Org Chem **64**: 684
90. White JD, Sundermann KF, Carter RG (1999) Improved Synthesis of Epothilone B Employing Alkylation of an Alkyne for Assembly of Subunits. Org Lett **1**: 1431
91. White JD, Sundermann KF, Wartmann M (2002) Synthesis, Conformational Analysis, and Bioassay of 9,10-Didehydroepothilone D. Org Lett **4**: 995
92. a) White JD, Carter RG, Sundermann KF, Wartmann M (2001) Total Synthesis of Epothilone B, Epothilone D, and cis- and trans-9,10-Dehydroepothilone D. J Am Chem Soc **123**: 5407; b) White JD, Carter RG, Sundermann KF, Wartmann M (2003) Erratum: J Am Chem Soc **125**: 3190

93. Ermolenko MS, Potier P (2002) Synthesis of Epothilone A and B from D-Glucose. Tetrahedron Lett **43**: 2895

94. Martin N, Thomas EJ (2001) Total Syntheses of Epothilones B and D: Applications of Allylstannanes in Organic Synthesis. Tetrahedron Lett **42**: 8373

95. Bijoy P, Avery MA (1998) Synthetic Studies Directed Towards Epothilone A: Enantioselective Synthesis of a C7–C15 Carboxaldehyde Segment. Tetrahedron Lett **39**: 209

96. Panicker B, Karle JM, Avery MA (2000) An Unusual Reversal of Stereoselectivity in a Boron Mediated Aldol Reaction: Enantioselective Synthesis of the C1–C6 Segment of the Epothilones. Tetrahedron **56**: 7859

97. Valluri M, Hindupur RM, Bijoy P, Labadie G, Jung J-C, Avery MA (2001) Total Synthesis of Epothilone B. Org Lett **3**: 3607

98. Hindupur RM, Panicker B, Valluri M, Avery MA (2001) Total Synthesis of Epothilone A. Tetrahedron Lett **42**: 7341

99. Jung J-C, Kache R, Vines KK, Zheng Y-S, Panicker B, Valluri M, Avery MA (2004) Total Syntheses of Epothilones B and D. J Org Chem **69:** 9269

100. Taylor RE, Haley JD (1997) Towards the Synthesis of Epothilone A: Enantioselective Preparation of the Thiazole Sidechain and Macrocyclic Ring Closure. Tetrahedron Lett **38**: 2061

101. Taylor RE, Galvin GM, Hilfiker KA, Chen Y (1998) A Formal Total Synthesis of Epothilone A: Enantioselective Preparation of the C1–C6 and C7–C12 Fragments. J Org Chem **63**: 9580

102. Taylor RE, Chen Y (2001) Total Synthesis of Epothilones B and D. Org Lett **3**: 2221

103. Claus E, Pahl A, Jones PG, Meyer HM, Kalesse M (1997) Synthesis of the C1–C9 Segment of Epothilones. Tetrahedron Lett **38**: 1359

104. Gerlach K, Quitschalle M, Kalesse M (1998) Synthesis of the C(7)–C(17) Segment of Epothilones by a 10-Membered Ring Closing Metathesis Reaction. Synlett: 1108

105. Kalesse M, Quitschalle M, Claus E, Gerlach K, Pahl A, Meyer HH (1999) The Formal Total Synthesis of Epothilone A. Eur J Org Chem: 2817

106. Quitschalle M, Kalesse M (1999) Improved Synthesis of the Northern Hemisphere of Epothilone A by a Sharpless Asymmetric Dihydroxylation. Tetrahedron Lett **40**: 7765

107. Gerlach K, Quitschalle M, Kalesse M (1999) Synthesis of the Northern Hemisphere of Epothilone A by a Ten-Membered Ring Closing Metathesis Reaction. Tetrahedron Lett **40**: 3553

108. Chandrasekhar S, Reddy CR (2002) Towards a Synthesis of Epothilone A: Asymmetric Synthesis of C(1)–C(6) and C(7)–C(15) Fragments. Tetrahedron Asym **13**: 261

109. Ramachandran PV, Prabhudas B, Pratihar D, Chandra JS, Reddy MVR (2003) Stereoselective Synthesis of the C7–C21 Segment of Epothilone A *via* Asymmetric Alkoxyallyl- and Crotylboration. Tetrahedron Lett **44**: 3745

110. Ramachandran PV, Prabhudas B, Chandra JS, Reddy MVR, Brown HC (2004) Preparative-scale Synthesis of Both Antipodes of B-γ,γ-Dimethylallyldiisopinocampheylborane: Application for the Synthesis of C1–C6 Subunit of Epothilone. Tetrahedron Lett **45**: 1011

111. a) De Brabander J, Rosset S, Bernardinelli G (1997) Towards a Synthesis of Epothilone A. Rapid Assembly of the C(1)–C(6) and C(7)–C(12) Fragments. Synlett: 824; b) De Brabander J, Rosset S, Bernardinelli G (1998) Towards a Synthesis of Epothilone A. Rapid Assembly of the C1–C6 and C7–C12 Fragments [Erratum to document cited in CA127:234203]. Synlett: 32

112. Scheid G, Ruijter E, Konarzycka-Bessler M, Bornscheuer UT, Wessjohann LA (2004) Synthesis and Resolution of a Key Building Block for Epothilones: A Comparison of Asymmetric Synthesis, Chemical and Enzymatic Resolution. Tetrahedron Asym **15**: 2861
113. Bekish AV, Isakov VE, Kulinkovich OG (2005) A Cyclopropanol Approach to the Synthesis of the C13–C21 Fragment of Epothilones from Diethyl (S)-Malate. Tetrahedron Lett **46**: 6979
114. Reiff EA, Nair SK, Reddy BSN, Inagaki J, Henri JT, Greiner JF, Georg GI (2004) Practical Syntheses of the C12–C21 Epothilone Subunit *via* Catalytic Asymmetric Reductions: Itsuno-Corey Oxazaborolidine Reduction and Asymmetric Noyori Hydrogenation. Tetrahedron Lett **45**: 5845
115. Shioji K, Kawaoka H, Miura A, Okuma K (2001) Synthesis of C1–C6 Fragment for Epothilone A *via* Lipase-Catalyzed Optical Resolution. Synthetic Comm **31**: 3569
116. Tanimori S, Tanimoto K, Kirihata M (1998) Simple Synthesis of Both Enantiomers of the C7–C12 Segment of Epothilones. Biosci, Biotech BioCh **62**: 2428
117. Dong SD, Sundermann K, Smith KMJ, Petryka J, Liu F, Myles DC (2004) Rapid Access to Epothilone Analogs *via* Semisynthetic Degradation and Reconstruction of Epothilone D. Tetrahedron Lett **45**: 1945
118. Höfle G, Glaser N, Leibold T, Karama U, Sasse F, Steinmetz H (2003) Semisynthesis and Degradation of the Tubulin Inhibitors Epothilone and Tubulysin. Pure Appl Chem **75**: 167
119. Chakraborty TK, Dutta S (1998) Radical-Induced Opening of Trisubstituted Epoxides: Application in the Synthesis of C1–C12 Segment of Epothilones. Tetrahedron Lett **39**: 101
120. Lee HW, Hong YD, Lee I-YC (1999) Synthesis Toward Epothilone A: Coupling Reaction Between the Sulfone of C1–C10 and the Allylic Bromide of C11–C21. Bull Kor Chem Soc **20**: 403
121. Lee HW, Lee I-YC, Hong YD (2000) Synthesis of the C11–C21 and C13–C21 Fragments of Epothilones from D-Glucose. Bull Kor Chem Soc **21**: 1177
122. Lalic G, Galonic D, Matovic R, Saicic RN (2202) A Model Study of Epothilone Synthesis: An Alternative Synthetic Approach to the C1–C7 Fragment. J Serb Chem Soc **67**: 221
123. Guzman A, Alfaro R, Diaz E, Quintero A, Solano J (2002) Synthesis and Cytotoxic Evaluation of Some Structural Fragments of Epothilone A. Revista de la Sociedad Quimica de Mexico **46**: 235
124. a) Zhang HS, Zhong CF, Bao XP (2003) Studies on the Total Synthesis of Epothilone B and D: A Facile Synthesis of C7–C14 and C15–C21 Fragments. Chin Chem Lett **14**: 115; b) Hong YD, Lee HW, Lee IYC, Jung J-K (2007) Synthesis of the C11–C21 Fragments of Epothilones A and B Using Ring-Closing Metathesis. Bull Kor Chem Soc **28**: 367
125. Broadrup RL, Sundar HM, Swindell CS (2005) Total Synthesis of 12,13-Desoxyepothilone B (Epothilone D). Bioorg Chem **33**: 116
126. Nicolaou KC, Hepworth D, King NP, Finlay MRV, Scarpelli R, Pereira MMA, Bollbuck B, Bigot A, Werschkun B, Winssinger N (2000) Total Synthesis of 16-Desmethylepothilone B, Epothilone B10, Epothilone F, and Related Side Chain Modified Epothilone B Analogues. Chem Eur J **6**: 2783
127. Nicolaou KC, Pratt BA, Arseniyadis S, Wartmann M, O'Brate A, Giannakakou P (2006) Molecular Design and Chemical Synthesis of a Highly Potent Epothilone. Chem Med Chem **1**: 41

128. Sinha SC, Sun J, Miller GP, Wartmann M, Lerner RA (2001) Catalytic Antibody Route to the Naturally Occurring Epothilones: Total Synthesis of Epothilones A–F. Chem Eur J **7**: 1691
129. Nicolaou KC, Sasmal PK, Rassias G, Reddy MV, Altmann K-H, Wartmann M, O'Brate A, Giannakakou P (2003) Design, Synthesis, and Biological Properties of Highly Potent Epothilone B Analogues. Angew Chem Int Ed **42**: 3515
130. Nicolaou KC, Ritzen A, Namoto K, Buey RM, Diaz JF, Andreu JM, Wartmann M, Altmann K-H, O'Brate A, Giannakakou P (2002) Chemical Synthesis and Biological Evaluation of Novel Epothilone B and *trans*-12,13-Cyclopropyl Epothilone B Analogues. Tetrahedron **58**: 6413
131. Nicolaou KC, Namoto K, Li J, Ritzen A, Ulven T, Shoji M, Zaharevitz D, Gussio R, Sackett DL, Ward RD, Hensler A, Fojo T, Giannakakou P (2001) Synthesis and Biological Evaluation of 12,13-Cyclopropyl and 12,13-Cyclobutyl Epothilones. Chem BioChem **2**: 69
132. Nicolaou KC, Scarpelli R, Bollbuck B, Werschkun B, Pereira MMA, Wartmann M, Altmann K-H, Zaharevitz D, Gussio R, Giannakakou P (2000) Chemical Synthesis and Biological Properties of Pyridine Epothilones. Chem Bio **7**: 593
133. Nicolaou KC, Namoto K, Ritzen A, Ulven T, Shoji M, Li J, D'Amico G, Liotta D, French CT, Wartmann M, Altmann K-H, Giannakakou P (2001) Chemical Synthesis and Biological Evaluation of *cis*- and *trans*-12,13-Cyclopropyl and 12,13-Cyclobutyl Epothilones and Related Pyridine Side Chain Analogues. J Am Chem Soc **123**: 9313
134. Klar U, Buchmann B, Schwede W, Skuballa W, Hoffmann J, Lichtner RB (2006) Total Synthesis and Antitumor Activity of ZK-EPO: The First Fully Synthetic Epothilone in Clinical Development. Angew Chem Int Ed **45**: 7942

5. Semisynthetic Derivatives of Epothilones

Karl-Heinz Altmann

Department of Chemistry and Applied Biosciences, Institute of Pharmaceutical Sciences, Swiss Federal Institute of Technology (ETH), Zürich, Switzerland

Contents

5.1. Introduction

Semisynthetic derivatives of natural products traditionally occupy a prominent space in natural-product-based drug discovery (*1*, *2*). As many biologically active natural products exhibit a high degree of structural complexity (*3*), the chemical derivatization of material isolated from natural sources often represents the only feasible means (or at least the only economically viable approach) to explore structure-activity-relationships (SAR) and to produce analogs with more favorable pharmacokinetic and pharmacological properties than the natural product lead. Examples of clinically important drugs that are semisynthetic derivatives of natural products exist in virtually all disease areas (*1*, *2*); in the treatment of cancer this includes compounds such as etoposide or teniposide (derived from podophyllotoxin) (*4–6*), irinotecan and topotecan (derived from camptothecin) (*7–9*), or docetaxel (derived from 10-deacetylbaccatin III) (*10*, *11*). Even for taxol (*11*), which is a natural product (*12*), the sustained supply of sufficient quantities of material for widespread clinical use could only be secured through the development of a semisynthetic production process from another natural product, namely, 10-deacetylbaccatin III (*13*). In light of these facts, it is not surprising that semisynthesis approaches have also featured prominently in the elucidation of the SAR for epothilones and in the discovery of a number of clinical development candidates. As will be shown in subsequent chapters, epothilones have proven not to be subject to the same chemical limitations as many other natural products and much of their SAR has been derived from fully synthetic analogs, many of which would not have been accessible through semisynthesis (see Chapter 6). Nevertheless, out of seven epothilones that have entered clinical evaluation in humans so far (including the natural product epothilone B (Epo B)), only one is produced by total chemical synthesis (ZK-Epo (Sagopilone); see Chapter 4). This bias towards semisynthetic derivatives most likely reflects the technical (fewer chemical steps) and economic (cost of goods) advantages still associated with natural product derivatization.

Obviously, the most fundamental provision for the generation of semisynthetic derivatives of a natural product is a sufficient supply of the natural product starting material itself. In an initial exploration phase, which is largely characterized by SAR work at the *in vitro* and, to a more limited extent, the *in vivo* level, the availability of the material as such is the most important aspect. However, at later stages of drug development, a cost-effective production process is an indispensable requirement, in order to ensure a sustained supply of material of a semisynthesis derivative for clinical studies and eventual commercialization. In the case of the epothilones, only few groups have had access to fermentatively produced starting materials as the basis for semisynthesis work, which reflects the complexities associated with the fermentation process for the production of natural epothilones (see Chapter 3). Thus, most of the semisynthesis work reported in literature originates either from the group of *Höfle* at the former "Gesellschaft für Biotechnologische Forschung" in Braunschweig, Germany (GBF, now "Helmholtz Centre for Infection Research"), one of the discoverers of epothilones, or the group at Bristol-Myers-Squibb (BMS). These groups have entered into a broad-based research collaboration in 1997 (*14*). Semisynthesis work on epothilones, although more limited in scope, has also been reported by the group at Novartis and, more recently, at Kosan. The work at Kosan is based on a process that allows the production of natural epothilones through heterologous expression of the epothilone gene cluster in organisms other than *Sorangium cellulosum* (*15*).

This chapter will provide an overview on the semisynthetic work that has been reported for epothilones in public literature over the last decade. The discussion will mostly focus on the organic chemistry of the system, while aspects of biological activity will only be addressed in a more general way. In order to ensure a coherent discussion of the epothilone SAR, the biological activities of important semisynthetic epothilone derivatives and the conclusions derived from their analysis are discussed in detail in Chapter 6 together with the data for fully synthetic analogs. In order to facilitate cross-referencing between chapters, the current chapter is structured by analogy to Chapter 6, *i.e.*, modifications are grouped according to their particular location in the overall molecular framework in largely the same way as in Chapter 6.

5.2. The O16–C8 Sector ("Polyketide Sector")

5.2.1. Modification of the Ester Moiety

One of the most obvious and seemingly straightforward modifications of the epothilone scaffold is the cleavage of the ester bond to produce the corresponding *seco*-acid. This transformation has been achieved by treatment of Epo A with NaOH/MeOH, which gave *seco*-acid **V-1** in 65% yield together with 23% of the retro aldol product **V-2** (*16*). Ester hydrolysis can also be accomplished with pig liver esterase (PLE) as a catalyst, which leads to a more selective course of the reaction (*14*).

Seco-acid **V-1** can be further degraded to ester **V-3** and diol **V-4** *via* acid-catalyzed epoxide hydrolysis followed by esterification with

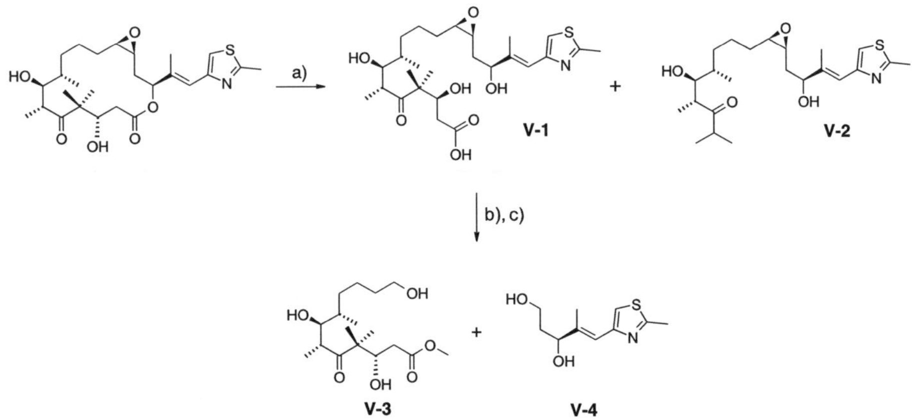

Scheme 1. a) NaOH/MeOH, RT, 5 min, 65% (**V-1**) and 24% (**V-2**). b) 1. H_2SO_4, THF/H_2O, RT, 5 h. 2. TMS-CH_2N_2, 2.5 h, 54%. c) 1. $NaIO_4$, MeOH/H_2O, RT, 45 min; 2. BH_4^--ion exchange resin, 41% (**V-3**) and 49% (**V-4**)

TMS-diazomethane, periodate cleavage of the resulting diol, and finally borohydride-based reduction of the mixture of cleavage products. Compounds **V-3** and **V-4** may be used as chiral building blocks for the synthesis of other complex target structures. Although no primary data on the biological activity of **V-1** have appeared in the literature, the Novartis group in several presentations has characterized the corresponding Epo B derivative as several thousand-fold less active in cell proliferation assays than Epo B. This is in line with the observation by the BMS group that the preincubation of epothilones with mouse plasma, which causes rapid ester cleavage, leads to significantly diminished cytotoxicity *in vitro* (*17*). While the situation is more complex *in vivo* and the plasma stability of Epo B is unlikely to be an issue in humans (due to the much higher stability of the compound in human plasma), these findings have inspired the BMS group to pursue lactam-based epothilone analogs as metabolically more stable alternatives to the natural macrolactones (*17*, *18*). In this process, the group has developed a very elegant and creative approach for the preparation of the macrolactam analogs of Epo A and B, which exploits the allylic nature of the ester group in epothilones and involves Pd(0)-catalyzed ring opening and trapping of the ensuing Pd-π complex with a nitrogen nucleophile. Thus, the treatment of Epo A or B with NaN_3 in the presence of $Pd(PPh_3)_4$ leads to azido acids **V-5** and **V-6**, with full retention of configuration at C15 (Scheme 2).

Reduction of the azide group either through catalytic hydrogenation or with trimethyl phosphine, followed by ring closure under amide-bond-forming conditions then furnished the lactam analogs of Epo A and B, **V-9** and **V-10**. Building on this chemistry, the BMS group has

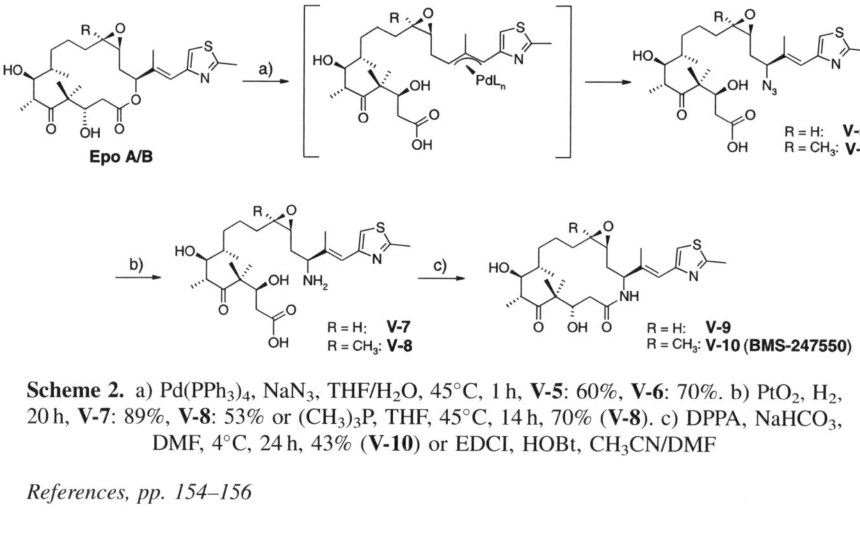

Scheme 2. a) $Pd(PPh_3)_4$, NaN_3, THF/H_2O, 45°C, 1 h, **V-5**: 60%, **V-6**: 70%. b) PtO_2, H_2, 20 h, **V-7**: 89%, **V-8**: 53% or $(CH_3)_3P$, THF, 45°C, 14 h, 70% (**V-8**). c) DPPA, $NaHCO_3$, DMF, 4°C, 24 h, 43% (**V-10**) or EDCI, HOBt, CH_3CN/DMF

also developed a highly efficient one-pot process for the conversion of Epo B into lactam derivative **V-10**, which involves the above Pd(0)-catalyzed ring-opening reaction, reduction of the azide with trimethyl phosphine and macrolactam formation with EDCI/HOBt (*17*) and provides the desired lactam in 45% overall yield. Epo B lactam **V-10**, although less potent than Epo B, is still a highly active antiproliferative agent that was selected by BMS for clinical development (as BMS-247550) and has recently been approved by the US FDA for clinical use in cancer patients (s. Chapter 7).

5.2.2. Modification in the C2–C8 Region

In addition to lactam-based analogs, the BMS group, in collaboration with *Höfle's* group at the GBF, has also investigated modifications at the 3-position of the epothilone macrocycle. Based on chemistry initially developed at the GBF, Epo A and B could be converted into the α,β-unsaturated lactones **V-13** and **V-14**, through bis-formylation of the C3- and C7–OH groups and subsequent selective based-induced elimination of formic acid across the C2–C3 bond (followed by removal of the C7–O–formyl group with NH_3/MeOH; Scheme 3) (*19*). **V-13** and **V-14** were then further elaborated into (3*S*)-3-deoxy-3-cyano derivatives **V-17** and **V-18** (Scheme 3). The *Michael* addition of cyanide ion to **V-15** and **V-16** proceeded without detectable selectivity, thus furnishing the addition products as 1/1 mixtures of (3*S*) and (3*R*) isomers. These mixtures could be separated by flash chromatography and the individual

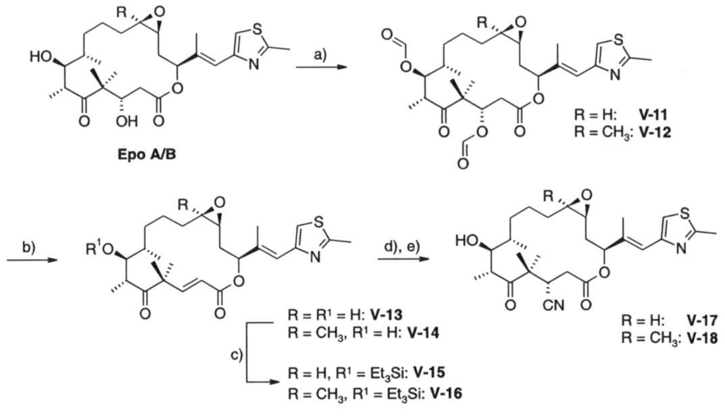

Scheme 3. a) HCO_2H, TEA, DMAP, H_2O. b) DBU, CH_2Cl_2; then NH_3, MeOH, **V-11**: 73%, **V-12**: 95%. c) Et_3SiCl, TEA, DMAP, CH_2Cl_2, 45°C, **V-13**: 96%, **V-14**: 90%. d) KCN, 18-C-6, DMF, **V-15**: 38%, **V-16**: 65%. e) AcOH/THF/H_2O, 83% (**V-17**), 41% (**V-18**)

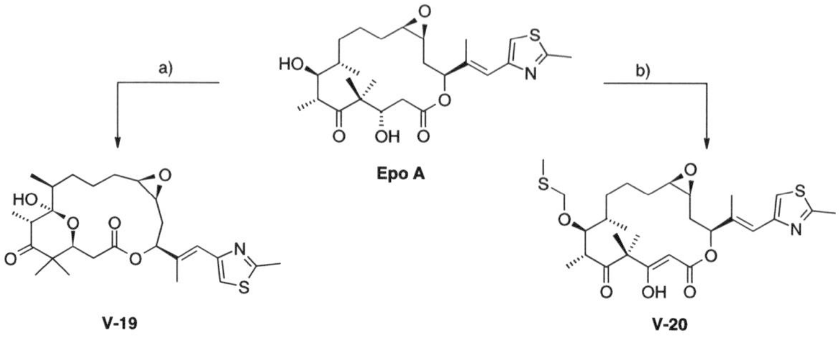

Scheme 4. a) PDC, CH_2Cl_2, RT, 36 h, 80%. b) $(CH_3)_2S$, $(PhCO_2)_2$, CH_3CN, 0°C → RT, 24 h, 30%

isomers were separately deprotected with acetic acid to provide **V-17** and **V-18** (and also their corresponding (3*R*) isomers). Analogs **V-13/V-17** and **V-14/V-18** each are *ca.* one order of magnitude less potent in cytotoxicity assays than Epo A or B, respectively (*19*); the *in vitro* activity of Epo B derivatives **V-14** and **V-18** thus appears to be comparable with that of taxol. In contrast, the (3*R*) isomers of **V-17** and **V-18** exhibit significantly reduced activity (*19*).

Oxidative and reductive transformations in the C2–C8 region of Epo A have been investigated by the GBF group, who established that the 7-hydroxyl group can be selectively oxidized with neutral pyridinium dichromate (PDC) in CH_2Cl_2 or DMF to provide acetal **V-19** in 80% yield (Scheme 4) (*20*).

In contrast, oxidation under *Swern* conditions or with basic PDC was either less selective or produced complex reaction mixtures (*20*). Selective oxidation of the 3-hydroxyl group is more difficult, but could be accomplished by treatment of Epo A with a mixture of dimethylsulfide and dibenzoylperoxide (*21*), which furnished the 3-oxo derivative as its C7-methylthiomethyl (MTM) ether **V-20** in 30% yield and as a 2/3 mixture of keto/enol isomers (*20*). In addition, the reaction produced the C7-mono-MTM and the C3,C7-bis-MTM ethers of Epo A each in about 30% yield. Treatment of Epo A with $NaBH_4$ in a THF/*pH* 7 buffer system gave a 1/1 mixture of diastereomeric C5-alcohols (*ca.* 70%), while unbuffered reaction conditions led to rapid cleavage of the ester bond (*20*). Interestingly, treatment of **V-19** with $NaBH_4$ in MeOH results in the selective reduction of the C5-keto group without any effect on the C7-ketone, in spite of the fact that *ca.* 3% of the latter is present as the keto alcohol (rather than the cyclic acetal) (*20*). Acetal **V-19** has been reported to retain some activity against the mouse leukemia cell line L929 (IC_{50} of *ca.* 400 n*M* *vs.* 8 n*M* for Epo A (*22*)), while oxidation at C3 or reduction at C5 led to IC_{50} values in excess of 1 μM (*20*).

5.3. Modification of the Epoxide Moiety

In light of the multitude of transformations that are conceivable for an oxirane ring and the potential for further elaboration of the initial reaction products, it is not surprising that modifications of the epoxide moiety have been an important trait of the semisynthetic work on epothilones from the very beginning. The earliest contributions to this area again originate from the GBF group, who investigated the possible transformation of epothilones A, B, and C into a variety of C12/C13 modified derivatives (23). Thus, treatment of Epo A with HCl in aqueous THF or with 1 *M* HCl produced chlorohydrins **V-21** and **V-22** in 60–80% overall yield with a 2–4:1 preference for the C12-chloro isomer **V-21** (*23*).

Rather unexpectedly, the corresponding bromo- and iodohydrins can be obtained with bromine or iodine in $CCl_4/CHCl_3$, and, as for the formation of the chlorohydrin, a *ca.* 3:1 preference for the C12-halo regioisomer is observed in both cases (*23*). Preferential (but not completely selective) opening of the epoxide moiety at position 12 upon treatment of Epo A with a variety of nucleophiles (HCl, $MgBr_2 \cdot Et_2O$, NaI/TMS-I, LiN_3, $Mg(OMe)_2$) has also been reported by the Novartis group (*24*). In contrast, the reaction of Epo A with $MgBr_2 \cdot Et_2O$ in CH_2Cl_2 at −20 to −5°C was found by the BMS group to produce the

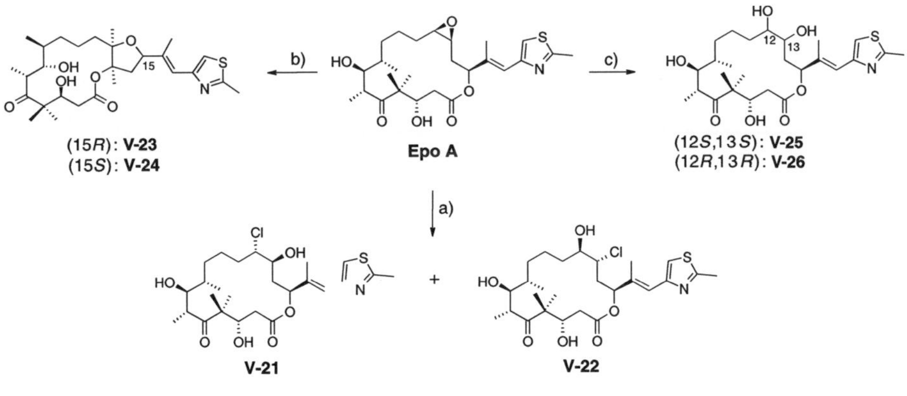

Scheme 5. a) THF/aqueous HCl or 1 *M* HCl, RT, 20 min, 60–80%, **V-21**:**V-22**, 2:1–4:1. b) 0.65 *M* CF_3COOH, acetone, 50°C, 10 h, 50% (**V-23**) and 35% (**V-24**). c) 0.65 *M* CF_3COOH, H_2O, 23°C, 48 h, 55% (**V-25**) and 15% (**V-26**)

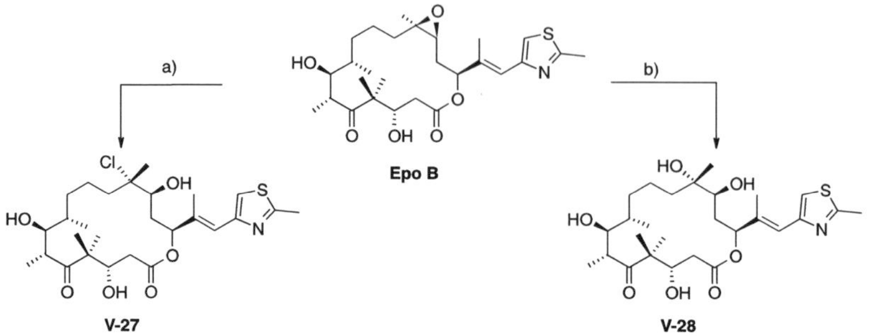

Scheme 6. a) THF/aqueous HCl or 1 *M* HCl, RT, 20 min, > 80%. b) CF_3COOH/H_2O or H_2SO_4/H_2O, 23°C, 45% (H_2SO_4) or 75% (CF_3COOOH)

C13-bromo isomer preferentially, with less than 2% of the C12-regioisomer being formed (*19*, *25*) (*vide infra*). As opposed to Epo A, treatment of Epo B with HCl gives chlorohydrin **V-27** as the *only* regioisomer in >80% yield, which is a direct reflection of the greater stability of the C12 over the C13 carbocation in a S_N1-type reaction (Scheme 6) (*23*).

Treatment of Epo A with non-nucleophilic *Brønsted* acids or *Lewis* acids in general produces relatively complex reaction mixtures, including rearranged products **V-23** and **V-24**. The latter were obtained in 85% combined yield and as the only isolable products upon treatment of Epo A with CF_3COOH in acetone (Scheme 5) (*23*). In contrast, exposure of Epo A or B to non-nucleophilic acids in the presence of water leads to diols **V-25/V-26** (Scheme 5) and **V-28** (Scheme 6). By analogy to halohydrin formation, nucleophilic attack of the epoxide moiety in Epo A preferentially, but not exclusively, occurs at position 12 (leading to isomer **V-25** as the major product), while **V-28** is the only isomer formed with Epo B. The GBF group has also used fermentatively produced Epo C (deoxyepothilone A) to prepare *cis*-diols **V-29** and **V-30** through OsO_4-catalyzed dihydroxylation (Scheme 7); these compounds were

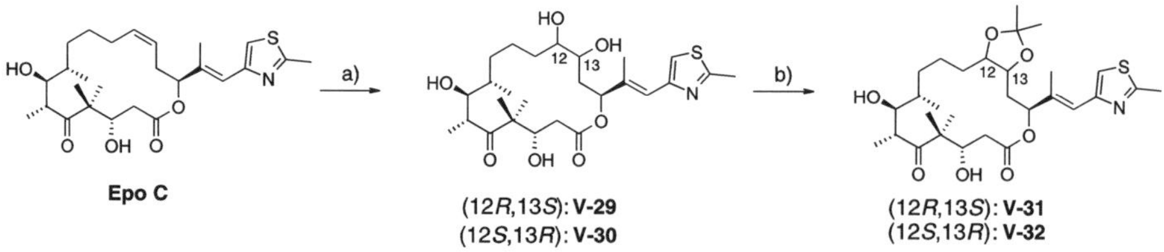

Scheme 7. a) OsO_4 cat., NMO, *t*-BuOH, THF/H_2O, RT, 75 min, 62%, **V-29**:**V-30**, 2:1 (inseparable mixture). b) Acetone, *p*-TsOH, RT, 2 h, 90% (for separable mixture of isomers)

References, pp. 154–156

subsequently converted into acetonides **V-31** and **V-32** (as were diols **V-25** and **V-26**) (*23*, *24*). The biological activity of these derivatives is discussed in Chapter 6.

The C12–C13 double bond in Epo C can be selectively reduced (over the C16–C17 double bond) with *in situ* generated diimide to provide an analog with a completely saturated macrocyclic core structure in 60% yield (98% based on recovered starting material) (*23*). This compound proved to be several hundred-fold less active than Epo A (*23*), and the same is true for the rearranged products **V-23** and **V-24** (*23*).

Approaches have also been developed for the chemical conversion of Epo A or B into Epo C and D (deoxyepothilones A and B). In the case of Epo A this can be achieved through epoxide deoxygenation either with 3-methyl-2-selenoxo-benzothiazole in CH_2Cl_2/CF_3COOH (40%) (*24*) or with $TiCp_2Cl_2$/Mg in THF (80%, Scheme 8) (*26*).

The Epo B → Epo D conversion seems to be best accomplished with WCl_6 and *n*-BuLi in THF, which gave Epo D in 78% yield (Scheme 9) (*26*).

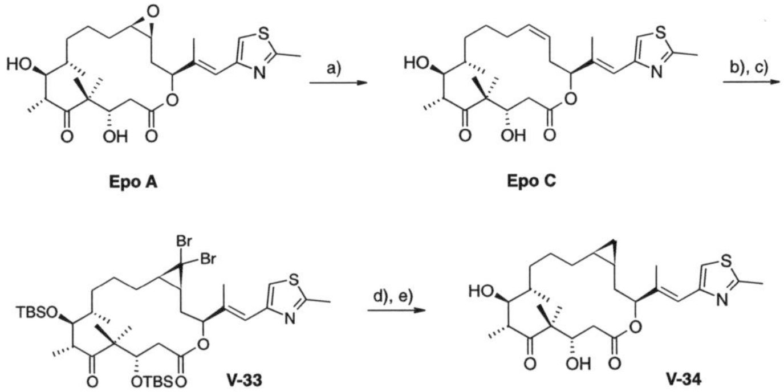

Scheme 8. a) $TiCp_2Cl_2$, Mg(s), THF, 80%. b) TBSOTf, CH_2Cl_2, lutidine, 0°C, 69%. c) Benzyltriethyl-ammonium chloride, 50% NaOH (aqueous), $CHBr_3$, 45°C, 12%. d) Bu_3SnH, AIBN, C_6H_{12}, 70°C, 76%. e) 20% CF_3COOH/CH_2Cl_2, −15°C, 90%

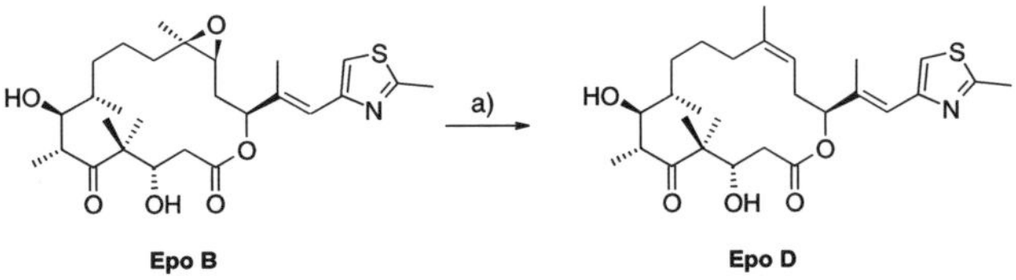

Scheme 9. a) WCl_6, *n*-BuLi, THF, 78%

The methodology developed by the BMS group for the reduction of Epo A and B to (*Z*)-olefins Epo C and D, respectively, provided the basis for the semisynthesis of cyclopropane-based analogs of epothilones, which is outlined in Scheme 8 for 12,13-cyclopropyl-Epo A **V-34** (*26*). While direct cyclopropanation of Epo C under *Simmons-Smith* conditions did not produce any practical amounts of the desired product, treatment of TBS-protected Epo C with $CHBr_3$/50% NaOH in the presence of benzyl-triethyl-ammonium chloride at 45°C gave dibromocyclopropane **V-33** as a single diastereoisomer in 12% yield. Although still far from satisfactory, this process gave access to sizable quantities of the target structure **V-34** (after reduction and deprotection, Scheme 8). Biological evaluation of **V-34** and the corresponding Epo B analog proved these compounds to be equally active as Epo A and B, respectively, thus invalidating prior literature reports, which had suggested that the replacement of the epoxide moiety in epothilones by a cyclopropane ring would result in a complete loss of activity.

In addition to cyclopropyl-epothilones, the BMS group has also devised a strategy for the conversion of Epo A into a whole range of

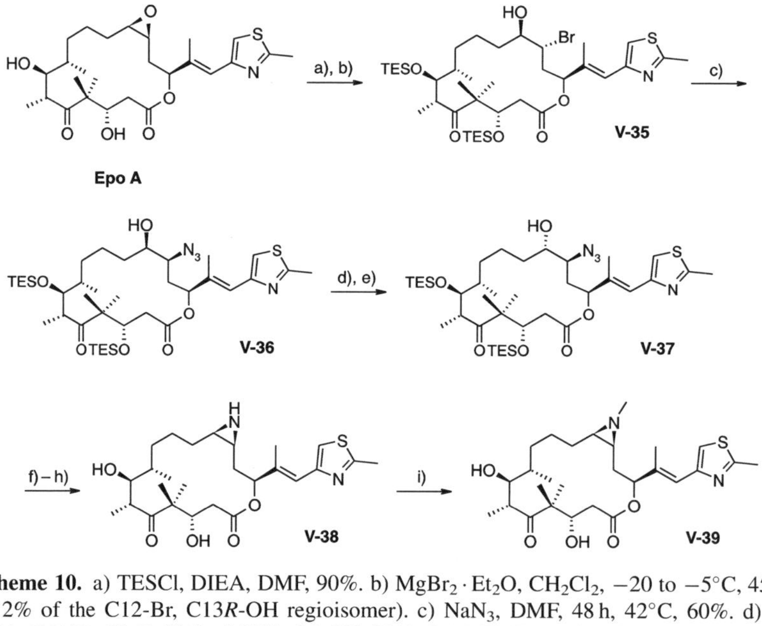

Scheme 10. a) TESCl, DIEA, DMF, 90%. b) $MgBr_2 \cdot Et_2O$, CH_2Cl_2, −20 to −5°C, 45% (<2% of the C12-Br, C13*R*-OH regioisomer). c) NaN_3, DMF, 48 h, 42°C, 60%. d) *p*-NBA, DEAD, Ph_3P, 99%. e) NH_3, MeOH, 88%. f) MsCl, Et_3N, CH_2Cl_2, 98%. g) Me_3P, THF/H_2O, 98%. h) 10% CF_3COOH/CH_2Cl_2, −10°C, 90%. i) Me_2SO_4, proton sponge, THF, 31%

analogs incorporating a (substituted) aziridine ring in place of the epoxide moiety (*25*).

As illustrated in Scheme 10, the optimized route to these compounds involves regioselective epoxide ring-opening with $MgBr_2 \cdot Et_2O$ (*vide supra*) followed by bromide displacement with azide ion, *Mitsunobu*-based inversion of configuration at C12, activation of the 12-hydroxyl group as its mesylate and finally aziridine ring formation through azide reduction under *Staudinger* conditions. The resulting N-unsubstituted 12,13-aziridinyl-Epo A **V-38** has been converted into a series of N-substituted derivatives *via* alkylation, acylation, carbamoylation, or sulfonylation. Several of these derivatives show antiproliferative activities that are comparable with or even superior to that of Epo A (*25*). These findings are discussed in more detail in Chapter 6. Bromohydrin **V-35** (Scheme 10) also served as an intermediate for the synthesis of 12-cyano-Epo A **V-41**, which was conceived as a more acid-stable analog of Epo B (Scheme 11) (*19*).

As outlined in Scheme 11, **V-35** was converted into bromo ketone **V-40** by oxidation with pyridinium chlorochromate (PCC), which was followed by cyanohydrin formation with KCN and concomitant epoxide ring closure. Deprotection with CF_3COOH then provided the target structure **V-41**, whose antiproliferative activity is comparable with that of Epo A. Interestingly, and in contrast to the facile formation of the epoxide ring from the cyanohydrin derived from **V-40**, chlorohydrins **V-21**, **V-22**, and **V-27** (Schemes 5 and 6) have been reported not to undergo epoxide formation even under strongly basic conditions (*23*).

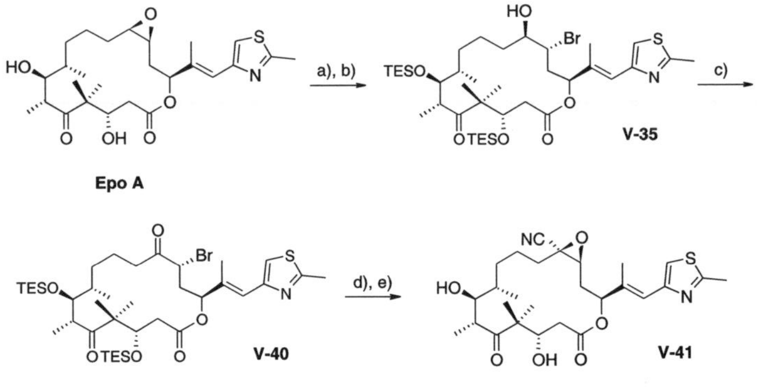

Scheme 11. a) TESCl, DIEA, DMF, 90%. b) $MgBr_2 \cdot Et_2O$, CH_2Cl_2, −20 to −5°C, 45%. c) PCC, pyridine, CH_2Cl_2, 92%. d) KCN, 18-C-6, THF, 49%. e) 10% CF_3COOH/CH_2Cl_2, −10°C, 98%

5.4. Side Chain Modifications

5.4.1. *Modifications of the C16/C17 Bond and the Thiazole Moiety*

The heterocycle-bearing side chain of epothilones has been targeted for semisynthesis in a number of different ways, in order to produce derivatives with modifications in the heterocycle, in the vinyl linker between the heterocycle and the macrolactone ring, or in both. As shown by the GBF group in one of their earliest studies, the treatment of Epo A with an excess of *n*-BuLi in THF at −90°C and subsequent quenching with different electrophiles mostly produces C19-substituted derivatives in modest yields (Scheme 12) (*27*).

In some cases, reaction occurs also at C21 (in addition to C19), albeit to a lower extent. As the only example, reaction with EtI provided C20-desmethyl-C20-propyl-Epo A exclusively, with no alkylation taking place at the C19-position, but the compound was isolated only in 8% yield (40% based on recovered starting material). Notably, electrophilic fluorination at C19 could not be achieved with a range of fluorinating agents. All of the C19-modified Epo A analogs were found to be significantly less active than the parent natural product in cancer cell proliferation assays (*27*).

While treatment with strong base followed by electrophilic quenching does not provide access to C21-substituted epothilone derivatives, mod-

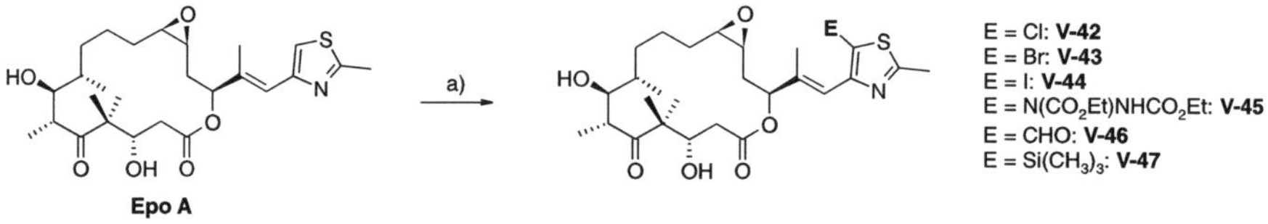

Scheme 12. a) *n*-BuLi (5 equ), THF, −90°C, electrophile (CCl_4, NBS, I_2, EtO_2CN=N-CO_2Et, $HCON(CH_3)_2$, $(CH_3)_3SiCl$). **V-42**: 50%; **V-43**: 15%; **V-44**: 25%; **V-45**: 50%; **V-46**: 30%; **V-47**: 35%

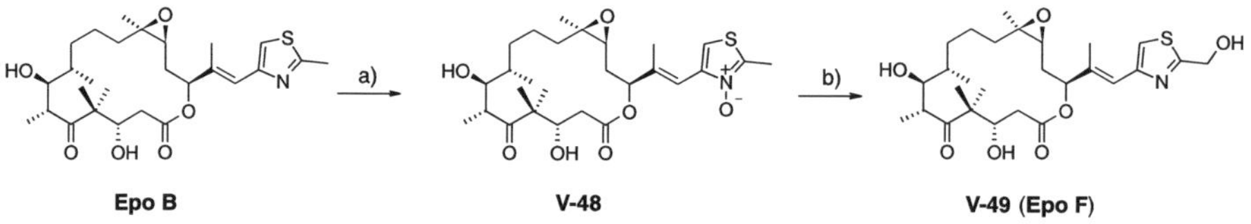

Scheme 13. a) *m*-CPBA, CH_2Cl_2, RT, 3 h, 48%. b) 1. $(CF_3CO)_2O$, 2,6-lutidine, CH_2Cl_2, 75°C, 10 min; 2. THF/25% NH_3, 45°C, 10 min, 78%

ification of the C21 position can be achieved in a highly efficient manner through conversion of Epo A or B into Epo E or F, respectively, and further transformation of the ensuing hydroxymethyl group attached to C20.

As illustrated in Scheme 13, Epo B can be converted into Epo F through a *Polonovsky*-type (*28*) rearrangement (*22*). Thus, treatment of Epo B with *m*-chloroperbenzoic acid (*m*-CPBA) produces the thiazole *N*-oxide **V-48**; upon reaction with trifluoroacetic anhydride, **V-48** rearranges to C21-trifluoroacetoxy-Epo B, from which Epo F (**V-49**) is obtained through cleavage of the trifluoroacetyl group with aqueous. NH_3 in THF in 38% yield (for the 3-step sequence from Epo B). In a completely analogous way, Epo A can be converted into Epo E (= C20-desmethyl-C20-hydroxymethyl-Epo A) (*22*). As exemplified in Scheme 14, Epo F (and also Epo E) can be elaborated into different C21-modified derivatives (*29*).

The conversion of the C21-methyl group into other small or even moderately sized substituents generally produces highly potent analogs and these compounds are discussed in more detail in Chapter 6. The most important C21-modification to date involves the incorporation of a primary amino group to produce C21-amino-Epo B **V-51**, which has been selected by BMS for clinical development (as BMS-310705) and which is currently undergoing Phase I clinical studies (*30*). As outlined in Scheme 14, BMS-310705 (**V-51**) is obtained from Epo F through conversion of the C21-hydroxyl group into an azide moiety with diphenylphosphoryl azide (DPPA) followed by *Staudinger* reduction with $P(CH_3)_3$ (*30*). The initiation of clinical development has also been

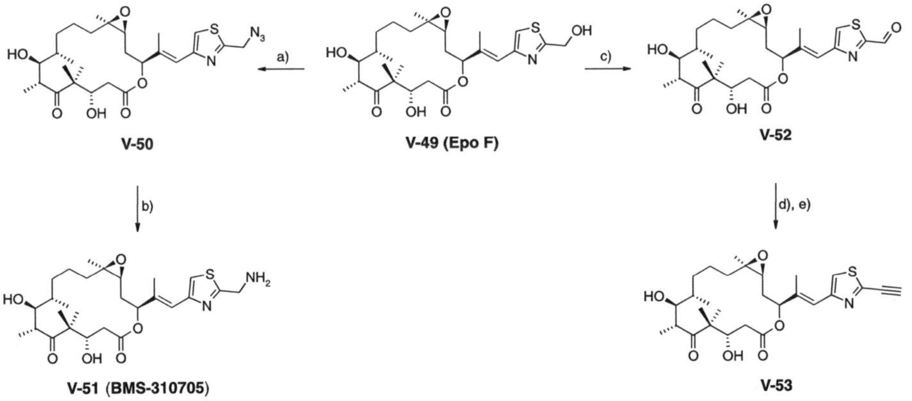

Scheme 14. a) DPPA, DBU, THF, 94%. b) $P(CH_3)_3$, THF/H_2O, 91%. c) MnO_2. d) $ICH_2P(CH_3)_3I$, *n*-BuLi. e) *n*-BuLi

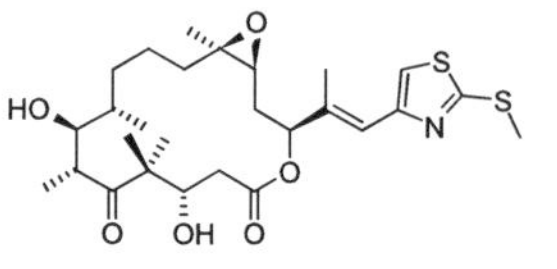

V-54 (ABJ879)

Fig. 1. Structure of ABJ879

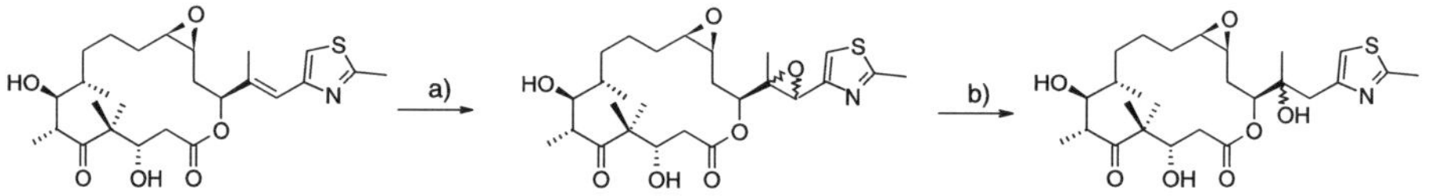

Epo A V-55 V-56

Scheme 15. a) *m*-CPBA, CH_2Cl_2, RT, 1 h, 10% or DMDO, acetone, 0°C → RT, 5 h, 25% (3/2 mixture of isomers). b) 10% Pd-C, EtOH, RT, 90 h, quant

reported for C20-desmethyl-C20-methylsulfanyl-Epo B **V-54** (ABJ879; Novartis), which, like BMS-310705 (**V-51**), is a semisynthetic Epo B derivative (*31*). However, so far, no details for the preparation of this compound have been disclosed in literature.

As indicated above, semisynthetic modifications of the epothilone side chain are not limited to changes in the substituent pattern on the heterocycle, but also include manipulation of the C16–C17 double bond. The latter can be epoxidized with *m*-CPBA or dimethyl dioxirane DMDO in low to moderate yield to furnish **V-55** as a 3/2 mixture of diastereoisomers, which can be converted to a mixture of epimeric C16-alcohols **V-56** by hydrogenolysis over Pd/C in EtOH in 90% yield (*i.e.*, hydrogenation occurs exclusively at the benzyl-like position α to the thiazole ring, Scheme 15) (*20*).

5.4.2. Cleavage/Restitution of the C16/C17 Bond

The C16–C17 double bond can be completely cleaved by means of ozonolysis (*20*); the resulting ketone **V-57** ((in the case of Epo A) was immediately reacted with trimethylsilyl (TMS) chloride to provide bis-TMS ether **V-58** in 70% overall yield (Scheme 16).

Quite remarkably, **V-58** proved to be highly resistant to *Wittig*-type and other olefination reactions and could only be converted to methylene derivative **V-59** in a modest overall yield of 15% (after deprotection). Attempts to re-introduce the natural thiazole side chain or to create a phenyl-based Epo A analog were unsuccessful (*20*, *32*). In contrast,

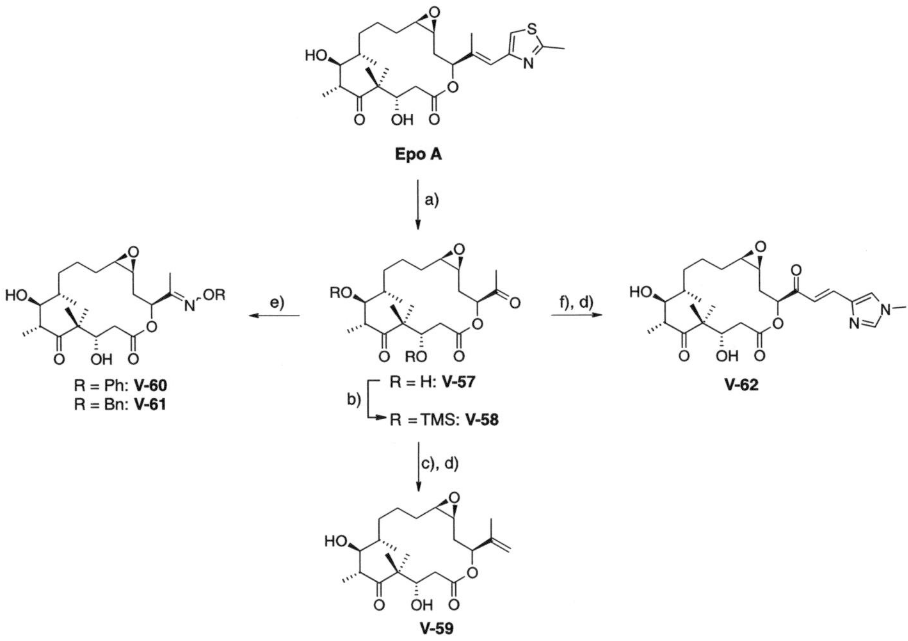

Scheme 16. a) O_3, CH_2Cl_2, −70°C, 10 min, then $(CH_3)_2S$. b) TMSCl, Et_3N, DMAP, CH_2Cl_2, RT, 18 h, 70% (2 steps). c) $Ph_3P{=}CH_2$, THF. d) Citric acid, MeOH, 65°C, 1 h, 15% (2 steps). e) **V-57**, $RONH_3Cl$, EtOH, pyridine, RT, 18 h, 50–60%. f) 1. LiTMP, THF, −70°C; 2. *N′*-methyl-imidazole-4-carboxaldehyde

transformation of **V-57** into O-substituted oximes **V-60** and **V-61** proved to be unproblematic (*20*) as was its elaboration into the imidazole-containing analog **V-62** through a base-catalyzed aldol reaction (*32*) (Scheme 16). All analogs depicted in Scheme 16 were found to be significantly less active in proliferation or tubulin assays than Epo A (*20*, *32*).

In order to overcome the limitations imposed by the low reactivity of ketone **V-58** under a variety of olefination conditions and to enable the replacement of the thiazole ring by other (hetero)aryl moieties, *Höfle* and co-workers have developed an alternative strategy for the reconstruction of the aryl-vinyl part of the epothilone side chain or modified versions thereof (Scheme 17).

The approach is based on the conversion of ketone **V-58** into boronic acid **V-64**. In contrast to olefination methods that require strongly basic reaction conditions, the reaction of **V-58** with the less basic bisboryl methyl lithium reagent **V-63** produced **V-64** in excellent yield (83%), albeit as a 7/3 mixture of (*E*)/(*Z*) diastereomers (*32*). However, this mixture could be separated by preparative HPLC and the pure (*E*)-isomer

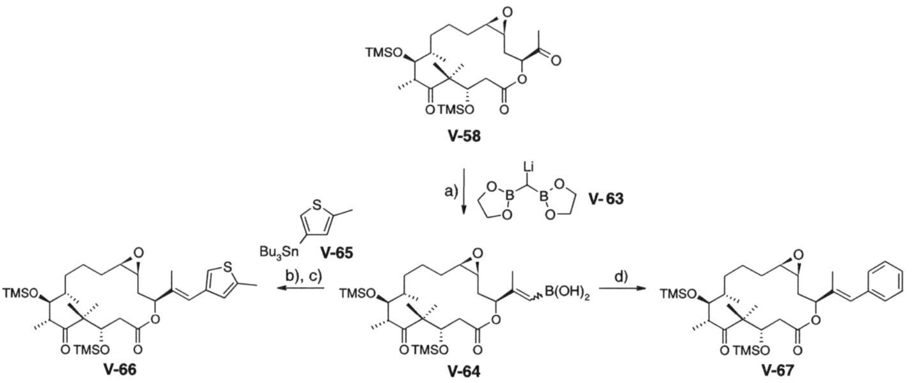

Scheme 17. a) 1. **V-63**, CH_2Cl_2/THF; 2. hydrolysis, 83% (*E*/*Z*, 7/3). b) *N*-iodo succinimide. c) **V-65**, $Pd(PPh_3)_4$. d) PhI, $Pd(PPh_3)_4$, TlOEt, EtOH

could be converted into the protected phenyl-based Epo A analog **V-67** through *Suzuki* coupling. Alternatively, reaction of *E*-**V-64** with *N*-iodosuccinimide (NIS) gave the corresponding vinyl iodide, which underwent smooth *Suzuki* coupling with stannane **V-65** to give **V-66** (*32*). No activity data have been reported for the (deprotected version of) thiophene derivative **V-66**; the activity of other heterocycle-modified epothilone analogs is discussed in Chapter 6.

5.5. Removal/Incorporation of the C13–O16 Segment

Semisynthesis-based modifications of the epothilone side chain have also been achieved *via* intermediates obtained through excision of the entire C13–O16 segment (including the pendant side chain), which could be (re)elaborated into modified versions of the original structure. Thus, employing *Grubbs-Hoveyda* catalyst **V-68**, the GBF group has used ring-opening olefin metathesis (ROM) with ethylene for the conversion of Epo C into bis-olefin **V-69** in 73% yield (*33*) (Scheme 18).

Diene **V-69** was then further elaborated into protected acid **V-71** through TBS-protection (at −20°C) followed by ester cleavage with an excess of TBSOTf (at 30°C). Esterification of **V-71** with alcohol **V-72** provided diene **V-74**, which underwent smooth ring-closing olefin metathesis (RCM) with *Grubbs* catalyst **V-73** to reconstitute the macrocyclic epothilone framework in the form of 12/13-deoxy analog(s) **V-75** (obtained as a 1/1 mixture of (*E*)/(*Z*)-isomers in 52% yield). Deprotection of **V-75** with CF_3COOH led to a separable mixture of (*E*)- and (*Z*)-isomers, each of which could be isolated in 43% yield. Epoxidation of

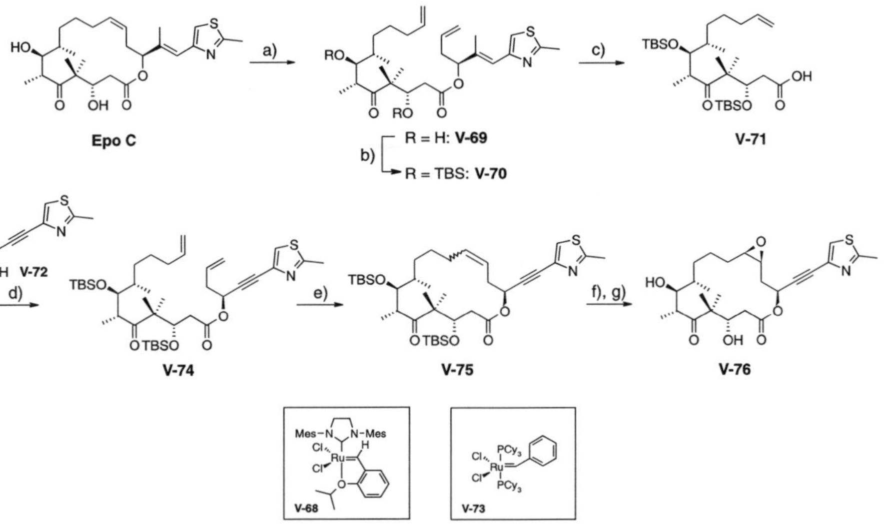

Scheme 18. a) $CH_2{=}CH_2$, **V-68** (0.15 equ), 44 h, 73% (83% based on recovered starting material). b) TBSOTf, 2,6-lutidine, CH_2Cl_2, −20°C, 24 h, 75%. c) TBSOTf, 2,6-lutidine, 30°C, 2 h, 94%. d) **V-72**, DCC, DMAP, 0°C → RT, 16 h, 91%. e) **V-73** (0.2 equ), CH_2Cl_2, RT, 48 h, 52% (1/1 mixture of *E*/*Z*-isomers). f) CF_3COOH/CH_2Cl_2 6/1, 0°C, 2 h, 43% (+ 43% of (12,13*E*)-isomer). g) DMDO, CH_2Cl_2/acetone, −20°C, 2 h, 61% (+ 32.5% (12*S*,13*R*) isomer)

the (*Z*)-isomer with DMDO gave Epo A analog **V-76** in 61% yield (together with 33% of its (12*S*,13*R*) isomer). Unfortunately, analog **V-76** (as well as its 12,13-deoxy derivative) proved to be significantly less active against the mouse fibroblast cell line L929 than Epo A (IC_{50} values >1 μ*M* *vs.* 8 n*M* for Epo A (*22*)) (*33*). It should be noted that **V-76** could conceivably be accessible from ketone **V-58** and this was also the approach originally pursued by *Höfle* and co-workers (*33*). However, they were unable to convert the C15-acetyl group in **V-58** into an alkyne moiety (*33*), which could have served as a substrate for *Sonogashira* couplings with a variety of halides.

An approach similar to the one above by *Höfle* and co-workers has been developed more recently by *Dong et al.* at Kosan, who have converted fermentatively produced Epo D into keto ester **V-77** (Scheme 19) (*34*). In contrast to *Höfle*'s approach, cleavage of the 12,13-double bond in Epo D in this case was not achieved by ROM, but involved stoichiometric OsO_4-mediated dihydroxylation followed by oxidative cleavage of the resulting 12,13-diol with $Pb(OAc)_4$. Treatment of the crude keto aldehyde thus obtained with K_2CO_3 led to facile elimination of the C1–C12 segment as the free carboxylic acid, which was converted

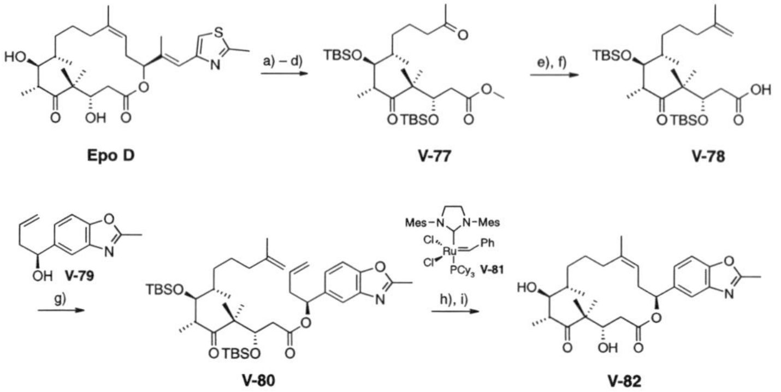

Scheme 19. a) TBSOTf, NEt_2, CH_2Cl_2, −78°C, 85%. b) OsO_4, TMEDA, −78°C, then $NaHSO_3$, 65°C, 75%. c) $Pb(OAc)_4$, benzene, then K_2CO_3, MeOH. d) $TMSCHN_2$, MeOH, toluene, 81% (2 steps). e) $Cp_2Ti(CH_3)_2$, toluene, 80°C, 74%. f) LiOH, *i*-PrOH/H_2O, 86%. g) **V-79**, EDC, DMAP, CH_2Cl_2, 54–82%. h) **V-81**, CH_2Cl_2, refl., 8 h, 60% (1/1 mixture of *E*/*Z*-isomers). i) CF_3COOH, CH_2Cl_2, 0°C → RT or HF · pyridine, THF, 0°C → RT, 32–59%

into ester **V-77** with TMS-diazomethane. *Petasis* olefination followed by ester saponification and reesterification of the resulting acid **V-78** with alcohol **V-79** then provided diene **V-80**. The latter underwent RCM in the presence of *Grubbs* second-generation catalyst **V-81**, to provide the desired macrolactone as a 1/1 mixture of (*E*)/(*Z*) isomers about the C12/C13 bond in 60% yield. Isomer separation was possible after deprotection of the mixture with CF_3COOH. No biological data were reported in Ref. (*34*) for analog **V-82**, but the activity of closely related analogs is discussed in Chapter 6.

Employing PLE-catalyzed hydrolysis of the lactone group and cleavage of the C12/C13 double bond by ozonolysis, the BMS group has reported the transformation of Epo C into ester **V-83** (Scheme 20).

Oxidation of the aldehyde group in **V-83** with $NaClO_2$ gave the corresponding carboxylic acid which could be further elaborated into C12–C13 amide-based analogs **V-86** and **V-87** through coupling with amines **V-84** or **V-85**, respectively, ester saponification, *Yamaguchi*-type macrolactonization and final deprotection with CF_3COOH (Scheme 20) (*18*). Alternatively, **V-83** was combined with amines **V-84** or **V-85** *via* reductive amination to provide the corresponding secondary or tertiary amines; employing the same sequence of reactions that led to **V-86** and **V-87**, the reductive amination products were then converted into 13-aza epothilones **V-88** and **V-89** (Scheme 20). Unfortunately, none of the derivatives **V-86**, **V-87**, **V-88**, or **V-89** showed any significant tubulin-polymerizing or growth-inhibitory activity.

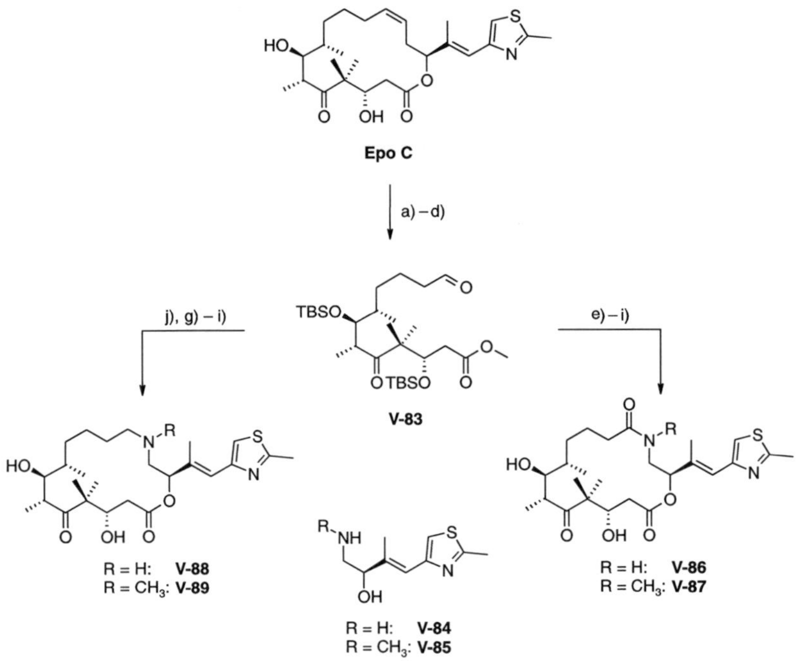

Scheme 20. a) PLE. b) $TMSCHN_2$, 25% (2 steps, plus recovered starting material). c) TBSOTf, 44%. d) O_3, Ph_3P, 86%. e) $NaClO_2$, NaH_2PO_4, 43%. f) **V-84** or **V-85**, EDC, HOBt, Et_3N, 79%. g) LiOH, 86%. h) 2,4,6-trichlorobenzoyl chloride, 37%. i) CF_3COOH, 56%. j) **V-84** or **V-85**, $NaBH(OAc)_3$, 74% (with **V-85**). (Yields for steps f) – i) are for intermediates leading to **V-87**)

5.6. Conclusions

The work summarized in this chapter has defined the chemistry associated with the epothilone molecular framework in terms of possible functional group transformations and acid/base stability in significant detail. This has enabled the preparation of a plethora of structurally modified, semisynthetic epothilone derivatives that have contributed significantly to our current understanding of the epothilone SAR (as discussed in more detail in Chapter 6). Most importantly, however, these efforts have led to the discovery of three derivatives that are currently undergoing clinical trials in humans. Thus, in spite of the wealth of fully synthetic analogs that have been prepared over the last decade and many of which exhibit very attractive biological profiles (at least *in vitro*), semisynthesis has had the more profound impact on the clinical advancement of the epothilone class of microtubule stabilizers than total synthesis. While fully synthetic analogs other than Bayer-Schering's ZK-Epo (Sagopilone) (see Chapters 4, 6.4, and 7.7) may be advanced to clinical trials in the future, for the time being the prevalence of semi-synthetic derivatives or fermentatively produced epothilone variants among clinical development compounds still reflects the higher technical hurdles associated with molecules that have to be prepared through multi-step chemical syntheses.

On the other hand, the potential of semisynthesis for the creation of structurally modified epothilone analogs is far from being exhausted. Many additional epothilone derivatives with a diverse range of structural features are conceivable that should be chemically accessible from natural epothilones as starting materials and that might exhibit interesting biological properties. However, the further exploration of semisynthetic approaches towards new epothilone analogs is limited by access to the fermentatively produced starting materials. It is unclear at this point, whether those groups with a sufficient supply of natural epothilones (BMS, GBF (now *Helmholtz* Centre for Infection Research), Novartis) continue to be active in the area of epothilone semisynthesis. Time will tell.

References

1. Butler MS (2004) The Role of Natural Product Chemistry in Drug Discovery. J Nat Prod **67**: 2141
2. Hamann MT (2003) Enhancing Marine Natural Product Structural Diversity and Bioactivity Through Semisynthesis and Biocatalysis. Curr Pharm Design **9**: 879
3. Henkel T, Brunne RM, Müller H, Reichel F (1999) Statistical Investigation into the Structural Complementarity of Natural Products and Synthetic Compounds. Angew Chem Int Ed **38**: 643
4. Kuhn M, Keller-Juslén C, von Wartburg A (1969) Mitosis-inhibiting Natural Substances. XXII. Partial Synthesis of 4′-Demethylepipodophyllotoxin. Helv Chim Acta **52**: 944
5. Keller-Juslén C, Kuhn M, von Wartburg A, Stähelin H (1971) Mitosis-inhibiting Natural Products. 24. Synthesis and Antimitotic Activity of Glycosidic Lignan Derivatives Related to Podophyllotoxin. J Med Chem **14**: 936
6. You Y (2005) Podophyllotoxin Derivatives: Current Synthetic Approaches for New Anticancer Agents. Curr Pharm Design **11**: 1695
7. Sawada S, Yokokura T, Miyasaka T (1995) Synthesis and Antitumor Activity of A-ring or E-lactone Modified Water-Soluble Prodrugs of (20*S*)-Camptothecin, Including Development of Irinotecan Hydrochloride Trihydrate (CPT-11). Curr Pharm Design **1**: 113
8. Kingsbury WD, Boehm JC, Jakas DR, Holden KG, Hecht SM, Gallagher G, Caranfa MJ, McCabe FL, Faucette L, Johnson RK, Hertzberg RP (1991) Synthesis of Water-Soluble (Aminoalkyl) Camptothecin Analogs: Inhibition of Topoisomerase I and Antitumor Activity. J Med Chem **34**: 98

9. Mathijssen RHJ, Loos WJ, Verweij J, Sparreboom A (2002) Pharmacology of Topoisomerase I Inhibitors Irinotecan (CPT-11) and Topotecan. Curr Cancer Drug Targets **2**: 103
10. Guenard D, Gueritte-Voegelein F, Potier P (1993) Taxol and Taxotere: Discovery, Chemistry, and Structure-activity Relationships. Acc Chem Res **26**: 160
11. Rowinsky EK (1997) The Development and Clinical Utility of the Taxane Class of Antimicrotubule Chemotherapy Agents. Ann Rev Med **48**: 35
12. Wani MC, Taylor HL, Wall ME, Coggon P, McPhail AT (1971) Plant Antitumor Agents. VI. Isolation and Structure of Taxol, a Novel Antileukemic and Antitumor Agent from *Taxus brevifolia*. J Am Chem Soc **93**: 2325
13. Holton RA, Biediger RJ, Boatman D (1995) Semisynthesis of Taxol and Taxotere. In: Suffness M (ed.) Taxol: Science and Applications, p. 97. CRC, Boca Raton, FL
14. Höfle G, Reichenbach H (2005) Epothilone, a Myxobacterial Metabolite with Promising Antitumor Activity. In: Cragg GM, Kingston DGI, Newman DJ (eds.) Anticancer Agents from Natural Products, p. 413. Taylor & Francis, Boca Raton, FL
15. Arslanian RL, Tang L, Blough S, Ma W, Qiu R-G, Katz L, Carney JR (2002) A New Cytotoxic Epothilone from Modified Polyketide Synthases Heterologously Expressed in *Myxococcus xanthus*. J Nat Prod **65**: 106
16. Niggemann J, Michaelis K, Frank R, Zander N, Höfle G (2002) Natural Product-derived Building Blocks for Combinatorial Synthesis. Part 1. Fragmentation of Natural Products from Myxobacteria. JCS Perkin Trans: 2490
17. Borzilleri RM, Zheng X, Schmidt RJ, Johnson JA, Kim SH, DiMarco JD, Fairchild CR, Gougoutas JZ, Lee FYF, Long BH, Vite GD (2000) A Novel Application of a Pd(0)-Catalyzed Nucleophilic Substitution Reaction to the Regio- and Stereoselective Synthesis of Lactam Analogues of the Epothilone Natural Products. J Am Chem Soc **122**: 8890
18. Vite GD, Borzilleri RM, Kim SH, Regueiro-Ren A, Humphreys WG, Lee FYF (2001) Epothilones A and B: Springboards for Semisynthesis of Promising Antimitotic Agents. In: Ojima I, Vite GD, Altmann K-H (eds.) Anticancer Agents – Frontiers in Cancer Chemotherapy. ACS Symposium Series 796, p. 148. American Chemical Society, Washington DC
19. Regueiro-Ren A, Leavitt K, Kim SH, Höfle G, Kiffe M, Gougoutas JZ, DiMarco JD, Lee FYF, Fairchild CR, Long BH, Vite GD (2002) SAR and pH Stability of Cyano-Substituted Epothilones. Org Lett **4**: 3815
20. Sefkow M, Kiffe M, Schummer D, Höfle G (1998) Oxidative and Reductive Transformations of Epothilone A. Bioorg Med Chem Lett **8**: 3025
21. Medina JC, Salomon M, Kyler KS (1988) A Mild Method for the Conversion of Alcohols to Methylthiomethyl Ethers. Tetrahedron Lett **29**: 3773
22. Höfle G, Glaser N, Kiffe M, Hecht H-J, Sasse F, Reichenbach H (1999) *N*-Oxidation of Epothilone A-C and *O*-Acyl Rearrangement to C-19- and C21-Substituted Epothilones. Angew Chem Int Ed **38**: 1971
23. Sefkow M, Kiffe M, Höfle G (1998) Derivatization of the C12–C13 Functional Groups of Epothilones A, B and C. Bioorg Med Chem Lett **8**: 3031
24. Altmann K-H, Bold G, Caravatti G, End N, Flörsheimer A, Guagnano V, O'Reilly T, Wartmann M (2000) Epothilones and Their Analogs – Potential New Weapons in the Fight Against Cancer. Chimia **54**: 612
25. Regueiro-Ren A, Borzilleri RM, Zheng X, Kim SH, Johnson JA, Fairchild CR, Lee FY, Long BH, Vite GD (2001) Synthesis and Biological Activity of Novel Epothilone Aziridines. Org Lett **3**: 2693

26. Johnson J, Kim SH, Bifano M, DiMarco J, Fairchild C, Gougoutas J, Lee F, Long B, Tokarski J, Vite GD (2000) Synthesis, Structure Proof, and Biological Activity of Epothilone Cyclopropanes. Org Lett **2**: 1537
27. Sefkow M, Höfle G (1998) Substitutions at the Thiazole Moiety of Epothilone. Heterocycles **48**: 2485
28. Huisgen R, Kolbeck W (1965) *N*-Acyloxyammonium Salts. Tetrahedron Lett **6**: 783
29. Höfle G, Glaser N, Leibold T, Karama U, Sasse F, Steinmetz H (2003) Semisynthesis and Degradation of the Tubulin Inhibitors Epothilone and Tubulysin. Pure Appl Chem **75**: 167. See also: Glaser N (2001) Semisynthese Seitenketten-modifizierter Epothilone, Doctoral Thesis, Technical University of Braunschweig.
30. Borzilleri RM, Vite GD (2003) Epothilones: New Tubulin Polymerization Agents in Preclinical and Clinical Development. Drugs of the Future **27**: 1149
31. Wartmann M, Loretan J, Reuter R, Hattenberger M, Muller M, Vaxelaire J, Maira S-M, Flörsheimer A, O'Reilly T, Nicolaou KC, Altmann K-H (2004) Preclinical Pharmacological Profile of ABJ879, a Novel Epothilone B Analog with Potent and Protracted Anti-tumor Activity. Proc Am Assoc Cancer Res **45**: Abstract #5440
32. Höfle G, Glaser N, Leibold T, Sefkow M (1999) Epothilone A-D and Their Thiazole-modified Analogs as Novel Anticancer Agents. Pure Appl Chem **71**: 2019
33. Karama U, Höfle G (2003) Synthesis of Epothilone 16,17-Alkyne Analogs by Replacement of the C13–C15(O)-Ring Segment of Natural Epothilone C. Eur J Org Chem: 1042
34. Dong SD, Sundermann K, Smith KMJ, Petryka J, Liu F, Myles DC (2004) Rapid Access to Epothilone Analogs via Semisynthetic Degradation and Reconstruction of Epothilone D. Tetrahedron Lett **45**: 1945

6. Preclinical Pharmacology and Structure-Activity Studies of Epothilones

Karl-Heinz Altmann

Department of Chemistry and Applied Biosciences, Institute of Pharmaceutical Sciences, Swiss Federal Institute of Technology (ETH), Zürich, Switzerland

Contents

6.1. Introduction

Microtubule inhibitors are an important class of anticancer agents (*1*), with clinical applications in the treatment of a variety of cancer types, either as single agents or as part of different combination regimens (*2*, *3*). Microtubule-interacting agents can be grouped into two distinct functional classes, namely, compounds that inhibit the assembly of tubulin heterodimers into microtubule polymers ("tubulin polymerization inhibitors") and those that stabilize microtubules under normally

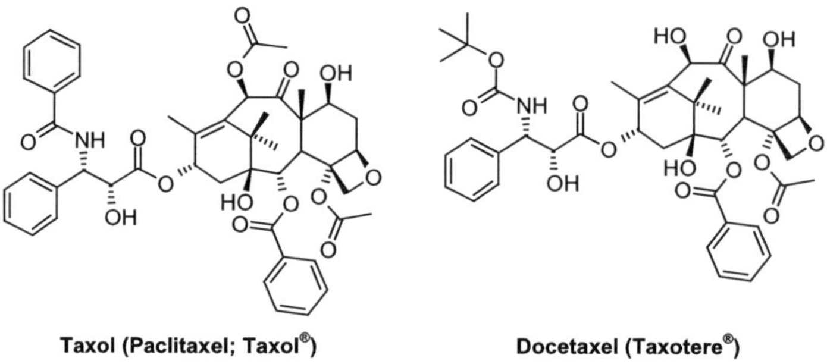

Fig. 1. Molecular structures of taxol and its semi-synthetic analog docetaxel

destabilizing conditions ("microtubule stabilizers") (*4*). The latter will also *promote* the assembly of tubulin heterodimers into microtubule polymers and, as will be discussed later, the induction of tubulin polymerization is often used as a biochemical readout for a quantitative assessment of the interaction of microtubule-stabilizing agents with tubulin. Tubulin polymerization inhibitors such as vincristine and vinblastine have been employed in cancer therapy for more than 40 years (vincristine and vinblastine received FDA approval in 1963 and 1965, respectively). In contrast, the clinical history of microtubule-stabilizing agents is significantly shorter and its beginning dates back no more than 15 years. The first microtubule stabilizer to be introduced into clinical practice was the natural product taxol (paclitaxel; Taxol®) in 1993 (Fig. 1), with FDA approval for the semi-synthetic taxol analog docetaxel (Taxotere®) following three years later. Both compounds rapidly became an important part of today's armamentarium in the battle against cancer, which marks the emergence of microtubule-stabilizing anticancer drugs as an important milestone in the development of modern cancer chemotherapy (*5*).

While efficient inhibition of tubulin polymerization can be effected by a number of small synthetic molecules (apart from complex natural products like vincristine or vinblastine) (*6*), it is intriguing to note that all potent microtubule-*stabilizing* agents identified to date are natural products or natural product-derived (for recent reviews *cf.* (*7–9*)). Historically, more than a decade passed after the elucidation of taxol's mode of action in 1979 (*10*) before other microtubule-stabilizing agents with non-taxol-like structures were discovered. Most prominent among these new microtubule stabilizers is a group of bacteria-derived macrolides that were discovered in 1987 by *Reichenbach* and *Höfle* (*11*) and have been termed "epothilones" by their discoverers (Fig. 2) (*11*, *12*).

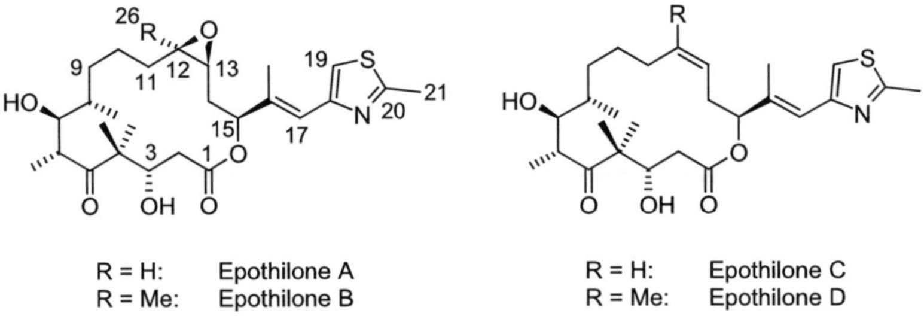

Fig. 2. The epothilone family

The taxol-like mechanism of action of epothilones (*i.e.* their microtubule-stabilizing properties) was not immediately recognized, however, and was only discovered 8 years later by a group at Merck Research Laboratories (*13*).

The major products originally isolated from the myxobacterium *Sorangium cellulosum* So ce 90 are epothilone A and epothilone B (Epo A and B), but numerous related members of this natural products family have subsequently been recovered as minor components from fermentations of myxobacteria (*14*). The relative and absolute stereochemistry of Epo B was determined by *Höfle* and co-workers in 1996 based on a combination of X-ray crystallography and chemical degradation studies (*15*). The availability of this information in combination with their (at that time) unusual mechanism of action made Epo A and B attractive and widely pursued targets for total synthesis and, in particular, turned them into important lead structures for anticancer drug discovery. Although not within the scope of this book, it is also worth noting in this context that a growing number of additional natural products have been recognized over the last few years to be microtubule stabilizers (for reviews *cf.* (*7–9*)), thus providing a whole new set of diverse lead structures for anticancer drug discovery.

While they exert their antiproliferative activity through interference with the same molecular target, a major distinction between taxol and epothilones is the latter's ability to inhibit the growth of multidrug-resistant cancer cell lines at concentrations similar or only slightly lower than those required for growth inhibition of drug-sensitive cancer cells (*13*, *16–18*). Epothilones have also been shown to be active *in vitro* against cancer cell lines, whose taxol resistance is mediated by specific tubulin mutations (*16*, *19*). At the same time, epothilones possess more favorable biopharmaceutical properties than taxol (*e.g.*, improved water-solubility (*15*)), which enables the use of clinical formulation vehicles less problematic than Cremophor-EL®. (For a discussion of clinical side effects of Taxol® believed to originate in this particular formulation vehicle *cf.*, *e.g.*, (*5*).) Epo B and a number of its analogs have been demonstrated to possess potent *in vivo* antitumor activity and up to this point at least seven compounds based on the epothilone structural scaffold have entered clinical evaluation in humans. These include Epo B itself (EPO906, patupilone; developed by Novartis), Epo D (deoxyEpo B, KOS-862; Kosan/Roche), BMS-247550 (ixabepilone, the lactam analog of Epo B; BMS), BMS-310705 (C21-amino-Epo B; BMS), ABJ879 (C20-desmethyl-C20-methylsulfanyl-Epo B; Novartis), 9,10-didehydro-Epo B (KOS-1584 (Kosan/Roche)), and the fully synthetic analog ZK-Epo (Sagopilone) (Schering). These compounds will be discussed in more detail below.

As discussed in Chapter 4, epothilones have been (and continue to be) attractive targets for total chemical synthesis and numerous syntheses of Epo A and B have been published in literature since the first disclosure of their absolute stereochemistry in 1996 (*15*) (for reviews *cf.* (*20–26*)). At the same time, the methodology developed in the course of those studies has been exploited for the synthesis of a host of synthetic analogs (reviewed in (*20*, *21*, *24*, *27–31*)), which have allowed the empirical elucidation of the most important structural parameters required for biological activity. The chemistry developed for the preparation of some of these analogs should even allow the production of amounts of material sufficient for clinical trials (*21*, *32*), thus highlighting the difference in structural complexity (which is reflected in synthetic accessibility) between epothilone-type structures and taxol, for which an industrial scale synthesis is clearly out of reach.

The chemistry, biology, and SAR (structure-activity-relationship) of epothilones have been extensively discussed in recent review articles (*17*, *20*, *21*, *24*, *27–31*). This chapter will summarize the most pertinent aspects of the *in vitro* and *in vivo* pharmacology of Epo B and its most important analogs and it will review the most significant findings and conclusions that have arisen from extensive SAR studies of a multitude of epothilone variants. This will, however, not include any discussion of the organic chemistry underlying the synthesis of these compounds, which is presented in detail in Chapters 4 and 5 for fully synthetic analogs and semisynthetic derivatives, respectively. Likewise, the impressive advances in the elucidation of epothilone biosynthesis and the development of heterologous expression systems, which form the basis for the production of some of the analogs discussed below, are summarized in Chapter 3.

6.2. *In vitro* Pharmacology of Epo B

The basic biology and pharmacology of Epo B, which is the most potent (and, therefore, most extensively studied) natural epothilone have been reviewed in a number of previous accounts (*11*, *17*, *20*, *27–31*, *33*, *34*). As indicated above, the biological effects of the compound are based on its ability to bind to microtubules, thereby altering the intrinsic stability and dynamic properties of these supramolecular structures.

Microtubules, the cellular targets of microtubule-stabilizing agents, are hollow filaments of *ca.* 240 Å outer diameter, which are composed of the 55 kD proteins α- and β-tubulin as the constituent subunits (Fig. 3). (For excellent recent reviews on microtubule structure and function *cf.* Refs. (*35–38*)). Together with actin polymers and intermediate filaments, microtubules are one of the constituent components of the cytoskeleton. As such they are of critical importance for a variety of cellular functions, including the development and maintenance of cell shape and motility as well as the intracellular transport of vesicles, mitochondria and other components. Of particular relevance in the context of cell proliferation is their role in cell division, when the entire microtubule network rearranges to form the mitotic spindle, thus providing the structural framework for the physical segregation of sister chromatids. In general, cellular microtubules are composed of 13 protofilaments, *i.e.* linear polymers with a head-to-tail arrangement of α- and β-tubulin subunits (Fig. 3). Structures with different numbers of protofilaments are also possible, but

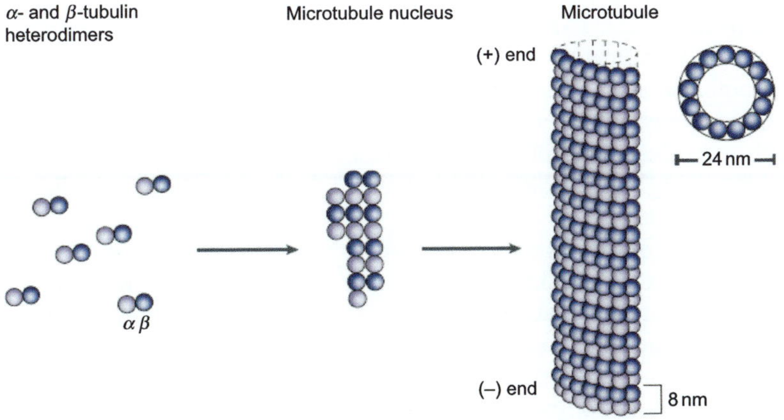

Fig. 3. Schematic representation of the α/β-tubulin/microtubule equilibrium and the structure of microtubules. Reprinted by permission from Macmillan Publishers Ltd: NATURE REVIEWS – CANCER, Vol. 4, p. 254, 2004

only with 13 protofilaments can these be aligned fully in parallel with the long axis of the microtubule cylinder, which may be required for uninterrupted transport of cellular cargo by motor proteins over long distances (*38*). One of the characteristic features of microtubules is their ability to shorten and lengthen in a stochastic fashion through loss or addition of α/β-tubulin heterodimers from microtubule ends, a phenomenon referred to as "dynamic instability" (*35–42*). For microtubules formed *in vitro* (*i.e.* outside of a cellular context) growth and shrinkage can occur from both ends of the cylinder, *i.e.* the plus-end, where β-tubulin is exposed to solvent, or the minus-end with solvent exposure of α-tubulin (Fig. 3) (*35–38*). In contrast, in cells, all microtubules originate from one microtubule-organizing center, from which the plus end is able to grow and shrink. Microtubule growth is coupled to GTP hydrolysis in the β-tubulin subunit (there is a second non-hydrolyzable GTP in the α-subunit). The dynamic properties of microtubules are a fundamental prerequisite for the proper assembly of the mitotic spindle and the subsequent movement of sister chromatids to the spindle poles, with spindle microtubules being significantly (*ca.* 4–100-fold) more dynamic than those forming the interphase cytoskeleton (*41*). It is thus not surprising that agents interfering with microtubule dynamics can have profound and detrimental effects on spindle formation and cell division.

Epothilones can prevent the Ca^{2+}- or cold-induced depolymerization of preformed microtubule polymers in cell-free *in vitro* systems (*16*). They can also *promote* the polymerization of soluble tubulin into microtubule-like polymers in the absence of either microtubule associated proteins (MAPs) and/or guanosine triphosphate (GTP), at temperatures significantly below 37°C, and in the presence of Ca^{2+} (*13, 16*). As indicated above, the induction of tubulin polymerization is frequently used as quantitative biochemical readout for the interaction of ligands with the tubulin/microtubule system. Epo B is a more potent tubulin-polymerizing agent than taxol, which in turn polymerizes tubulin with about equal efficiency to Epo A. (*E.g.*, EC_{50} values for the polymerization of microtubule protein by Epo A, Epo B, and taxol have been determined as 1.12, 0.67, and 1.88 μM (*43*).) It is important to note, however, that the exact magnitude of tubulin-polymerizing effects *in vitro* (absolute and even relative polymerization rates, extent of tubulin polymer formation) strongly depends on the assay conditions employed (*e.g.*, biological source and purity of tubulin, concentration of microtubule-stabilizing buffer components, and reaction temperature) (*44*).

Epothilones are able to displace (^{3}H)-taxol from microtubules with efficiencies similar or superior to those of unlabelled taxol or docetaxel

(*13*, *16*). As demonstrated by kinetic experiments, inhibition of taxol binding by epothilones is of the competitive type (with apparent K_i values of 1.4 μM (Epo A) and 0.7 μM (Epo B)), which implies that the microtubule binding sites of taxol and Epo A/B are largely overlapping or even identical. This notion has recently been confirmed by structural studies on a complex between Zn^{2+}-stabilized tubulin polymer sheets and Epo A (*45*) (*vide infra*). Most recently, the binding constants of Epo A and B to stabilized (crosslinked) microtubules *in vitro* have been determined as $2.93 \times 10^7\ M^{-1}$ (Epo A) and $6.08 \times 10^8\ M^{-1}$ (Epo B) (37°C) using a fluorescence-based displacement assay (*46*).

Microtubule stabilization by epothilones at low temperatures has also been demonstrated in cells (*27*), but the effect (like cancer cell growth inhibition, *vide infra*) is observed at strikingly lower concentrations than those required for the induction of tubulin polymerization *in vitro*. This apparent discrepancy has been resolved by careful uptake experiments in HeLa cells (*17*, *43*), which have shown that Epo A and B, like taxol (*47*), accumulate several-hundred fold inside cells over external medium concentrations (Fig. 4). Similar findings have been reported for a close analog of Epo B in MCF-7 cells (*48*).

Early experiments investigating the effects of epothilones on microtubule bundling in intact cells had demonstrated that treatment of cultured cells (RAT1, HeLa, Hs578T, Hs578Bst, PtK_2) with high (10^{-6}–$10^{-4}\ M$) concentrations of epothilones resulted in the formation of

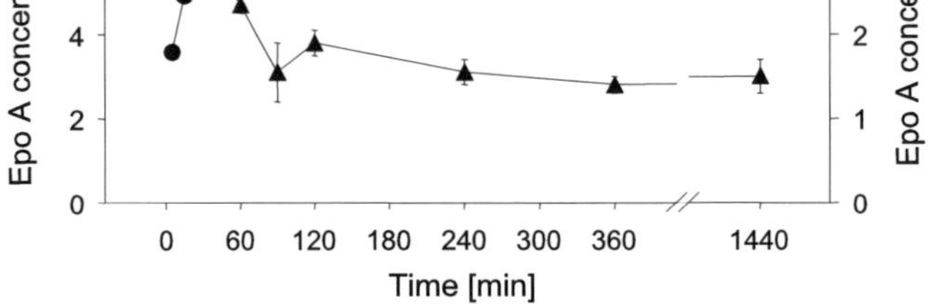

Fig. 4. Time-dependent accumulation of Epo A in HeLa cells. Initial medium concentration of Epo A ($t = 0$) was 10 n*M*. Filled circles indicate cell-associated concentrations of Epo A (μM), filled triangles are medium concentrations (n*M*). Concentrations were determined by μ-HPLC-MS. (*M. Wartmann et al.*, unpublished data; see also Ref. (*43*))

characteristic, extensive microtubule bundles (lateral association of microtubules) throughout the cytoplasm of interphase cells (*13*, *16*). In contrast, lower epothilone concentrations (10^{-7}–10^{-8} *M*) were reported to have little effect on interphase microtubule arrays, with the primary effect occurring on cells entering mitosis (*13*). While the more relevant effects of epothilones at low n*M* concentrations are unquestionably associated with the spindle apparatus, it has also been demonstrated (employing live fluorescence microscopy of HeLa cells ectopically expressing mouse β6-tubulin fused to enhanced green fluorescent protein (EGFP)) that bundling of interphase microtubules occurs even at low n*M* concentrations of Epo B after 24 h of drug exposure (*17*).

It is generally assumed that the growth inhibitory effect of epothilones (and also other microtubule-interacting agents) is a consequence of the suppression of microtubule dynamics rather than an overall increase in microtubule polymer mass due to massive induction of tubulin polymerization (*35*, *40*). The concentration-dependent inhibition of the dynamics of interphase microtubules by Epo B has been experimentally demonstrated by *Kamath* and *Jordan* (*49*) by means of time-lapse microscopy in MCF-7 cells stably transfected with GFP (green fluorescent protein)-α-tubulin. The effects on microtubule dynamics correlated with the extent of mitotic arrest (*vide infra*), with microtubule dynamics remaining unaffected at concentrations that did not lead to a G2/M block (0.2 n*M* Epo B). On the other hand, dynamicity (*i.e.*, the length grown and shortened by an individual microtubule divided by its total life span) was reduced by 62% at the IC_{50} for mitotic arrest after 20 h (3.5 n*M*). Similar results were obtained for taxol and it should be noted that the calculated reduction in dynamicity only pertains to those cells for which dynamic microtubules could still be observed. In fact, at 3.5 n*M* Epo B no dynamic microtubules could be detected in the majority of cells (over a 2 min observation period). These data provide direct evidence for the suppression of cellular microtubule dynamics by Epo B, but the question remains how effects on interphase microtubules relate to the dynamics of *spindle* microtubules (which could not be measured) und thus to mitotic arrest. Likewise, it is unclear, how mitotic entry and the associated rearrangement of the entire microtubule network can occur from a largely static state of this network (in the presence of Epo B). However, mitotic signals may lead to an increase in inherent microtubule dynamics in spindle microtubules, which would then be more difficult to suppress. (As pointed out above, microtubule dynamics is significantly increased in spindle microtubules (over interphase microtubules) (*49*).)

Treatment of human cancer cells with low n*M* concentrations of Epo B leads to profound growth inhibition and cell death (Table 1). In

Table 1. Human cancer cell growth inhibition by Epo A and B and taxol (IC_{50} n*M*)[a]

	Cell line						
	HCT-116 (colon)	PC-3M (prostate)	A549 (lung)	MCF-7 (breast)	MCF-7/ ADR[b]	KB-31 (cervix)	KB-8511[c]
Epo A	2.51	4.27	2.67	1.49	27.5	2.1	1.9
Epo B	0.32	0.52	0.23	0.18	2.92	0.19	0.19
Taxol	2.79	4.77	3.19	1.80	9105	2.31	533

[a] Cells were exposed to drugs for 3–5 days, allowing for at least two population doublings. Cell numbers were estimated by quantification of protein content of fixed cells by methylene blue staining. Data from Ref. (*43*). [b] Multiple resistance mechanisms/ MDR. [c] P-gp overexpression/MDR.

line with the effects on tubulin polymerization *in vitro*, Epo B is a more potent growth inhibitor than Epo A, which in turn is approximately equipotent with taxol. Epothilone treatment produces aberrant mitotic spindles, results in cell cycle arrest in mitosis, and eventually leads to apoptotic cell death (*13*, *16*). Work at Novartis with HCT-116 human colon carcinoma cells has shown that a 4 h exposure period to Epo B produces virtually the same growth inhibitory effect as a continuous 72 h exposure (Fig. 5) (*17*). In contrast, for Epo A as well as taxol exposure times significantly above 4 h are required to produce effects equal to

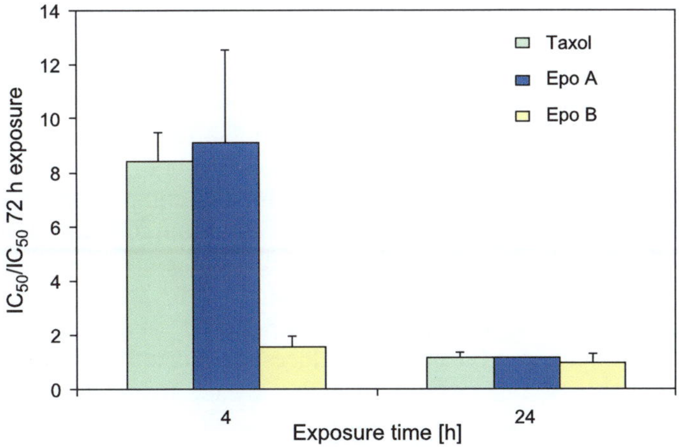

Fig. 5. Effect of exposure time on anti-proliferative activity in HCT-116 cells: Epothilones and paclitaxel. Reprinted from Biochim. Biophys. Acta – Reviews on Cancer, 1470, K.-H. Altmann, M. Wartmann, T. O'Reilly, Epothilones and related structures – a new class of microtubule inhibitors with potent in vivo antitumor activity, p. M83, 2000, with permission from Elsevier

those observed after 72 h of continuous exposure; IC_{50}'s after a 4 h exposure period are 8–9-fold higher than those observed after 72 h continuous exposure. These findings suggest that Epo B is retained inside cells more effectively than either Epo A or taxol.

Apoptosis is often assumed to be a direct consequence of G2/M arrest, but *Horwitz* and co-workers have recently shown that the situation is clearly more complex (*50*, *51*). Thus, low concentrations of Epo B (and also taxol or discodermolide) produce a large aneuploid cell population in A549 lung carcinoma cells in the *absence* of a mitotic block. These cells arise from aberrant mitosis after formation of multipolar spindles, are arrested in the G1 phase of the cell cycle, and will eventually undergo apoptosis. In contrast, higher drug concentrations lead to a protracted mitotic block from which the cells exit without division, thus forming tetraploid G1 cells (*51*). Both types of G1 cell populations are characterized by a similar pattern of up-regulated gene expression, which includes *p53*, *p21* (an inhibitor of cyclin-dependent kinases 2 and 4, which are required for G1-S progression), *CD95* (Fas ligand) and *BTG2* as well as survival genes *PTGF-β* and *GADD45*. At the same time, PARP cleavage products *p*24 and *p*85 are observed in both the aneuploid and the tetraploid G1 cells, with PARP cleavage being one of the earliest markers for apoptosis. Surprisingly, mitotic cells showed little, if any PARP cleavage, indicating the absence of apoptosis for cells blocked in G2/M. Consistent with this finding, mitotic cells express significantly elevated levels of the apoptosis-inhibiting protein survivin (*51*). None of the above differential behavior was observed for tubulin polymerization inhibitors such as colchicine, nocodazole, or vinblastine, which do *not* give rise to aneuploid cells, but always lead to mitotic arrest followed by apoptosis (*50*). This suggests that the suppression of microtubule dynamics, which is common to both types of tubulin-interacting agents, cannot fully account for the complex array of biological effects displayed by Epo B or other microtubule-stabilizing agents. The notion of cell death occurring from a G1-arrested state in Epo B-induced aneuploid cells is consistent with the observation that sequential treatment of cancer cells with Epo B and flavopiridol, an agent which enhances G1 arrest, produces significantly more apoptosis in MB-468 breast cancer cells than the reverse treatment regimen or either agent alone (*53*). In summary, the results from the *Horwitz* lab demonstrate that *entry* of cells into mitosis is a fundamental prerequisite for cell killing by microtubule-stabilizing agents, but that cell death itself does not necessarily require prior mitotic arrest. At low concentrations of Epo B cells may simply undergo mitotic slippage (aberrant mitosis) and subsequent cell cycle arrest in G1.

References, pp. 210–220

The *Horwitz* lab has also investigated the effects of Epo B (and other microtubule-interacting agents) on activation of the MAPK signaling pathways, using *trans*-reporting systems for the phosphorylation of nuclear transcription factors Elk-1 and c-jun (*52*). In these experiments, both Epo B and taxol were found to induce *trans*-activation of Elk-1 in CHO-cells *via* the Erk- and JNK-pathways, but not the p38 pathway. At the same time, both drugs induced mitotic arrest in HeLa cells which was coupled with phosphorylation of Bcl-2 and raf-1. A more detailed analysis of MAPK signaling was only performed for taxol, but the results of these experiments should also be relevant for Epo B. Most significantly, taxol was demonstrated to induce activation of Erk and JNK in CHO cells and of Erk, JNK, and p38 in HeLa cells. In HeLa cells time-dependent activation of Erk is accompanied by an increase in apoptosis markers (PARP cleavage, phosphatidylserine externalization) and accumulation of cells in G2/M. However, Erk-activation merely coincides with, but is not causally involved in apoptosis induction, as combination treatment of cancer cells exhibiting taxol-induced Erk activation (A549, HeLa) with taxol and MEK inhibitor U0126 resulted in additive cytotoxicity. Apart from effects on signal transduction, Epo B has also been demonstrated to induce a conformational change of Bax protein, which is required for its localization to the mitochondrial membrane and to trigger the mitochondrial pathway for apoptosis (*53*). Furthermore, inhibition of microtubule dynamics by Epo B produces significant nuclear localization of p53, which induces transcription of a variety of apoptosis related genes (*54*). On the other hand, the cytotoxic effects of Epo B in non-small cell lung cancer (NSCLC) cells (as measured by the appearance of cells with a hypodiploid DNA content) have been reported to be independent of both caspase inhibition or blockage of the death receptor pathway, thus raising questions about apoptosis as the predominant mechanism of cell killing by Epo B (*55*). In contrast, Epo B-mediated cell kill could be fully blocked with a specific inhibitor of the cysteine protease cathepsin B (Cat B), thus implicating Cat B as a central player in the execution of cell death in response to Epo B treatment of NSCLC cells (*56*). Consistent with these findings, Epo B was shown to induce release of pro-cathepsin B from the lysosomal compartment, thus allowing for cytoplasmic activation of the protease and subsequent cleavage of cytoplasmic substrates. It should be emphasized here that, although caspase activity does not seem to be an essential prerequisite for Epo B-induced cell kill, activation of these proteases still occurs after prolonged exposure of NSCLC cells to Epo B, as indicated by PARP cleavage, chromatin condensation, and phosphatidylserine externalization (*55*). Furthermore, it should be noted that the overall effects described by *Bröker et al.* (*55*, *56*) are very similar to those reported by the *Horwitz* group (*50*, *51*). This includes the insensitivity of Epo B-induced cell kill to caspase inhibition, which had been previously observed in HeLa cells by *Horwitz* and coworkers (*52*). It remains to be seen to what extent the above findings for NSCLC cells, in particular with regard to the role of cathepsin B, can be extended to other types of cancer cells.

As is generally the case for anticancer drugs, the cellular response to microtubule-stabilizing agents can be modulated by adaptive changes of the cell which lead to acquired drug resistance. Alternatively, cells may be inherently protected from the antiproliferative effects of cytotoxic agents by a variety of mechanisms. As indicated above, Epo A/B, in contrast to taxol and other standard cytotoxic anticancer agents, are largely non-susceptible to permeation glycoprotein-170 (Pgp)-mediated drug efflux and thus retain full (or almost full) antiproliferative activity against the corresponding multidrug-resistant cell lines *in vitro* (*13*, *16*, *17*, *43*) (Table 1). This may represent a distinctive advantage of epothilones over current taxane-based therapy, provided that Pgp-mediated drug resistance is indeed of practical relevance clinically. The latter question, however, is a matter of significant debate (*57*). At the same time, it is now well established that cancer cells can become resistant to epothilones through alternative (*i.e.* non-Pgp-dependent) mechanisms, such as tubulin mutations. For example, *Wartmann* (*27*) has isolated an epothilone-resistant subline of the KB-31 cervix carcinoma cell line (termed KB-31/C5), in which Thr274, which maps to the taxol/epothilone binding site on tubulin (*45*, *58*), is mutated to Pro, resulting in 14-fold resistance to Epo A and 3.4-fold resistance to Epo B. Similar findings have been independently reported by *Giannakakou et al.*, who have produced an Epo A-resistant cell line, 1A9/A8, in which Thr274 is mutated to Ile rather than Pro (*59*). Consistent with the fact of a shared binding site between epothilones and taxol both 1A9/A8 as well as KB-31/C5 cells are cross-resistant to taxol, albeit to varying degrees.

More recently, *Horwitz et al.* have generated three different Epo-resistant cell lines, A549B40, HeLa.EpoA9, and HeLa.EpoB1.8, each of which is characterized by a specific β-tubulin mutation, namely, Gln292Glu in A549.EpoB40 cells, Pro173Ala in HeLa.EpoA9 cells and Tyr422Cys in HeLa.EpoB1.8 cells (*60*). The highest level of resistance was associated with A549.EpoB40 cells, which were 95-fold resistant to Epo B and also exhibited marked cross-resistance with other microtubule-stabilizing agents, with the notable exception of discodermolide. Induction of tubulin polymerization in cell extracts prepared from A549.B40 cells is markedly decreased at low concentrations of Epo B

(<100 n*M*) in comparison with extracts derived from wild-type A549 cells. No difference is observed, however, in the overall extent of tubulin polymerization at higher compound concentrations. While exhibiting significant resistance to epothilones, A549.EpoB40 cells are *hyper*sensitive towards tubulin polymerization inhibitors and this behavior may provide a clue as to the effects of the Gln292Glu mutation on microtubule stability (*vide infra*). Through further selection of A549.EpoB40 cells, *Horwitz* and co-workers have subsequently generated an even more resistant cell line, A549.EpoB480 (*61*), which is *ca.* 900-fold resistant to Epo B. In addition to the βGln292Glu mutation the EpoB480 line is characterized by mutations of β60 from Val to Phe and of α195 (α-tubulin) from Leu to Met. As βVal60 is located at the end of the H1-S2 loop, which is assumed to be involved in M-loop contacts between protofilaments, amino acid changes at this position could affect lateral contacts between protofilaments and thus lead to less stable microtubules. This hypothesis is supported by the fact that EpoB480 cells are again hypersensitive to the action of vinblastine and, rather intriguingly, also depend on (low concentrations of) Epo B for survival (*61*).

The above Gln292Glu mutation was also identified by *Verrills et al.* (*62*) in a highly resistant subline of the human T cell acute leukemia cell line CCRF-CEM in combination with a second mutation at position 231 of β-tubulin (Thr $\rightarrow$ Ala). This cell line (termed dEpoB300) is 307-fold resistant to Epo D (which was the selecting agent) and exhibits 77-fold and 467-fold cross-resistance with Epo B and taxol, respectively. dEpoB300 cells contain significantly ($>60\%$) decreased levels of polymerized tubulin (compared to wild-type) and the polymerized tubulin fraction shows greatly reduced affinity to taxol. Most strikingly, Epo D failed to induce any measurable tubulin polymerization in cell lysates prepared from dEpoB300 cells at 8 μM compound concentration, thus illustrating that impaired growth inhibition is indeed paralleled by diminished effects on tubulin polymerization.

In summary, all tubulin mutations identified to date in epothilone-resistant cells are found in regions of the tubulin structure that are predicted to be important for tubulin polymerization and/or microtubule stability (including those which may additionally affect drug binding). Thus, these mutations may not only affect drug-target interactions, but they may also (or alternatively) impair intrinsic tubulin functions in a way that could result in hypostable microtubules (*27*, *60*, *61*). This hypothesis, which has been previously put forward by *Cabral et al.* based on observations on taxol-resistant CHO cells (*63*), is supported by the fact that all the above cell lines are hypersensitive to tubulin-depolymerizing agents, such as vincristine or colchicine.

Tubulin mutations arising in response to taxol treatment of human cancer cell lines do not necessarily lead to cross-resistance with epothilones (*19*) and again this may enable epothilone-based treatment of Taxol®-resistant tumors. However, any such predictions must be treated with great caution, as the clinical significance of individual resistance mechanisms identified *in vitro* has not been established.

6.3. *In vivo* Pharmacology of Epo B

The *in vivo* effects of Epo B have been investigated in some detail by a group at the Sloan-Kettering Cancer Center as well as the group at Novartis. While early experiments by the Sloan-Kettering group in xenograft models of human leukemia (CCRF-CEM and CCRF-CEM/VBL (MDR)) in CB-SCID mice (drug-sensitive as well as multidrug-resistant tumors) had suggested promising antitumor activity (*64*), the compound was found in subsequent experiments to have only limited effects on tumor growth in human MX-1 breast or SKOV-3 ovarian tumors in mice and to exhibit considerable toxicity. These data led to the conclusion that Epo B might simply be too toxic to become a clinically useful anticancer agent (*65*). In contrast to these findings, studies by the Novartis group have demonstrated potent antitumor activity of Epo B in a number of drug-sensitive human tumor models (nude mice) upon *i.v.* administration (*17*, *66*). Activity was observed in models encompassing all four major types of solid human tumors (lung, breast, colon, prostate) and was manifest either as profound growth inhibition (stable disease) or significant tumor regression. In addition, Epo B was found to be a potent inhibitor of tumor growth in P-gp-overexpressing multidrug-resistant human tumor models. Regressions were observed in two such models (KB-8511 (cervix carcinoma) (*17*) and HCT-15 (colon carcinoma) (*67*)), where tumors were either poorly responsive or completely non-responsive to treatment with Taxol®. Although Epo B treatment was frequently accompanied by significant body weight loss, therapeutic effects could generally be achieved at tolerated dose levels. As demonstrated in a recent study by *Pietras et al.*, the antitumor effect of Epo B in a model of human anaplastic thyroid carcinoma can be potentiated by co-administration with the tyrosine kinase inhibitor STI571 (Gleevec®) without any overt decrease in tolerability (*68*). The enhanced antitumor activity of the combination is presumed to be a consequence of a selective enhancement in drug uptake by the tumor due to inhibition of PDGF-R (platelet-derived growth factor receptor) by STI571.

References, pp. 210–220

The exact reasons for the disparate results of *in vivo* experiments by the Sloan-Kettering and the Novartis groups are unknown at this point. It should be noted, however, that the data are not necessarily incompatible, but may simply reflect differences in the experimental set-ups, such as tumor models, formulation, and/or dosing regimens.

6.4. Epothilone Analogs and SAR Studies

As indicated in the introductory section of this chapter, the chemistry of epothilones has been extensively explored and a wealth of SAR information has been generated for this family of structures over the last several years based on synthetic analogs and semi-synthetic derivatives. Most of the synthetic work was performed by the groups of *Nicolaou* (*cf.*, *e.g.*, (*20*)) and *Danishefsky* (*cf.*, *e.g.*, (*21*)) and to a lesser extent also the groups at Novartis (*Altmann* and colleagues, now at the ETH Zürich) (*31*, *43*, *69*) and Schering AG (*70*, *71*). Semi-synthesis work has been reported by the groups at the former "Gesellschaft für Biotechnologische Forschung" in Braunschweig, Germany (GBF, now "*Helmholtz* Centre for Infection Research"; *cf.*, *e.g.*, (*72*, *73*)) and Bristol-Myers Squibb (BMS; *cf.*, *e.g.*, (*74*)). The SAR data that have emerged from this research have been summarized in several recent review articles (*11*, *20*, *21*, *27–31*). In the following, the most significant results and conclusions from these SAR studies will be discussed, such that modifications will be grouped according to their particular location in the overall structural framework. For the largest part, this discussion will be based on publications that have appeared in scientific journals; however, as epothilones are also the subject of a large body of patent literature, reference will made in some cases to relevant patent applications (although most of these applications do not contain any specific biological data). In addition, it is important to emphasize that any data related to pharmacological effects (*i.e.* tubulin polymerization, *in vitro* and *in vivo* antiproliferative activity), independent of the type of literature source used, may not always be directly comparable when originating from different laboratories, due to differences in the experimental conditions employed.

6.4.1. The O16–C8 Sector ("Polyketide Sector")

6.4.1.1. Lactam-based Analogs

One of the most important and productive scaffold modifications in epothilones that has been investigated so far involves the replacement of

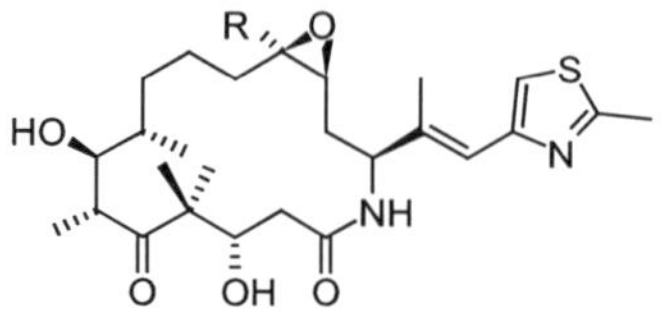

R = CH_3: **VI-1a**
R = H: **VI-2a**

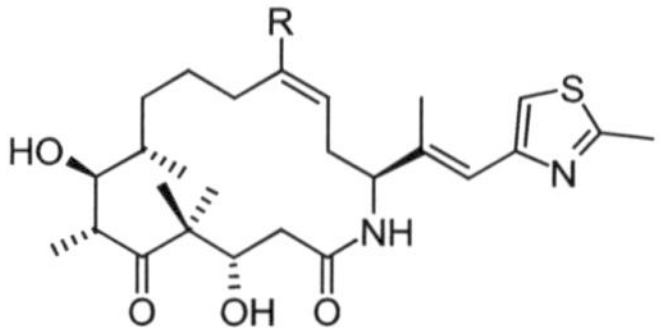

R = CH_3: **VI-1b**
R = H: **VI-2b**

Fig. 6. Lactam-based analogs of epothilones

the lactone oxygen by nitrogen, *i.e.* the conversion of the macro*lactone* into a macro*lactam* ring (*75, 76*). The work on lactam-based analogs was spearheaded by the group at BMS and it has led to the identification of the lactam analog of Epo B (**VI-1a** = BMS-247550, ixabepilone; Fig. 6) as a highly promising antitumor agent, which has undergone extensive clinical evaluation by BMS and has recently been approved by the US FDA for clinical use in humans (see Chapter 7).

Lactam-based analogs of epothilones were conceived by the BMS group as a means to stabilize the lactone-based natural products against metabolic degradation, as it had been observed in rodent plasma. It is worth noting, however, that in spite of its short plasma half-life in rodent species, Epo B shows potent antitumor activity in a variety of nude mouse human tumor models (*17*) and the same is true for Epo D (*vide infra*). Epo D has also been demonstrated to be significantly more stable in human than in rodent plasma (*77*), and similar observations were made at Novartis for Epo B. This species-dependent difference in metabolic stability is not surprising and reflects the well known differences in plasma esterase activity between humans and rats or mice.

BMS-247550 (**VI-1a**) is a potent inducer of tubulin polymerization, but its antiproliferative activity is *ca.* one order of magnitude lower than that of Epo B (*27, 75*) (*e.g.*, IC_{50} values against the human colon carcinoma cell line HCT-116 are 3.6 n*M* and 0.42 n*M*, for **VI-1a** and Epo B (*75*)). A similar trend holds for Epo A lactam **VI-2a**, as well as Epo D and C lactams **VI-1b** and **VI-2b** (Fig. 6), all of which are less active against drug-sensitive cancer cell lines than the corresponding parent compounds Epo's A, D, and C. Methylation of the lactam nitrogen in **VI-2b** results in a substantial loss in potency (*78*).

In contrast to Epo B, **VI-1a** is significantly less active against the P-gp overexpressing, multidrug-resistant KB-8511 variant of the KB-31 cell line than the drug-sensitive KB-31 parental line (IC_{50}'s of 2.85 n*M* and 128 n*M* against KB-31 and KB-8511 cells, respectively (*27*)). This indicates that the compound is a substrate for the P-gp efflux pump. Similar observations have been made by *Stachel et al.,* who have reported resistance factors of 1423 and 81.4 for **VI-1b** in the drug-sensitive/multidrug-resistant cell line pairs CCRF-CEM/CCRF-CEM/VBL_{100} and CCRF-CEM/CCRF-CEM/Taxol (*vs.* resistance factors of 6.1 and 3.1, respectively, for Epo B) (*79*).

Like other types of microtubule inhibitors BMS-247550 (**VI-1a**) was found by *Bhalla* and co-workers to induce G2/M arrest and apoptosis in human cancer cell lines (*80*). Interestingly, *Bhalla* and coworkers strongly emphasize that (BMS-247550 (**VI-1a**)-induced) "apoptosis of A2780-1A9 cells follows mitotic arrest, which is not associated with a marked increase in the levels of survivin" (*cf.*, however, recent work by *Horwitz* and coworkers (*50, 51*)).

Similar to Epo B (*53*) (*vide supra*), apoptosis induction by BMS-247550 (**VI-1a**) in MDA-MB-468 human breast carcinoma cells appears to be related to conformational changes in the Bax protein and its subsequent translocation to the mitochondrial membrane (*80*). In line with a Bax-dependent mechanism of cell death, BMS-247550 (**VI-1a**)-mediated apoptosis, but not G2/M arrest, is significantly delayed in the presence of increased levels of the anti-apoptotic protein Bcl-2 and this is paralleled by a delay in Bax conformational change. Interestingly, BMS-247550 (**VI-1a**)-induced cell death, at least in MDA-MB 468 breast carcinoma cells, appears to be caspase-*dependent* (*80*), contrary to what has been reported for Epo B in other cell lines (*52, 55, 56*). In a separate study in A2780-1A9 human ovarian carcinoma cells (*81*), *Bhalla* and coworkers found BMS-247550 (**VI-1a**) treatment to be associated with cytosolic accumulation of cytochrome c and smac (an antagonist of the IAP (inhibitors of apoptosis) family of proteins, which act as caspase inhibitors) as well as increased protein levels of death receptors DR4 and DR5. Consistent with these findings BMS-247550 (**VI-1a**) significantly increased the apoptotic effects of Apo-2L/TRAIL in A2780-1A9 cells when administered prior to Apo-2L/TRAIL. No such effect was observed upon simultaneous treatment with these agents or if Apo-2L/TRAIL treatment preceded treatment with BMS-247550 (**VI-1a**) (*81*).

BMS-247550 (**VI-1a**) exhibits antitumor activity similar to that of Taxol® in Taxol®-sensitive tumor models (*i.e.*, A2780 human ovarian carcinoma, HCT-116 and LS174T human colon carcinomas) when each drug is given at its optimal dose (*76*). In spite of its limited effects against highly multidrug-resistant cell lines *in vitro*, BMS-247550 (**VI-1a**) was also shown to be superior to Taxol® in Taxol®-resistant tumor models (*i.e.*, Pat-7 and A2780Tax human ovarian carcinomas, Pat-21 human breast carcinoma, Pat-26 human pancreatic carcinoma, M5076 murine sarcoma). Furthermore, the compound showed remarkable antitumor activity against Pat-7 ovarian and HCT-116 colon carcinoma xenografts following *oral* administration (*76*). BMS-247550 (**VI-1a**) has also been evaluated as a potential radio-sensitizer in a H460 human lung carcinoma model. Only a marginal potentiation of tumor growth delay (GD) was observed with an enhancement factor (EF) of 1.28 ($EF = (GD_{Drug + radiation\ treatment} - GD_{Drug})/GD_{radiation\ treatment}$) in this study (*82*). It is clear, however, that significantly more data will be required to allow solid conclusions with regard to the radiation-enhancing potential of BMS-247550 (**VI-1a**) or any other epothilone-type anticancer agent.

6.4.1.2. Modifications in the C2–C8 Region

Modifications at the 3-position of the epothilone macrocycle have been investigated by the BMS group, in collaboration with *Höfle*'s group at the GBF. These studies included 3-deoxy-2,3-didehydro derivatives **VI-3a** and **VI-3b** as well as 3-deoxy-3-cyano derivatives **VI-4a** and **VI-4b** (Fig. 7). Both types of analogs are readily accessible through semisynthesis from Epo A or B (*83*) (Chapter 5).

Analogs **VI-3** retain most of the activity of the parent natural products, with **VI-3b**, *e.g.*, being only 4-fold less potent than Epo B against the human colon carcinoma cell line HCT-116 (*83*). The same holds for 3-deoxy-(3*S*)-cyano derivative **VI-4a**, which is only two-fold less active than Epo A. In contrast, the corresponding (3*R*)-isomer shows a significant reduction in potency. More recently, *Altmann* and coworkers have also investigated *saturated* 3-deoxy derivatives of Epo A and B, which lack any direct conformational constraint about the C2–C3 bond (*84*). Quite remarkably, 3-deoxy-Epo B (**VI-5**) (Fig. 7) retains potent biological activity, which is manifested in IC_{50} values for human cancer cell growth inhibition in the low n*M* range. (*E.g.*, the IC_{50} values of **VI-5** against the human cervix carcinoma cell lines KB-31 and KB-8511 are 7.4 n*M* and 4.0 n*M*, respectively, *vs.* 0.29 n*M* and 0.22 n*M*, for Epo B (*84*).) These findings demonstrate that the presence of a

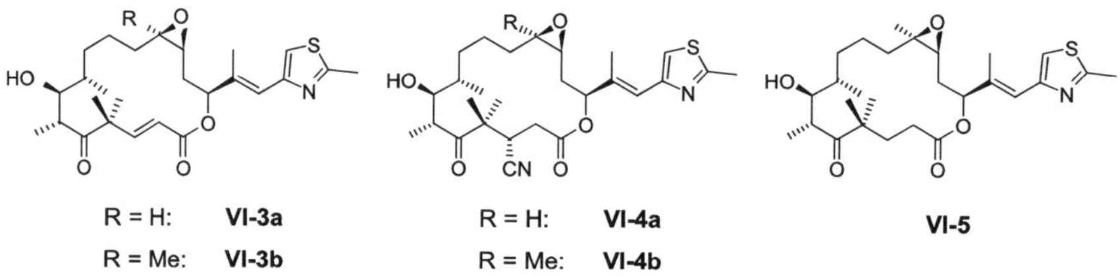

Fig. 7. Modifications at C2/C3

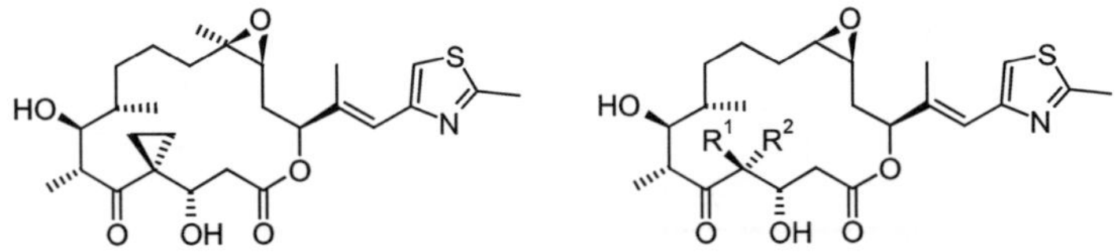

VI-6

R^1 = H, R^2 = CH_3: **VI-7a**
R^1 = CH_3, R^2 = H: **VI-7b**

Fig. 8. Modifications at C4

3-hydroxyl group in epothilones is not a crucial requirement for potent biological activity, even in the absence of a hard conformational constraint about the C2–C3 bond, as it is present in α,β-unsaturated lactones **VI-3**. It should be noted, however, that inversion of stereochemistry at C3 in Epo C has been reported to result in significantly reduced tubulin-polymerizing activity (*85*).

Analogs with a 3-membered ring on C4, such as **VI-6** (Fig. 8), have been reported by *Nicolaou et al.* to be devoid of any tubulin-polymerizing and antiproliferative activity (*85*). Interestingly, C4-monomethylated Epo A derivatives **VI-7a** and **VI-7b** (Fig. 8) have been reported to exhibit similar growth inhibitory activity against the mouse fibroblast cell line L929 as Epo A (IC_{50}'s of 14.6 n*M* and 42 n*M* *vs.* 8 n*M* for Epo A) (*86*). This finding is rather intriguing and indicates that the *gem*-dimethyl group at C4, contrary to intuitive belief, does not play a pivotal role in the stabilization of the bioactive conformation of epothilones.

As part of a general program on the synthesis and biological evaluation of aza-epothilones, *i.e.*, epothilone analogs in which one of the carbon atoms in the macrolide backbone has been replaced by nitrogen, *Altmann* and co-workers have investigated 4-aza analogs of Epo D that are characterized by the replacement of C4 by nitrogen and the presence of a C5/N4 amide group rather than a C5-ketone (**VI-8**, Fig. 9) (*87*). These analogs were inspired by the fact that one of the characteristic

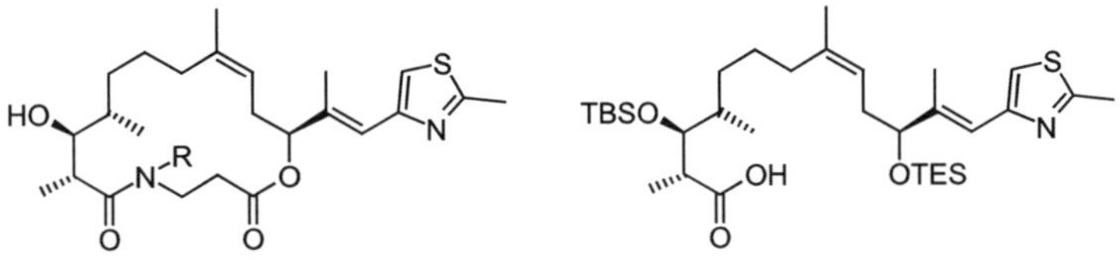

VI-8 **VI-9**

Fig. 9. 4-Aza-epothilones and structure of carboxylic acid **VI-9**

features of the tubulin-bound structure of Epo A (*88*) is the presence of a close to *syn-periplanar* conformation about the C4–C5 bond. The same geometry would be enforced in analogs of type **VI-8**, provided that the amide bond between N4 and C5 was present in a *cis* conformation. At the same time, preliminary modeling studies indicated that the presence of a *cis* amide bond in this position should allow replacement of the C1–C4 segment by various types of β-amino acids without causing significant distortions in the bioactive conformation of the C5–O16 segment. Apart from these structural considerations, structures of type **VI-8** also appeared attractive for chemical reasons, as they would lend themselves to an efficient combinatorial chemistry approach employing a single advanced intermediate (*i.e.*, a C5-carboxylic acid encompassing the C21–C5 fragment (**VI-9**); epothilone numbering). So far, only a limited number of examples of analogs of type **VI-8** have been investigated, all of which were found to lack any meaningful tubulin-polymerizing or antiproliferative activity (*87*). However, a larger number of these structures (incorporating different types of α-, β-, and γ-amino acids) will need to be investigated before allowing a final conclusion on the (pharmaceutical) validity of this modification approach.

Based on a wealth of patent literature, C6-modified epothilone analogs have been extensively studied by the group at Schering AG (now Bayer-Schering). However, only a fraction of the chemistry described in these patent applications has been published in the scientific literature and relatively few biological data are available for this class of analogs. In general, the replacement of the methyl group attached to C6 by other alkyl substituents appears to be reasonably well tolerated (*48*, *70*, *71*, *89*). Thus, IC_{50} values of 3.4 n*M* and 38 n*M* have been reported for the

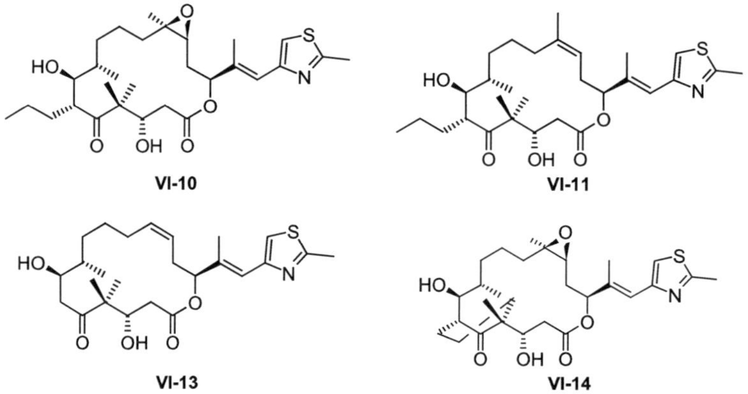

Fig. 10. Modifications at C6

C6-propyl analogs of Epo B (**VI-10**) and Epo D (**VI-11**) (Fig. 10), respectively, against the human breast cancer cell line MCF7 (*vs.* 0.6 n*M* and 19 n*M* for Epo B and Epo D); furthermore, these compounds retain full activity against the multidrug-resistant NCI/Adr variant of the MCF7 line (*48*).

Most significantly, a 6-allyl analog of a side-chain-modified derivative of Epo B developed by the group at Schering (ZK-Epo (Sagopilone)) is currently undergoing Phase II clinical trials (*70*). This compound will be discussed in somewhat more detail in the section on side chain modifications. 6-Desmethyl-6-propyl-Epo B (**VI-10**) was found to localize preferentially to the nuclear compartment of A431 or NCI/Adr cells (40–50%), while both taxol or 6-desmethyl-6-propyl-Epo D (**VI-11**) were primarily found in the cytoplasm (10–15% nuclear localization) (*48*). In addition, overall cellular uptake was significantly higher for **VI-10** than for **VI-11**. Based on these data, it was suggested that the difference in antiproliferative activity between epoxide-bearing epothilones and their deoxy analogs may be the consequence of differences in cellular accumulation and subcellular distribution (*cf.* also (*43*)). It is worth noting, however, that the enhanced nuclear accumulation of epothilones is not associated with any covalent modification of nuclear components such as DNA or nuclear proteins.

Hardt et al. have recently reported the 6-desmethyl analog of epothilone C (**VI-13**, Fig. 10) as a minor product in the fermentation of *Sorangium cellulosum* (*86*); this compound is 25-fold less active against the mouse fibroblast cell line L929 than Epo C. *Mulzer* and co-workers have described the synthesis of the conformationally constrained tricyclic analog **VI-14** (Fig. 10) (*90*), which was conceived based on the close proximity of the C4-(pro-*R*)- and the C6-methyl groups in the X-ray crystal structure of Epo B. The compound showed no biological activity, but it should be noted that the stereochemistry at C7 could not be rigorously established.

Modifications at C7 or C8 have generally been associated with reduced activity, although the exploration of structural space in this region has been far from exhaustive. Removal of the methyl group attached to C8 as in **VI-15** (Fig. 11) results in a >100-fold loss in activity against the CCRF-CEM leukemia cell line in comparison with Epo A (IC_{50} = 0.439 μM *vs.* 3 n*M*) (*91*). In addition, compounds **VI-16a** and **VI-16b** (Fig. 11) have been reported to exhibit a significantly decreased ability to induce tubulin polymerization *in vitro*, indicating that neither inversion of stereochemistry at C8 nor di-substitution is well tolerated (*85*). Likewise, the *simultaneous* inversion of stereochemistry at C6 and C7 in compound **VI-17** (Fig. 11) results in a loss of tubulin induction

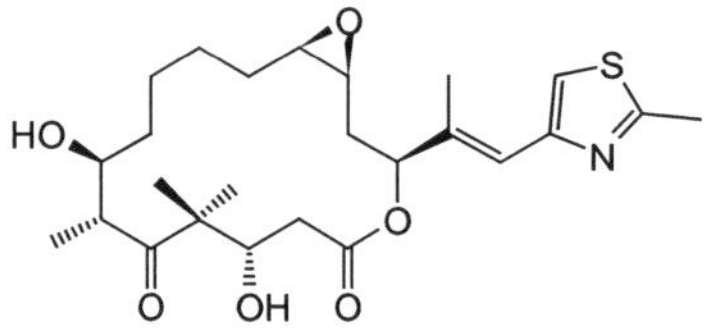

VI-15

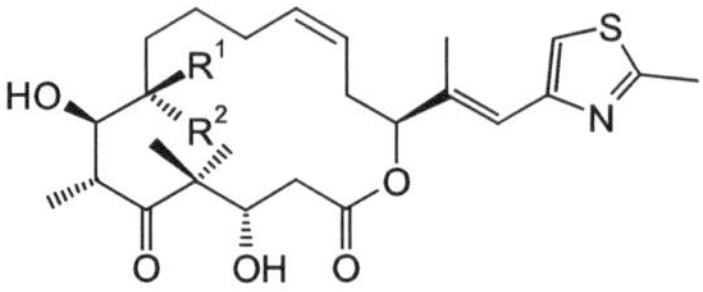

$R^1 = R^2 = CH_3$: **VI-16a**

$R^1 = CH_3$, $R^2 = H$: **VI-16b**

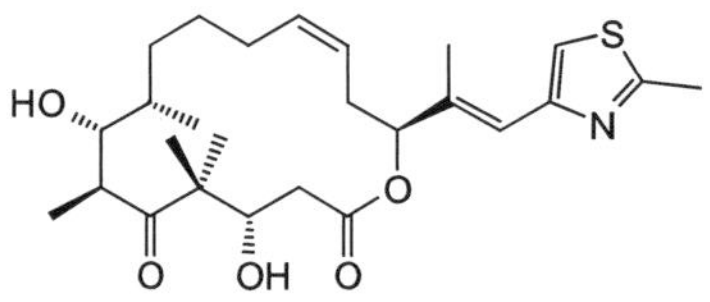

VI-17

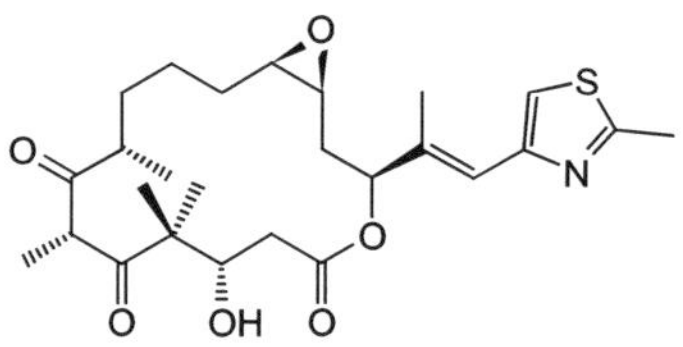

VI-18

Fig. 11. Modifications at C7 and C8

potential (25% *vs.* 76% for Epo A under identical experimental conditions (*85*)). Analogs with inverted stereochemistry at only C6 or C7 have not been reported in literature. *Sefkow et al.* have reported a number of oxidation and reduction products derived from epothilones, including the C7-ketone **VI-18** (Fig. 11) (*92*) (see Chapter 5). All of these products were reported to exhibit greatly reduced biological activity, but no specific biological data are available for most of these derivatives.

6.4.2. The C9–C13 Sector ("Epoxide Sector")

6.4.2.1. Modifications of the C9–C12 Segment

Initial structural modifications in the C9–C11 trimethylene fragment adjacent to the epoxide moiety were generally found to result in diminished biological activity. Thus, early work by the *Nicolaou* and *Danishefsky* groups showed that ring contraction or expansion *via* the removal of existing or the incorporation of additional CH_2-groups in the C9–C11 region both caused a substantial loss in biological potency (*64, 93*). An alternative approach pursued by the Novartis group for modifications situated in the Northern hemisphere of epothilones was based on an epothilone pharmacophore model, which had been derived from a comparative analysis of the X-ray crystal structure of Epo B (*15*) with those of taxol and discodermolide (*P. Furet, N. van Campenhout*, unpublished results; no structural data for a tubulin/Epo A complex were

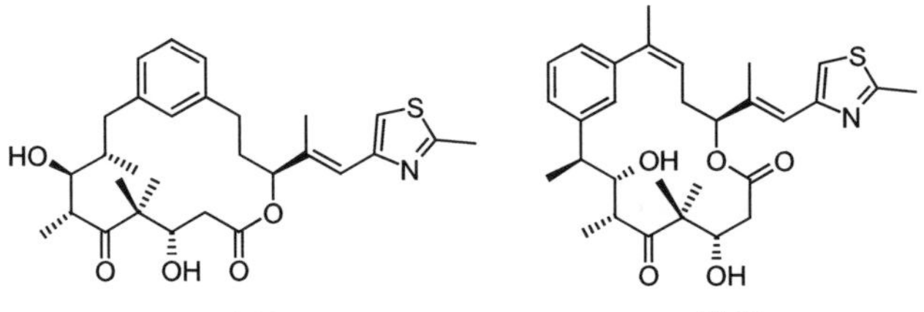

Fig. 12. Incorporation of *meta*-substituted phenyl rings in the C9 to C12 segment

available at the time of these investigations). Based on this model the three bonds between C8/C9, C9/C10, and C10/C11 were all predicted to adopt an *anti-periplanar* conformation in the bioactive state (*94*, *95*). As a consequence, the model suggested that the incorporation of *meta*-substituted phenyl rings in the C9 to C12 segment, such as in compounds **VI-19** or **VI-20** (Fig. 12) should lead to a stabilization of the bioactive conformation in this potentially rather flexible region of natural epothilones.

While the synthesis and biological evaluation of **VI-20** has been reported by the former Schering AG group (*71*), the Novartis group described the synthesis of analog **VI-19** (*94*). Unfortunately, **VI-19** and **VI-20** both are substantially less active than Epo B or D (**VI-19**; (*94*)) or to exhibit "reduced" activity (**VI-20**; (*71*). More specific data are not available from Ref. (*71*)).

In contrast to these disappointing early findings, a number of highly potent epothilone analogs with structural variations in the C9–C11 trimethylene segment were identified more recently and some of these exhibit highly favorable *in vivo* pharmacological properties. These analogs were obtained through a number of different approaches, including heterologous expression of the modified epothilone polyketide synthases in *Myxococcus xanthus* (*96*), total chemical synthesis (spearheaded by the *Danishefsky* group (*97–105*)), or biotransformation of Epo B (*106*, *107*). For example, the *in vitro* antiproliferative activity of **VI-21a** (epothilone 490, Fig. 13) is only 3–4-fold lower than that of the parent compound Epo D against the MCF7 breast, SF268 glioma, NCI-H460 lung cancer, and HL60 promyelocytic leukemia cell lines and the compound is equipotent with Epo D against the human T-cell leukemia cell lines CCRM-VEM and CCRM-VEM/VBL (*96*, *97*). These findings corroborate and extend earlier results reported by *Höfle* and co-workers for C10/C11 dehydro-Epo C (**VI-21b**, Fig. 13), which had been isolated as a minor fermentation product from cultures of the myxobacterium *S. cellulosum* (*14*). **VI-21b** is only 2-fold less active against the mouse

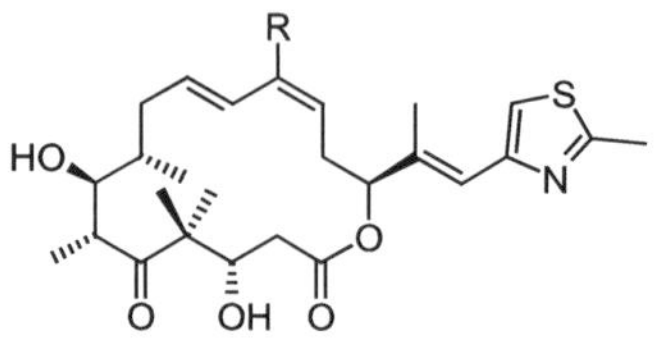

R = CH_3: **VI-21a**

R = H: **VI-21b**

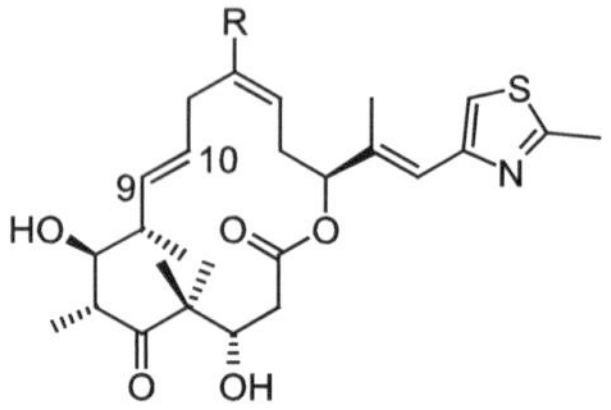

R = CH_3: **VI-22**

R = CF_3: **VI-23**

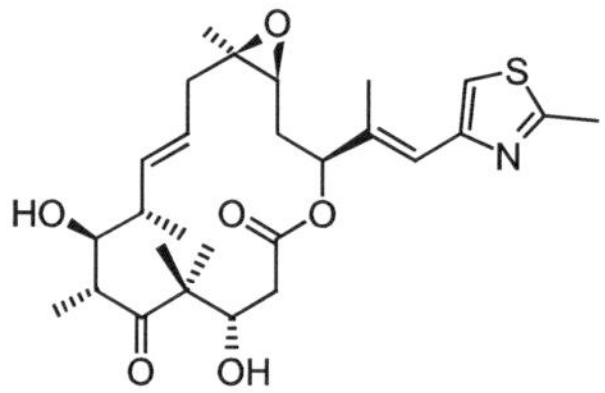

VI-24

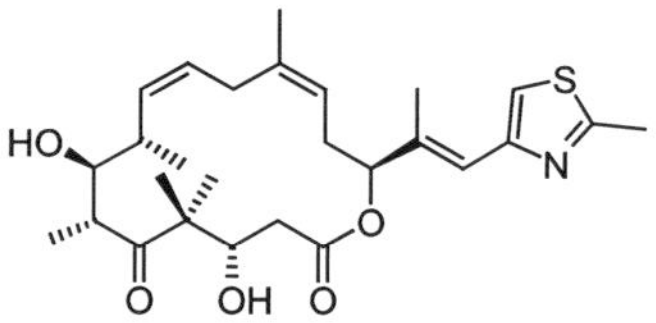

VI-25

Fig. 13. Modifications in the C9–C11 trimethylene segment

fibroblast cell line L929 than Epo C itself (*14*). The *Danishefsky* group has also reported additional analogs of **VI-21a**, which were obtained by dihydroxylation or cyclopropanation of the C10–C11 (*E*)-configured double bond (*97*). These compounds were found to be less active than **VI-21a**, although the cyclopropane derivative retained significant tubulin-polymerizing and antiproliferative activity.

Subsequent to these findings for epothilone 490 (**VI-21a**), *Danishefsky* and co-workers have shown that the presence of an (*E*)-configured double bond between C9 and C10 such as in **VI-22** (Fig. 13) results in a marked *increase* in antiproliferative activity over Epo D (the IC_{50} value of **VI-22** against the human leukemia cell line CCRF-CEM is 0.9 n*M* *vs.* 3.6 n*M* for Epo D) (*100*, *103*). Likewise, the C12/C13 epoxide corresponding to **VI-22**, *i.e.* **VI-24** (Fig. 13), is 3–4-fold more potent than Epo B (*101*). In contrast, (*Z*)-analog **VI-25** (Fig. 13) has been reported by *White et al.* to be *ca.* 30-fold less active than Epo D against the human cervix cancer cell line KB-31 (*108*). (Note that the compound assumed to be the (*E*)-analog **VI-22** in (*108*) later was found to be in fact the (*Z*)-isomer **VI-25** (*109*)). These bioactivity data are in agreement with recent spectroscopic studies (*88*), which indicate that the bioactive conformation of epothilones is characterized by *anti-periplanar* conformations about the C9/C10 and C10/C11 bonds. For analogs **VI-22** and **VI-24** it has also been suggested that the presence of a C9/C10 (*E*)-double bond favors the bioactive conformation of the macrocycle in the C5–C8 polyketide region (*101*). In light of these recent findings it appears likely that the lack of biological activity in analog **VI-20** (*vide supra*) is related to the increase in steric bulk associated with the presence of the phenylene moiety.

(*E*)-9,10-Didehydro-epothilone analogs **VI-22** and **VI-24** were found to possess markedly improved *in vivo* antitumor activity over their respective parent structures Epo D and Epo B in a mouse model of human breast cancer MX-1. For **VI-22** this effect was specifically ascribed to a combination of enhanced antiproliferative activity and improved plasma stability in mice (*100*, *102*), but, unfortunately, the compound is also associated with significantly enhanced toxicity (*102*). Nevertheless, **VI-22** (as KOS-1584) has been promoted to clinical development status and is currently undergoing Phase I clinical trials.

In contrast to **VI-22**, the corresponding C26-trifluoro derivative **VI-23** (termed "fludelone", Fig. 13) exhibits exquisite *in vivo* antitumor activity in the absence of unacceptable overt toxicity (*102*, *110*, *111*). The corresponding *in vivo* investigations included mouse models of human breast (MX-1), lung (A549), and colon (HCT-116) carcinomas as well as leukemia (CCRF) (*102*, *110*) and multiple myeloma (RPMI 8226 subcutaneous xenograft model and disseminated CAG MM model) (*111*). Sustained tumor remissions ("cures") were observed upon treatment with fludelone in several experiments and the compound proved to be clearly superior over taxol, but also over the non-fluorinated parent compound **VI-22**. **VI-23** showed similar activity against drug-sensitive and multidrug-resistant tumors (which were poorly responsive to Taxol®). In addition, comparable effects against MX-1 xenografts were observed after *i.v.* or oral administration of equal doses of fludelone, thus indicating that the compound has high oral bioavailability (*110*). The *in vitro* antiproliferative activity of **VI-23** is comparable with that of Epo D and the enhanced *in vivo* activity of the compound, as for the non-fluorinated analog **VI-22**, may be a consequence of improved pharmacokinetic properties, including enhanced metabolic stability in the liver (*110*). The discovery of fludelone could mark a major milestone in epothilone-based anticancer drug discovery and it represents the preliminary culmination of the extensive efforts of the *Danishefsky* group in this area.

Apart from the discovery of the potent *in vivo* activity of **VI-23**, work of the *Danishefsky* laboratory on C9–C11 modifications has also shown that the presence of a *trans* double bond between C10 and C11 allows the insertion of an additional methylene group between C11 and C12

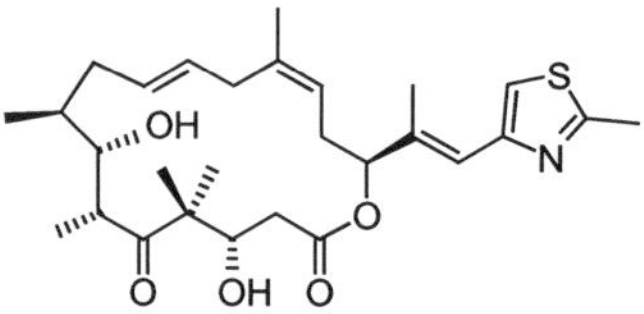

VI-26

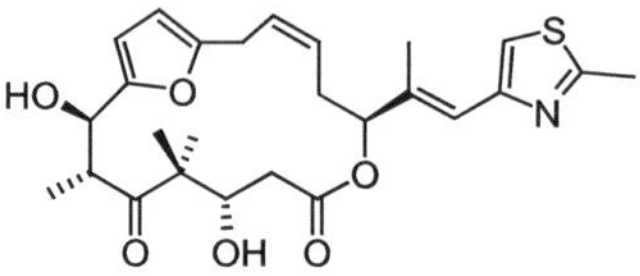

VI-27

Fig. 14. Structures of analogs with increased ring size and of furan-containing analog **VI-27**

(thus creating a 17-membered ring) without substantial loss in antiproliferative activity. Thus, in contrast to previously studied analogs with increased ring size (*vide supra*; (*93*)), compound **VI-26** (Fig. 14) is only 4-fold less active against the human leukemia cell line CCRF-CEM than the parent compound Epo D (*98*).

As for other modifications in the Northern part of the epothilone macrocycle, the replacement of C10 by oxygen (*112*, *113*) or the incorporation of a methyl group at position 10 of the macrolactone ring ((*S*)-10-methyl Epo C and its 12,13-(*E*)-isomer) (*114*) were found to be detrimental for biological activity. In contrast, the incorporation of a furan moiety, including C8 and C10, appears to be better tolerated (compound **VI-27**, Fig. 14) (*115*).

6.4.2.2. Modifications in the Epoxide Region (C12/C13)

Structural modifications of the epothilone macrocyclic framework in the C12/C13 (epoxide) region have been extensively investigated by several laboratories. These studies have resulted in a number of potent epothilone analogs, whose biological activity is comparable to that of Epo's A and B, with several of these compounds lacking the epoxide moiety present in the natural parent molecules. The fact that the epoxide moiety does not represent an absolute structural requirement for highly potent biological activity of epothilone-type molecules was first reported by the groups of *Danishefsky* (*64*, *116*, *117*) and *Nicolaou* (*85*, *118*) in the context of their early studies on Epo C and D, which incorporate a (*Z*) double bond between C12 and C13 in place of the oxirane ring system. (These compounds are also known as deoxyepothilones A and B; Fig. 2.) Thus, Epo C and D are virtually equipotent inducers of tubulin polymerization as Epo A and B. They are also potent inhibitors of human cancer cell growth *in vitro* (*17*, *43*, *64*, *65*, *85*, *116–118*), although they are somewhat less active than the corresponding parent epoxides. For example, Epo D inhibits the growth of the human cervix cancer cell line KB-31 and the leukemia cell line CCRF-CEM with IC_{50}

values of 2.7 n*M* (*43*) and 9.5 n*M* (*65*), respectively, *vs*. IC_{50}'s of 0.19 n*M* and 0.35 n*M* for Epo B. The reduced antiproliferative activity of Epo D compared to Epo B may be related to differences in cellular uptake between the two compounds (*48*). As for the corresponding epoxides, the deoxy compounds Epo's C and D are poor substrates for the P-gp efflux pump and, thus, are equally active against drug-sensitive and multidrug-resistant human cancer cell lines.

Epo D has been extensively characterized *in vivo* by the group at the Sloan-Kettering Cancer Center. Employing a specifically optimized *i.v.* dosing regimen (30 mg/kg, 6 h infusion, q2d $\times$ 5), the toxicity and efficacy of the compound was shown to be comparable to that of taxol when tested against MX-1 breast carcinoma and HT-29 colon tumors (*119*). However, in two multidrug-resistant models, MCF-7/Adr and CCRF-CEM/paclitaxel the activity of Epo D was far superior. Potent *in vivo* antitumor activity was observed in spite of the fact that Epo D exhibits a very short half-life in rodent plasma, which has also been reported for the parent compound Epo B (*75*) (*vide supra*). However, both Epo B and Epo D (*77*) are significantly more stable in human plasma (*in vitro*), thus indicating that plasma stability is an unlikely limitation for therapeutic applications of lactone-based epothilone analogs in humans. Finally, it should be noted that most of the available biological data on deoxyepothilones have been obtained with synthetic material, but these compounds are also produced as minor components in the fermentation of myxobacteria together with Epo's A and B and a host of additional analogs (*14*). In addition, heterologous expression systems have been developed more recently that allow the large scale production of Epo D by fermentation (*cf.* Chapter 3).

An alternative class of deoxyepothilones with alkyl groups attached to C13 rather than C12 has been described by *Sinha et al.* (*120*). The prototypical example for this series of analogs is compound **VI-28a** (Fig. 15), which is *ca.* 100-fold less potent than Epo D (IC_{50} against the human cervix cancer cell line KB-31 of 245 n*M* *vs.* 2.7 n*M* for Epo D). Replacement of the C13 methyl substituent by an ethyl group results in a

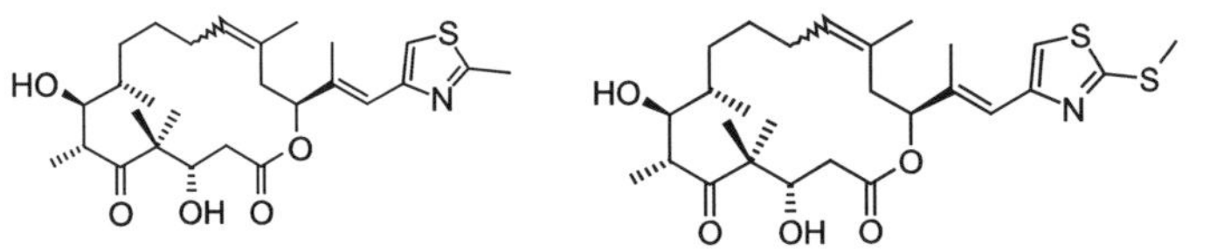

12,13-*Z*: **VI-28a**; 12,13-*E*: **VI-28b** 12,13-*Z*: **VI-29a**; 12,13-*E*: **VI-29b**

Fig. 15. C13-substituted deoxyepothilones

slight activity increase; more remarkable, however, is the observation that in contrast to Epo C and D (*vide infra*) the introduction of a C12/C13 *trans* double bond appears to result in improved tubulin polymerization and antiproliferative activity. The most potent analog reported in this series is the (*E*)-derivative **VI-29b** (Fig. 15), which incorporates a methylthio substituent in place of the natural methyl group at the 2-position of the thiazole moiety (78% tubulin polymerization and IC_{50} for growth inhibition of the KB-31 cell line of 4.0 n*M* *vs.* 93% and 2.77 n*M* for Epo D (*120*)). It should be noted that analogs related to **VI-28** and **VI-29** together with the corresponding epoxides have also been disclosed in a patent application from Schering AG (*121*), but no biological data have been published for these compounds so far.

Perhaps one of the most intriguing features of the epothilone SAR revealed during early SAR studies was the fact that even C12/C13 (*E*)-analogs of epothilones exhibit very potent tubulin-polymerizing as well as antiproliferative activity (*64*, *116–118*). The literature data on this phenomenon indicated that (*E*)-deoxyEpo A was only slightly less active than deoxyEpo A (Epo C), whereas in the B series the activity difference appeared to be more pronounced. At the same time, (epoxide-containing) (*E*)-Epo A was reported by *Nicolaou et al.* to be virtually equipotent with Epo A on an ovarian (1A9) and a breast cancer (MCF-7) cell line (*118*). However, the absolute stereochemistry of the active epoxide isomer was not reported and the *trans* analogs were obtained as minor components during the synthesis of the natural *cis* isomers. The absolute configuration of the active (*E*)-Epo A isomer was subsequently determined by the group at Novartis to be (12*S*,13*S*) (Fig. 16, **VI-30**) (*122*). This compound, which retains the natural stereochemistry at C13, was established to be a strong inducer of tubulin polymerization *in vitro* and to exhibit potent antiproliferative activity, whereas its (12*R*,13*R*)-isomer **VI-30a** is at least 500-fold less active (*122*). Inversion of stereochemistry at both epoxide carbons in Epo A results in a significant loss in tubulin-polymerizing activity (*85*); however, no data for cancer

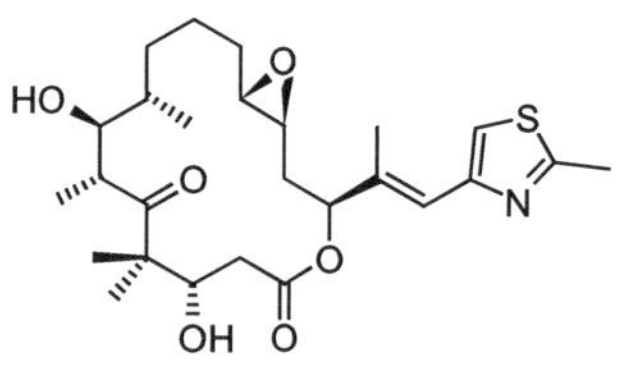

VI-30

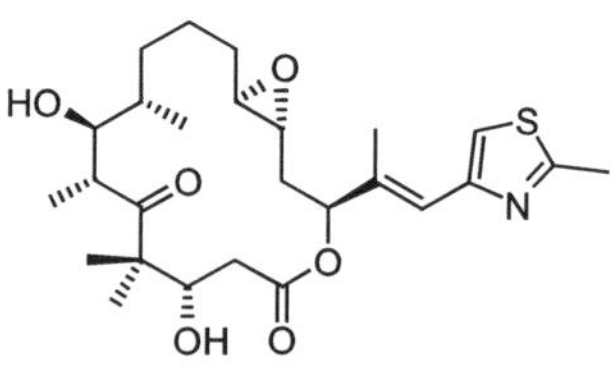

VI-30a

Fig. 16. Structures of *trans*-epothilones A

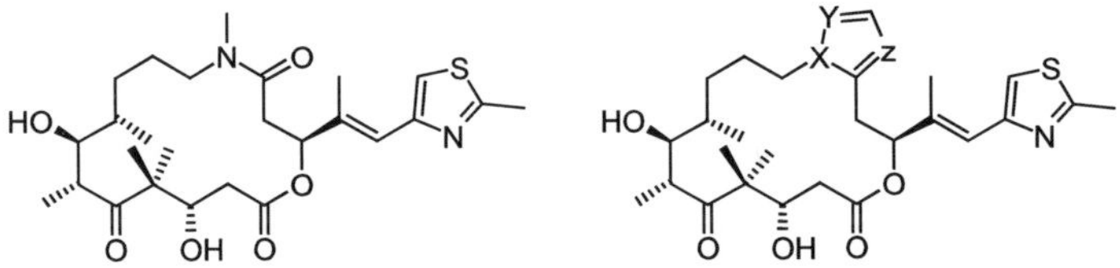

VI-31

X–Y = C=CH–CH, Z = CH: **VI-32**
X–Y = N–CH, Z = N: **VI-33**

Fig. 17. Replacement of the olefinic double bond in deoxyepothilones

cell growth inhibition by the (12*R*,13*S*)-isomers of either Epo A or B have been published (*cf.*, however, 6.4.3.1).

The average IC_{50} values across a panel of 7 human cancer cell lines have been reported as 2.72, 0.30, and 1.32 n*M* for Epo A, Epo B, and **VI-30** (*122*). These findings on the potent biological activity of the (*E*)-Epo A scaffold have recently been confirmed and expanded by *Nicolaou et al.* for a series of highly potent C12/C13 cyclopropane-based analogs of **VI-30** (*123*, *124*). In contrast, analogs of (*E*)-Epo B are generally less potent than the corresponding (*Z*)-isomers (*124*).

Attempts have been made to replace the olefinic double bond in deoxyepothilones by secondary or tertiary amide groups as in **VI-31** (Fig. 17) (*43*, *69*, *74*) or to make it part of a 1,2-disubstituted aromatic ring system, such as a phenylene (*126*) or an imidazole (*43*, *69*) moiety (**VI-32** and **VI-33**; Fig. 17). Structural units of this type were hypothesized to act as (*Z*)-configured double bond mimetics and were thus expected to induce a similar conformation of the macrocycle as for Epo D (assuming a preference of the C–N partial double bond in amide-based analogs for a *cis* conformation); as a consequence, the resulting analogs should exhibit similar antiproliferative activities as the parent deoxyepothilones. Unfortunately, contrary to these expectations, the biological activity of compounds **VI-31–VI-33** was found to be at least 100-fold lower than that of Epo D. Similar results were obtained by the BMS group for deoxyepothilone analogs incorporating C12–C13 amide bonds with the reverse orientation to that in **VI-31** (secondary and *N*-methyl amide: $–C^{12}(O)–N^{13}(R)–$, R=H, CH_3), which were also found to be inactive (*74*).

The underlying reasons for the lack of biological activity of such analogs have not been elucidated, but preliminary NMR studies with compound **VI-31** in DMSO/water showed a preference for the required *cis* conformation about the 12/13 *N*-methyl amide bond (*cis*/*trans* ratio ~4/1) (*69*), thus illustrating that the geometry about the 12–13 bond in **VI-31** was as desired. However, the data by *Sinha* on C13-modified

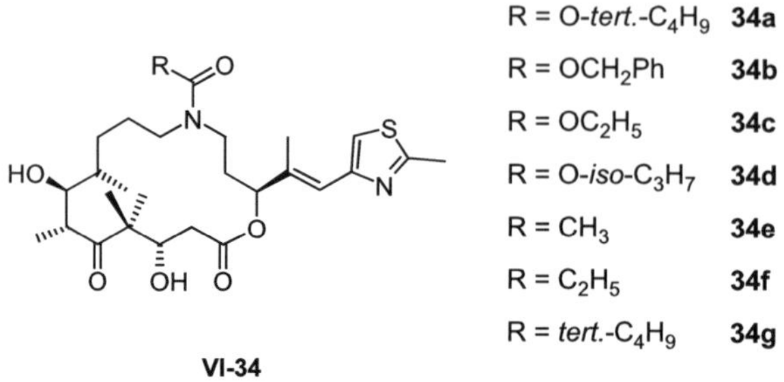

Fig. 18. Structures of 12-aza-epothilones

Epo C analogs (*120*) clearly demonstrate that increased steric bulk at C13 is associated with reduced potency and this may be a major reason for the strongly diminished potency of analogs such as **VI-31–VI-33**. In light of these findings *Altmann* and co-workers have continued to explore the potential utility of nitrogen incorporation at position 12 of the macrocycle, as a functional handle for further substitution, without concomitant modification of C13. This approach involved simple acylation of the 12-nitrogen atom, thus leading to amide- and carbamate-based analogs of type **VI-34** (Fig. 18), whose carbonyl oxygen could potentially assume the role of the epoxide oxygen in natural epothilones.

Analogs **VI-34** were tested for their ability to promote *in vitro* tubulin polymerization and their antiproliferative activity was assessed against the human cervix cancer cell lines KB-31 and KB-8511, which serve as representative examples of drug sensitive and P-gp-overexpressing, multidrug-resistant human cancer cell lines, respectively (*cf.*, *e.g.*, (*17*, *27*)). Compounds of type **VI-34** (which have been named "azathilones" (*127*)), although less active inhibitors of cancer cell growth than Epo A or B, can indeed be potent antiproliferative agents, with IC_{50} values against the human cervix carcinoma cell line KB-31 ranging from 30 n*M* (for **VI-34a**) to 300 n*M* (**VI-34b**, **VI-34d**) (*125*). Interestingly, however, some of these analogs were found to be significantly less active against the multidrug-resistant KB-8511 line than the drug-sensitive KB-31 parental line, thus indicating that compounds **VI-34** are better P-gp substrates than natural epothilones. The structural basis for this phenomenon is not understood at this point, but the finding is in line with a more general tendency for polar epothilone analogs (*e.g.*, compounds incorporating (other) amide bonds or additional hydroxyl groups) to exhibit increased resistance factors in the KB-31/KB-8511 cell line pair (*i.e.*, increased ratios of IC_{50} (KB-8511)/IC_{50} (KB-31) (*84*, *128*); *M. Wartmann*, *K.-H. Altmann*, unpublished observations). The most interesting analog investigated so far is **VI-34a** (Fig. 18), which is only *ca.* 15-fold less active against the drug-sensitive KB-31 line than Epo A (and thus roughly equipotent with Epo C) and is characterized by an only modest resistance factor of ~3. Guided by the SAR data for compounds **VI-34**, the azathilone scaffold has been further explored by the *Altmann* group in combination with a particular side chain modification (*127*) and these analogs will be discussed in more detail below. In contrast to analogs **VI-34**, the replacement of C13 by an unsubstituted NH or NCH_3 group (secondary and tertiary amines – $C^{12}H_2$–N^{13}(R)–, R=H, CH_3) has been reported to lead to inactive derivatives (*74*).

The strongest evidence that the oxirane ring system in epothilones merely serves to stabilize the proper bioactive conformation of the macrocyclic skeleton rather than acting as a reactive electrophile or a hydrogen bond acceptor comes from studies of cyclopropane-containing analogs **VI-35** and **VI-36** (Fig. 19). These studies have shown that the replacement of the oxirane ring in epothilones by a cyclopropane moiety is well tolerated and not associated with any loss in tubulin-polymerizing or antiproliferative activity (*123, 129, 130*). (*E.g.*, IC_{50}'s for **VI-35** and **VI-36** (Fig. 19) against the human colon carcinoma cell line HCT-116 are 1.4 n*M* and 0.7 n*M*, respectively (*129*).) As shown in a recent study by *Buey et al.* (*46*), the replacement of the epoxide moiety by a cyclopropane ring also produces enhanced binding to stabilized microtubules *in vitro*.

Bis-substitution of the oxygen-replacing carbon in **VI-36** by bulky bromine substituents leads to reduced activity, although the resulting analog is still more potent than Epo D against the HCT-116 cell line (IC_{50} against HCT-116 of 3.8 n*M* *vs.* 6.5 n*M* for Epo D (*129*)). Cyclopropyl-analogs **VI-35** and **VI-36** were first reported by the BMS group (*129*) and were obtained from Epo's A and B, through semisynthesis (see Chapter 5.2). *Nicolaou et al.* have subsequently reported the preparation of **VI-35** by total chemical synthesis together with the corresponding

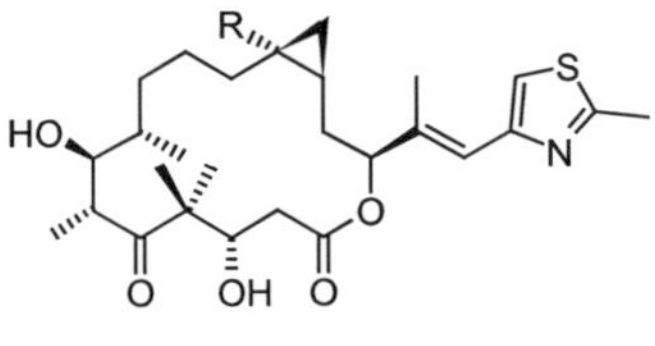

R = H: **VI-35**

R = CH_3: **VI-36**

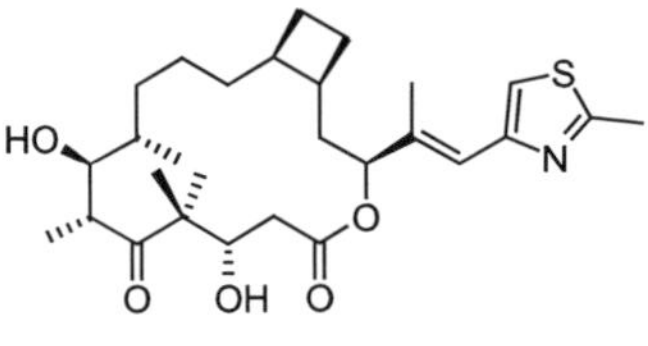

VI-37

Fig. 19. Structures of cyclopropane- and cyclobutane-containing epothilone analogs

C12/C13 cyclobutyl derivative **VI-37** (Fig. 19) (*123*, *130*). Cyclobutyl Epo A **VI-37** is less potent than cyclopropane-based analog **VI-35**, but the magnitude of the activity loss appears to be cell line-dependent (*123*). *Nicolaou et al.* (*123*) have also reported the cyclopropyl- and cyclobutyl-analogs of (*E*)-Epo A, which are essentially equipotent with the corresponding (*Z*)-isomers, thus confirming the observations discussed above for *cis*- and *trans*-epoxides.

Apart from cyclopropane analogs **VI-35/VI-36** and di-halogenated derivatives thereof, the BMS group has also prepared aziridine-based epothilone analog **VI-38** and N-substituted versions thereof (Fig. 20) (*131*). Like cyclopropane analog **VI-35**, the parent compound **VI-38** exhibits

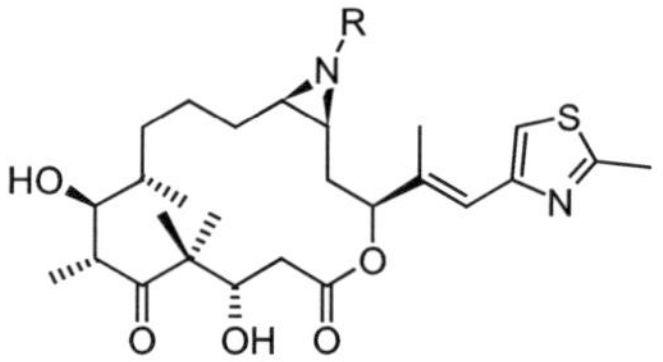

R = H: **VI-38**

R = CH_3: **VI-39**

Fig. 20. Structures of aziridine-based epothilone analogs

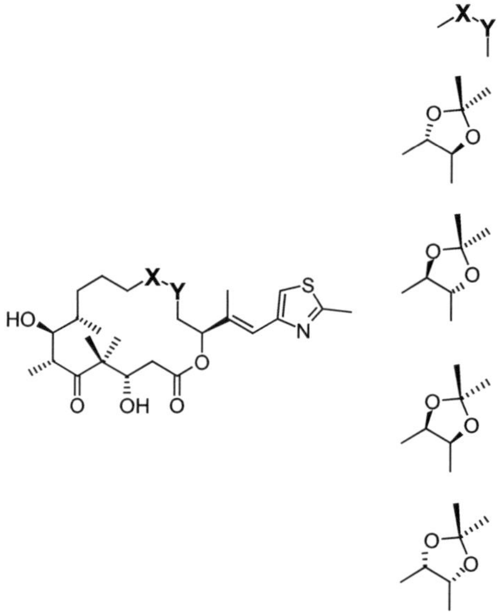

VI-40a

VI-40b

VI-41a

VI-41b

Fig. 21. Structures of analogs featuring C12/C13 acetonides

comparable activity to Epo A. Substitution of the aziridine nitrogen is well tolerated for a number of diverse groups, several of which lead to significantly improved potency over the parent compound **VI-38**. (*E.g.*, IC_{50}'s for **VI-38** and **VI-39** (Fig. 20) against the human colon carcinoma cell line HCT-116 are 2.7 n*M* and 0.13 n*M*.)

In addition to 12,13-cyclobutyl-epothilones, ring expanded analogs with a 5-membered cyclic acetal fused to the C12/C13 bond have been reported by *Sefkow et al.* (*73*) as well as *Altmann et al.* (43) (Fig. 21). Interestingly, acetonides **VI-40a** and **VI-41a** are only 10–15-fold less potent antiproliferative agents than Epo A itself (IC_{50} values against the KB-31 (KB-8511) line are 23 n*M* (10 n*M*) and 30 n*M* (17 n*M*) for **VI-40a** and **VI-41a**), while the respective diastereoisomers **VI-40b** and **VI-41b** are 30–100-fold less potent (*43*); *Sefkow et al.* (*73*) have reported **VI-40a** to have comparable antiproliferative activity with Epo C against the L929 mouse fibroblast cell line. These data suggest that for a tetrahedral geometry at C12 and C13 the size of the ring fused to the C12–C13 single bond can be significantly increased without substantial loss in biological potency (in contrast to analogs with a planar geometry of the C12–C13 bond; *vide supra*). In addition, the data for **VI-41a** and **VI-40a** also illustrate that, given the proper absolute stereochemistry at C12 and C13 and in analogy to 12,13-epoxide or cyclopropane-based analogs, activity is retained even upon moving from a *cis*- to a *trans*-fused system. It should be noted that the absolute configuration of compounds **VI-40a** or **VI-41a** (or the respective diastereoisomers) has not been rigorously established in literature, but is simply inferred from a comparison of the biological data with those obtained for Epo A/*epi*-Epo A (the inactive (12*S*,13*R*)-isomer of Epo A and (12*S*,13*S*)/(12*R*,13*R*)-(*E*)-Epo A (*vide supra*).

The *cis* and *trans* diol precursors of acetonides **VI-40** and **VI-41** (which can be derived from Epo A either by conversion into Epo C and subsequent *cis*-dihydroxylation or through hydrolytic epoxide opening (*trans* diols)), did not show any appreciable biological activity (IC_{50}'s for cancer cell growth inhibition $>1\ \mu M$) (*43, 73*). Interestingly, how-

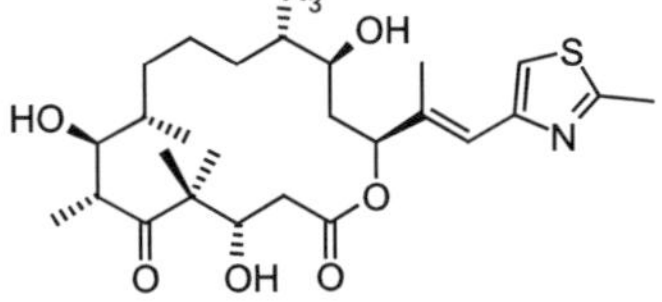

VI-42

Fig. 22. Structure of **VI-42** azido alcohol

ever, azido alcohol **VI-42** (obtained through epoxide ring opening with NaN_3, Fig. 22) was found to be significantly more potent (*e.g.*, IC_{50}'s of **VI-42** against the human cervix cancer cell lines KB-31 and KB-8511 are 61 n*M* and 64 n*M*) (*43*). This indicates that the loss in activity for the above diols cannot be simply ascribed to increased conformational flexibility. However, the interpretation of changes in cellular activity is not straightforward, as they may be caused by a combination of changes in target affinity, cellular penetration, and metabolic stability.

In addition to the changes in the epoxide structure itself a variety of modifications of the 26-methyl group in Epo B or D have been reported, mostly by the groups of *Danishefsky* and *Nicolaou*. These studies have shown that the replacement of one hydrogen atom of this methyl group by relatively small and apolar substituents such as F, Cl, CH_3, or C_2H_5 (Fig. 23, R = CH_2F, CH_2Cl, C_2H_5, n-C_3H_7), is well tolerated, thus producing analogs, which are only slightly less potent *in vitro* than Epo B or Epo D (*64*, *85*, *132*). In general, activity decreases with increasing size of the C26-substituents (*64*, *85*), but exceptions from this general theme have been reported. As an example, C26-(1,3-dioxolanyl)-12,13-Epo D **VI-43** (Fig. 23) is more active *in vitro* than Epo D (*133*). Unfortunately, the compound was found to be significantly less efficacious than Epo D *in vivo*. In contrast, C26-fluoro-Epo B **VI-44** (Fig. 23), which exhibits comparable *in vitro* antiproliferative activity to Epo B (*132*), was demonstrated to possess significantly better antitumor activity than taxol in a human prostate xenograft model when both compounds were administered at equitoxic doses (*134*). No comparison with Epo B was included in this

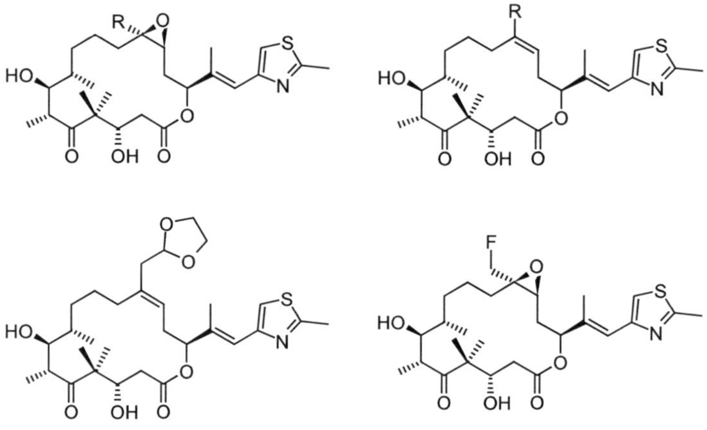

VI-43 **VI-44**

Fig. 23. Modifications of the 26-methyl group of Epo B and D

work, but data reported by the Novartis group indicate that the *in vivo* profile of C26-fluoro-Epo B is similar to that of Epo B itself (*135*).

6.4.3. The C14–C21 Sector

6.4.3.1. Side Chain Modifications

The heterocycle-bearing side chain of epothilones has been a frequent target for structural modifications, as these may allow modulation of the physico-chemical and perhaps also the pharmacokinetic properties of the natural products in a relatively straightforward way. Early work in this area has addressed the importance of the side chain as such, the presence of the olefinic linker element between the heterocycle and the macrocyclic skeleton, and the stereochemistry at C15 (Fig. 24).

Complete removal of the heterocycle as in **VI-45** (Fig. 24) results in complete loss in tubulin-polymerizing activity (1% *vs.* 72% for Epo C) and accordingly reduced antiproliferative activity (*136*). Unfortunately, the reported cellular data for this compound only indicate that **VI-45** is inactive at 100 n*M* concentration, but, based on the lack of tubulin polymerization activity its IC_{50}, would be expected to be in the μM range. Likewise, removal of the linker between the macrocycle and the heterocycle (Fig. 24; **VI-46**) reduces antiproliferative activity several hundred-fold (*64*). With regard to the importance of the configuration at C15, the early literature data are somewhat contradictory. While *Nicolaou et al.* found the (15*R*) isomer of Epo C (Fig. 24; **VI-47**) to be at

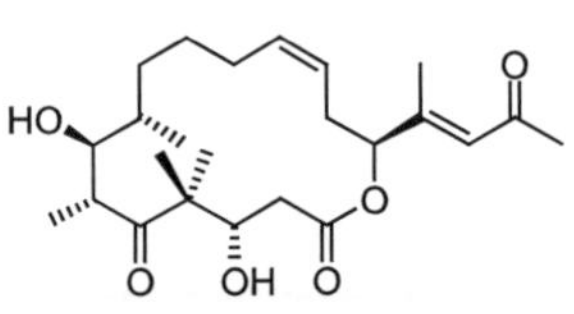

VI-45

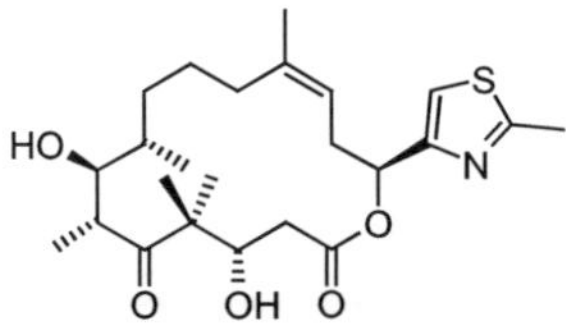

VI-46

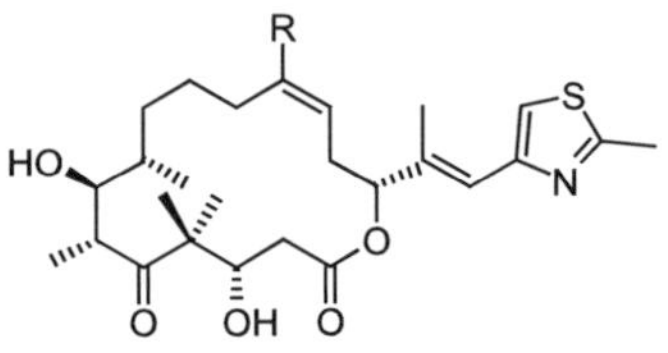

R = H: **VI-47**

R = C_2H_5: **VI-48**

Fig. 24. Modifications of the thiazolyl-vinyl moiety

least 25-fold less active than Epo C (*85*), Epo D analog **VI-48** (Fig. 24) was reported by *Su et al.* to be only 2.5-fold less active than the corresponding (15*S*) isomer (IC_{50}'s of 55 n*M* *vs.* 22 n*M* against the human leukemia cell line CCRF-CEM) (*64*). However, more recent work by *Nicolaou et al.* on a variety of different 12,13-cyclopropane analogs with either (15*S*) or (15*R*) stereochemistry has clearly established that the inversion of stereochemistry at C15 is generally associated with a profound loss in biological activity (*123*).

The allylic methyl group attached to C16 can be removed with only a minor change in biological activity (*43*). In contrast, a >10-fold loss in potency is incurred by extension of this methyl group to an ethyl substituent (*85*). Derivatives with larger substituents than ethyl attached to C16 have not been reported in the literature.

Numerous epothilone analogs with modifications of the thiazole ring at the 2- or 4-position have been investigated (positions 20 and 19 according to epothilone numbering, Fig. 2) (*85*, *124*, *137–139*). These compounds were obtained either through total synthesis or by means of semisynthesis from fermentatively produced epothilones (*137*, *138*) and a collection of structures with alternative substituents at the C20-position is depicted in Fig. 25. As illustrated by the data summarized in Table 2, the replacement of the 21-methyl group by relatively small substituents

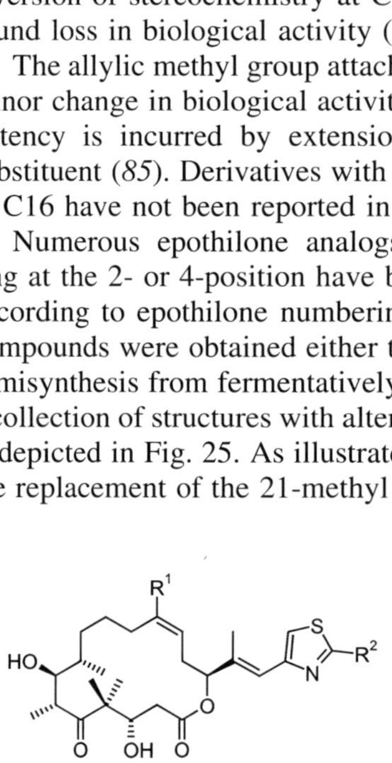

R^1 = H, R^2 = H: **VI-49a**
R^1 = H, R^2 = SCH_3: **VI-49b**
R^1 = H, R^2 = C_6H_5: **VI-49c**
R^1 = CH_3, R^2 = CH_2OH: **VI-49d**

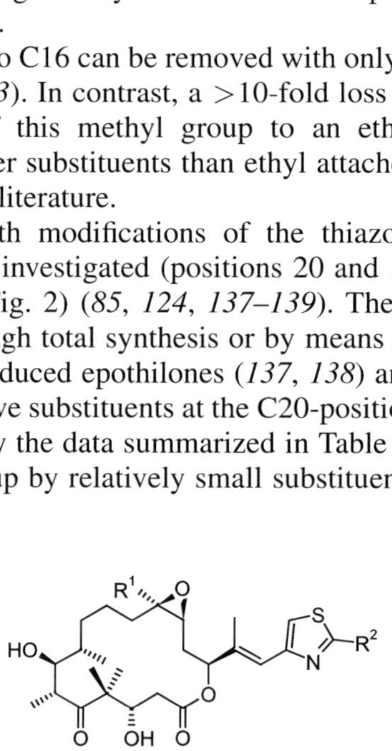

R^1 = CH_3, R^2 = CH_2CH_3: **VI-50a**
R^1 = CH_2F, R^2 = CH_2F: **VI-50b**
R^1 = H, R^2 = CH_2Cl: **VI-50c**
R^1 = CH_3, R^2 = *tert*-C_4H_9: **VI-50d**

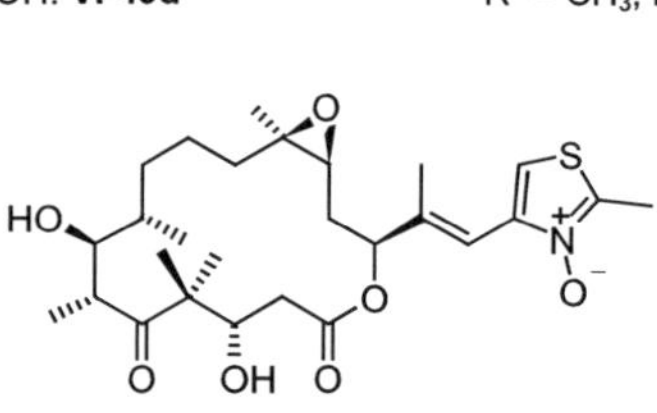

VI-51

Fig. 25. Epothilone analogs with modified C20-substituents and the structure of N-oxide **VI-51**

Table 2. Induction of tubulin polymerization and inhibition of human cancer cell growth by thiazole-substituted epothilone analogs **VI-49** and **VI-50**

Compound	% Tubulin polymerization[a]	Growth inhibition (IC_{50}/n*M*)			
		Cell line			
		1A9	PTX10	KB-31	KB-8511
Epo C	72	32	>100	24.7	9.9
VI-49a	50	>100	>100	32	23
VI-49b	92	9	22	–	–
VI-49c	25	–	–	590	257
Epo B	98	0.1	1.0	0.19	0.19
VI-50b	95	0.12	0.35	–	–
VI-50d	26	–	–	35.4	21.0

[a] Induction of tubulin polymerization at 2 *μM* or 5 *μM* test compound concentration *vs.* 0.5 M GTP or 25 *μM* Epo B. Data are taken from Ref. (*27*).

such as CH_2F, SCH_3, or CH_2CH_3 is well tolerated, although it should be noted that **VI-50c** has been reported to be substantially less active than Epo A (*137*). More bulky substituents result in a substantial loss in potency (Table 2; *cf.*, *e.g.*, **VI-49c** *vs.* Epo C or **VI-50d** *vs.* Epo B). Furthermore, the data for compound **VI-49a** suggest that a substituent at the 2-position of the thiazole ring is not required for potent biological activity. The activity of *N*-oxide **VI-51** (Fig. 25) is comparable with that of Epo B (*137*).

Epothilones with C20-substituents other than methyl have also been isolated from fermentations of myxobacteria (*86*). Most prominent among these derivatives are the C20-hydroxymethyl derivatives **VI-52a** and **VI-52b** (Epo E and Epo F, Fig. 26), which can also be obtained through microbial transformation of Epo A and B, respectively (*140*).

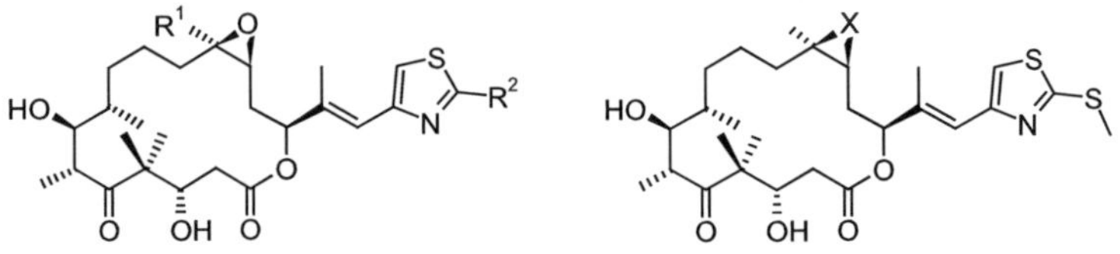

R^1 = H, R^2 = CH_2OH: **VI-52a** (Epo E)
R^1 = CH_3, R^2 = CH_2OH: CS**52b** (Epo F)
R^1 = CH_3, R^2 = CH_2NH_2: **VI-52c** (BMS-310705)

X = O: **VI-53** (ABJ879)
X = CH_2: **VI-54**

Fig. 26. Structures of Epo E and F and additional analogs with variations at the C20 position

Epo E and F show comparable potency to Epo A and B against drug-sensitive cell lines, but appear to be somewhat less potent against P-gp-overexpressing multidrug-resistant lines. (*E.g.* IC_{50} values of 0.23 n*M* and 0.94 n*M* have been reported for Epo F against the drug-sensitive and drug-resistant human cervix carcinoma cell lines KB-31 and KB-8511, respectively (*27*)). Similar findings have been reported by *Lee et al.* for deoxyEpo F (Fig. 25; **VI-49d**: IC_{50}'s of 2.7 n*M* and 47 n*M* against CCRF-CEM cells and multidrug-resistant CCRF-CEM/VBL100 cells, *vs.* 9.5 n*M* and 17 n*M*, respectively for Epo D) (*141*). DeoxyEpo F (C21-hydroxy-Epo D (**VI-49d**), Fig. 25) was also reported by the Sloan-Kettering group to exhibit comparable *in vivo* efficacy to Epo D (*79, 141*). Employing a 6 h continuous *i.v.* infusion regimen for both compounds, **VI-49d** was also found to show significantly superior antitumor effects over BMS-247550 (**VI-1a**) in a CCRF-CEM as well as a MX-1 tumor model (*77, 79*). However, these differences are difficult to interpret, as the 6 h continuous infusion schedule may be optimal for Epo D and **VI-49d**, but not necessarily for BMS-247550 (**VI-1a**). Due to its higher water solubility (compared with Epo D) **VI-49d** may be a more attractive drug development candidate than Epo D.

Epo F serves as a precursor for the semisynthesis of C21-amino-Epo B (**VI-52c** (BMS-310705), Fig. 26) (see Chapter 5.3), a compound which is currently undergoing clinical evaluation in humans (*29, 142, 143*). However, only limited biological data for this analog have appeared in literature so far. Thus, an IC_{50} value of 0.8 n*M* for growth inhibition of the human cervix cancer cell line KB-31 has been reported in a patent application (*144*) (*vs.* 1.2 n*M* for Epo B under comparable experimental conditions (*137*)) and the compound was demonstrated to induce substantial apoptosis in early passage, taxol- and platinum-refractive ovarian cancer cells (OC-2) at a concentration of 50 n*M* (*145*). Plasma concentrations of 50 n*M* of BMS-310705 (**VI-52c**) are clinically achievable at a dose of 10 mg/m^2, which is below the phase I MTD (maximum tolerated dose) for the compound (*29, 145*). BMS-310705 (**VI-52c**) exhibits improved water-solubility over BMS-247550 (**VI-1a**), which enables the use of clinical formulations not containing Cremophor-EL® (*29*). Variation of the C20-substituent has delivered a second clinical development candidate, C20-desmethyl-C20-methylsulfanyl-Epo B (**VI-53** (ABJ879), Fig. 26), which was identified through a collaborative effort between the *Nicolaou* group at the TSRI and the group at Novartis (*124, 146*). ABJ879 (**VI-53**) induces tubulin polymerization *in vitro* with slightly higher potency than Epo B. At the same time, the compound is a markedly more potent antiproliferative agent, with the average IC_{50} for growth inhibition across a panel of drug-sensitive human cancer cell lines being 0.09 n*M* *vs.* 0.24 n*M* for Epo B and 4.7 n*M* for taxol (*146*). The compound retains full activity against cancer cells overexpressing the drug efflux pump P-gp or harboring tubulin mutations.

The binding of **VI-53** to stabilized microtubules has been evaluated by *Buey et al.* (*46*), who found a difference in binding free energy at 35°C of −2.8 kJ/mol between ABJ879 (**VI-53**) and Epo B, corresponding to an increase in the binding constant from 7.5×10^8 for Epo B to 2.5×10^9 for ABJ879 (**VI-53**). Interestingly, the binding enthalpy is *less* favorable for ABJ879 (**VI-53**) than for Epo B and the increase in binding free energy is caused by an increase in binding entropy. Similarly, an increase in binding free energy is also observed for C12/C13-cyclopropane-based epothilone analogs and the energetic effects of the replacement of the C20-methyl group by a methylsulfanyl group and the substitution of a methylene group for the epoxide oxygen are in fact additive. Thus, analog **VI-54** (Fig. 26) binds to stabilized microtubules with 27.4-fold enhanced affinity over Epo B ($\Delta\Delta G^{35^\circ C} = -8.2$ kJ/mol) (*46*). Furthermore, this compound in some cases has been found to be a more potent antiproliferative agent *in vitro* than either Epo B or ABJ879 (**VI-53**) (*e.g.* IC_{50} values for growth inhibition of the human ovarian carcinoma cell line 1A9 are 0.3 (0.6) n*M*, 0.17 n*M*, and 0.10 n*M* for Epo B, **VI-53**, and **VI-54** (*124*, *147*)).

ABJ879 (**VI-53**) has demonstrated potent antitumor activity in experimental human tumor models in mice (*146*), where it produced transient regressions and inhibition of tumor growth of slow-growing NCI H-596 lung adenocarcinomas and HT-29 colon tumors. Inhibition of tumor growth was observed in fast-growing, difficult-to-treat NCI H-460 large cell lung tumors. Finally, single dose administration of ABJ879 (**VI-54**) produced long-lasting regressions and cures in a Taxol®-resistant KB-8511 cervix carcinoma model (*146*).

Apart from modifications of the thiazole moiety, a large part of the work on side chain modifications in epothilones has involved the replacement of the thiazole ring by other heterocycles (*64*, *85*, *148*) or a simple phenyl moiety (*64*, *149*). Again, this work was mainly driven by the *Danishefsky* and *Nicolaou* laboratories, who could independently demonstrate that the substitution of oxygen for sulphur in the heterocycle (thus producing oxazole-based epothilones) does not affect biological potency (*64*, *85*). Major contributions to the area of heterocycle modifications in epothilones have also come from the collaborative work of the *Nicolaou* group at the TSRI and the group at Novartis. Among other things, this research has shown that the pyridine-based analog **VI-55** (Fig. 27) and methyl-substituted variants thereof are basically equipotent cell growth inhibitors to Epo B (*148*) (Table 3). It is thus clear

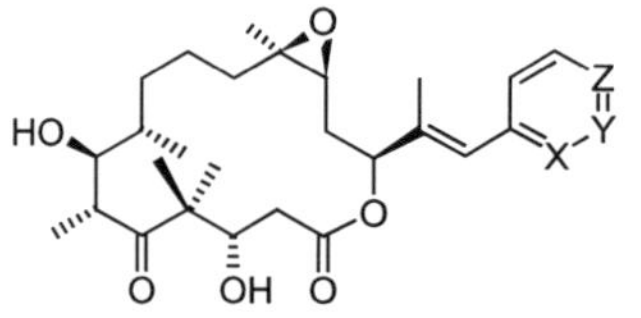

X, Y, Z = N, CH, CH: **VI-55**
X, Y, Z = CH, N, CH: **VI-56**
X, Y, Z = CH, CH, N: **VI-57**

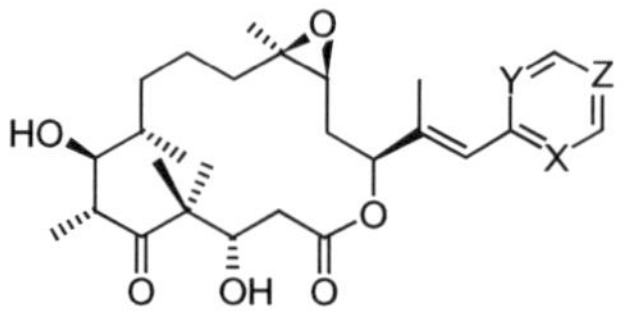

X, Y, Z = N, N, CH: **VI-58**
X, Y, Z = N, CH, N: **VI-59**
X, Y, Z = CH, CH, CH: **VI-60**

Fig. 27. Replacement of the thiazole ring by aromatic and heteroaromatic rings

Table 3. Pyridine and pyrimidine-based analogs of Epo B: importance of nitrogen positioning[a]

Compound	% Tubulin polymerization[a]	IC_{50} KB-31/n*M*[b]
VI-55	85	0.30
VI-56	41.7	4.32
VI-57	34.5	11.8
VI-58	74.4	8.78
VI-59	47.3	14.9
VI-60	28.0	2.88

[a] Induction of polymerization of porcine brain microtubule protein by 5 μM of test compound relative to the effect of 25 μM of Epo B, which gave maximal polymerization (85% of protein input). [b] IC_{50} values for growth inhibition of the human cervix carcinoma cell line KB-31. Data are from Ref. (*69*).

that the presence of a 5-membered heterocycle attached to C17 is not a prerequisite for highly potent biological activity. At the same time the occurrence of Epo B-*like* cellular activity in pyridine-based Epo B analogs (*i.e.*, sub-n*M* IC_{50}'s for growth inhibition) was found to require positioning of the ring N-atom *ortho* to the attachment point of the linker between the heterocycle and the macrocyclic skeleton. Transpositions of the ring N-atom such as in **VI-56** and **VI-57** (Fig. 27) lead to a significant decrease in cellular potency (*148*) (Table 3). Moreover, the incorporation of a second nitrogen atom either at the 3- or the 4-position of the 6-membered ring (**VI-58** or **VI-59**; Fig. 27) produces a profound decrease in antiproliferative activity, even with one of the N-atoms in the obligatory position (*69*). In fact, the corresponding analogs are even less potent than phenyl-derived analog **VI-60** (Fig. 27) (*69*). The underlying reasons for these differences have not been elucidated at this point and this will only be possible on the basis of detailed structural information on complexes between β-tubulin and various types of epothilone analogs. It should also be noted that studies with other types of side-

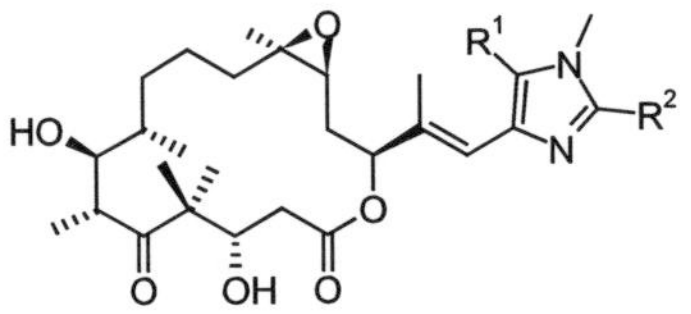

$R^1 = R^2 = H$: **VI-61a**
$R^1 = R^2 = SMe$: **VI-61b**

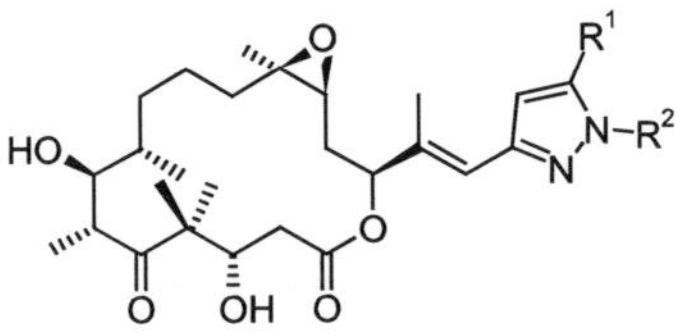

$R^1 = H, R^2 = Me$: **VI-62a**
$R^1 = H, R^2 = Ph$: **VI-62b**
$R^1 = SMe, R^2 = Me$: **VI-62c**

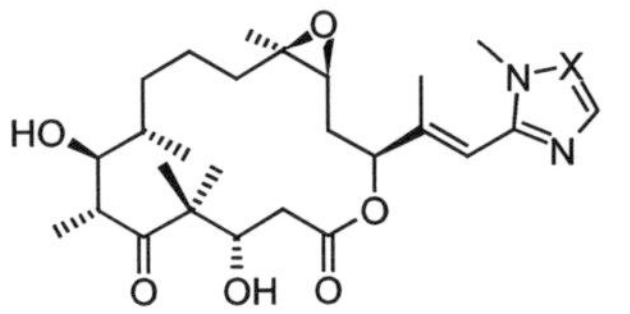

X = CH: **VI-63a**
X = N: **VI-63b**

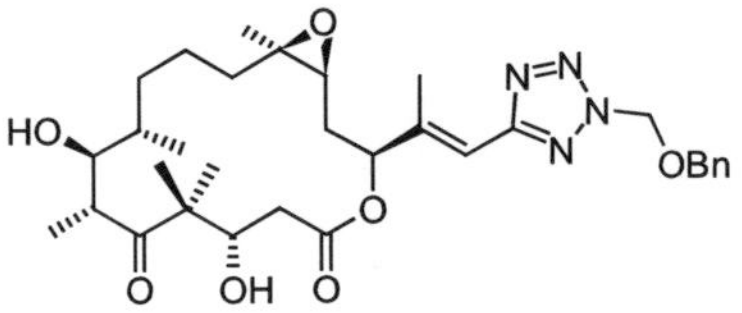

VI-64

Fig. 28. Replacement of the thiazole ring by alternative 5-membered heterocycles

chain-modified analogs have shown that potent cellular activity is also possible for structures with the N-atom in a "non-natural" position (*vide infra*).

In order to expand the structural scope of side chain modifications based on 5-membered heterocycles beyond simple thiazole or oxazole ring systems, the *Nicolaou* group has investigated Epo B analogs incorporating functionalized imidazole, pyrazole, triazole, and tetrazole rings (analogs **VI-61–VI-64**, Fig. 28) (*150*). Rather surprisingly, imidazole-containing side chains, such as in **VI-61a** or **VI-63a** caused a substantial loss in activity. This is rather intriguing, as structures of this type are part of the side chains of the natural product microtubule inhibitors eleutherobin or the sarcodictyins. In order to probe the importance of the electron density of smaller rings for biological activity, *Nicolaou*'s group investigated the replacement of the imidazole moiety by a triazole or a tetrazole ring (*e.g.*, analogs **VI-63b** and **VI-64**). While triazole derivative **VI-63b** showed no appreciable activity, the tetrazole-based analog **VI-64** proved almost equipotent with Epo B, in spite of the fact that earlier attempts at incorporating multiple nitrogen atoms into the heterocycle moiety (*e.g.*, the change from a pyridine to a pyrimidine ring (*27*, *69*)) had led to reduced potency. The activity of compound **VI-64** is particularly remarkable, as the analogous 2-substituted thiazole derivative is several hundred-fold less active than **VI-64** across the same panel of human cancer cell lines. Highly potent antiproliferative activity was also observed for pyrazole-based analogs, such as **VI-62**. Compound **VI-62a**, which incorporates a *N*-methyl pyrazole moiety and which may be regarded as a completely isosteric analog of Epo B, proved to be 2–18-fold more potent than the latter, depending on the cell lines investigated (*e.g.*, the IC_{50} of **VI-62a** against the human cervix carcinoma cell line KB-31 is 0.19 n*M* *vs.* 0.42 n*M* for Epo B). While phenyl-pyrazole derivative **VI-62b** was slightly less potent than Epo B, quite intriguingly, it showed enhanced activity against the Epo A-resistant 1A9/A8 cell line as compared with the parental Epo A-sensitive 1A9 line. The most potent pyrazole-based analog, however, proved to be the methylsulfanyl derivative **VI-62c**, which was found to be significantly more active than Epo B in all cell lines investigated. For example, the compound shows 17-fold enhanced activity over Epo B against both the human ovarian carcinoma cell line 1A9 and its Epo A-resistant variant 1A9/A8 (IC_{50} values of 0.06 n*M* and 0.6 n*M*, respectively, *vs.* 0.99 n*M* and 10.0 n*M* for Epo B) and its activity against the taxol-resistant 1A9/PTX10 line (IC_{50} = 0.1 n*M*) is 78-fold higher than that of Epo B. Based on these data, compound **VI-62c** represents the most potent epothilone analog identified to date.

In order to further probe the steric requirements of the epothilone binding pocket around the heterocycle moiety, a series of bicyclic aromatic thiazole replacements was synthesized and tested for antiproliferative activity (*150*). Analogs incorporating purine-based side chains, such as **VI-65a** and **VI-65b** (Fig. 29) exhibit very potent growth inhibitory activity, with methylsulfanyl derivative **VI-65a** being 4-fold more potent than Epo B against both the 1A9/PTX10 and the 1A9/A8 cell lines. Guided by previous data on Epo C analogs incorporating a thiazol-2-yl moiety (as opposed to the natural thiazol-4-yl group (*150*)) or a 2-quinoline moiety as a thiazole replacement, benzothiazole analog **VI-66** (Fig. 29) was designed and synthesized (*150*). As for **VI-65a**, this compound was found to be more potent than Epo B (up to 4-fold). Taken

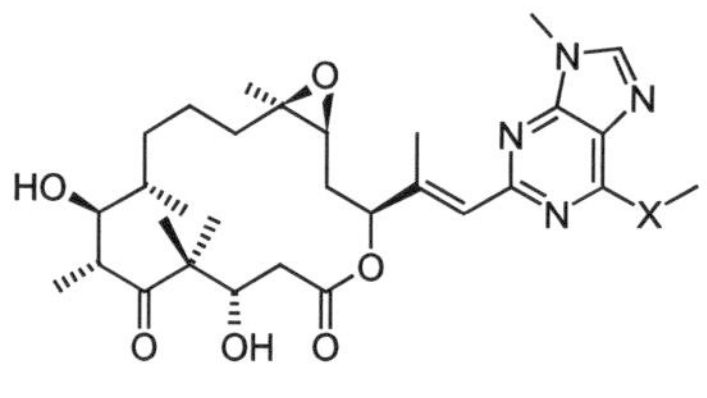

X = S: **VI-65a**
X = O: **VI-65b**

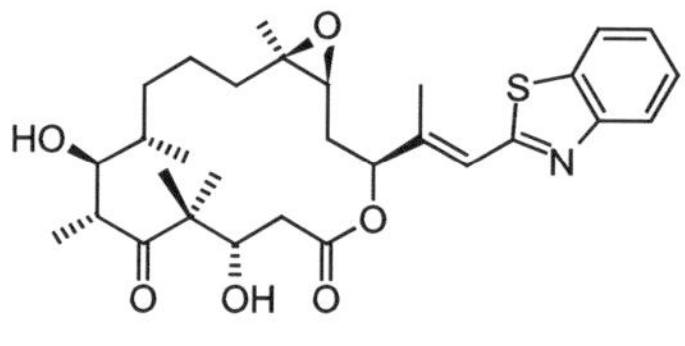

VI-66

Fig. 29. Epothilone analogs with purine- or benzothiazole-containing side chains

References, pp. 210–220

together, these findings suggest that the side chain binding pocket on tubulin can accommodate significantly expanded heterocycles, if the additional steric bulk is located in the northeastern quadrant of the structure. It is worth noting that this is contrary to the effects that have been observed for the incorporation of bulky substituents at the 2-position of the thiazole ring in Epo B, which have been reported to cause a substantial loss in biological potency (*27*, *151*).

In addition to analogs incorporating an olefinic double bond as a linker between the macrolactone ring and different types of heterocycles, the Novartis group has also studied a new family of side-chain-modified structures, which are characterized by rigidification of the entire side chain manifold (analogs **VI-67–VI-69**, Fig. 30) (*152*). The design of these structures was guided by the results of NMR studies on the bioactive (tubulin-bound) conformation of epothilones, which had indicated that the C16/C17 double bond and the aromatic C18-N bond were present in a *transoid* arrangement (corresponding to a ~180° C16-C17-C18-N22 torsion angle) (*88*). As illustrated by the data summarized in Table 4, epothilone analogs **VI-67–VI-69** are more potent inhibitors of human cancer cell proliferation than the respective parent compounds Epo D and Epo B (*152*) (*cf.* also Refs. (*84*, *128*)).

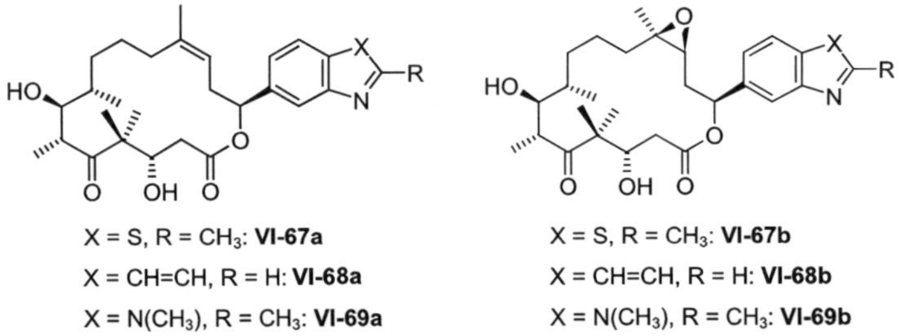

Fig. 30. Structures of epothilone analogs with conformationally constrained side chains

Table 4. Growth inhibition of the human epidermoid cancer cell lines KB-31 and KB-8511 by side-chain-modified epothilone analogs **VI-67–VI-69** (IC_{50}/nM)[a]

Compound	Cell line		Compound	Cell line	
	KB-31	KB-8511		KB-31	KB-8511
Epo D	2.70	1.44	Epo B	0.19	0.18
VI-67a	0.45	0.23	**VI-67b**	0.13	0.09
VI-68a	0.59	0.38	**VI-68b**	0.11	0.10
VI-69a	0.21	0.73	**VI-69b**	0.14	0.38

[a] Data are from Ref. (*152*).

The activity increase is more pronounced for analogs of the 12,13-deoxy type (**VI-67a**, **VI-68a**, **VI-69a**), which are up to 10-fold more potent antiproliferative agents than Epo D; in addition, these compounds are clearly more potent than Epo A. Interestingly, however, the observed increase in antiproliferative activity does not seem to be a consequence of more effective interactions with tubulin and this point will be discussed in more detail below. As an extension of the discussion on modifications of the epoxide moiety, it should be noted that the (12*S*,13*R*)-isomer of **VI-67b** is two orders of magnitude less potent than **VI-67b** (IC_{50}'s against the KB-31 and KB-85111 cell lines of 82 n*M* and 26 n*M*). This confirms that the simultaneous inversion of stereochemistry at both C12 and C13 causes a significant decrease in biological potency.

Side chain modifications of the above type have been independently investigated by the groups at Schering (*70*, *71*) and, more recently, at Kosan (*153*). The work at Schering has led to the identification of a C6-allylated benzothiazole-based analog, ZK-Epo (sagopilone) (**VI-70**) (Fig. 31), which is currently undergoing Phase II clinical trials and which is the first fully synthetic epothilone analog to have entered clinical studies (*70*).

ZK-Epo (sagopilone) (**VI-70**) is reported to be more potent *in vitro* than Epo B against a variety of drug-sensitive cancer cell lines and to retain full activity also against multidrug-resistant cancer cells (while Epo B exhibits somewhat reduced activity against certain types of MDR cell lines) (*70*). The compound also showed significant antitumor activity in a number of mouse xenograft models (stable disease), without inducing any profound body weight loss (*70*).

Based on the significant increase in cellular potency associated with a dimethyl-benzimidazole side chain within the structural framework of the Epo B/D macrocycle (Table 4) this modification has been further investigated by *Altmann* and co-workers as a potential potency-enhancing element in combination with a variety of macrocycle modifications (*84*, *128*, *154*, *155*). These studies revealed that the potency-enhancing effect of the dimethyl-benzimidazole moiety is rather general in nature

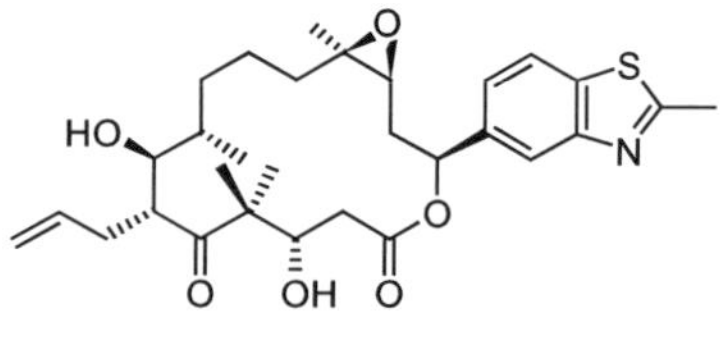

VI-70

Fig. 31. Structure of ZK-Epo (sagopilone)

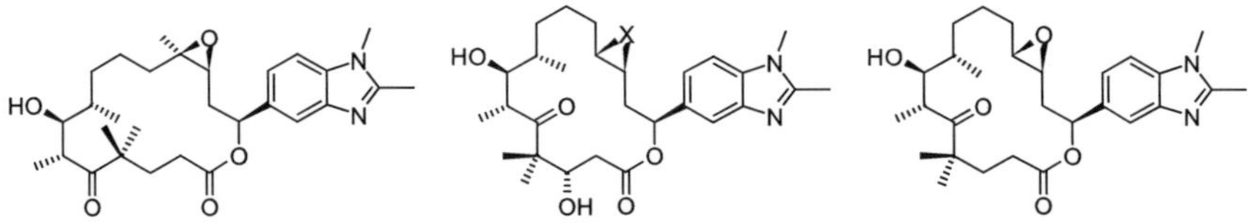

VI-71

X = O: **VI-72a**
X = CH_2: **VI-72b**

VI-73

Fig. 32. Dimethyl-benzimidazole-based epothilone analogs

and extends to the corresponding analogs of Epo A and C (*128*), of 3-deoxy-Epo B (**VI-71**) (*84*), and also of *trans*-Epo A (**VI-72a**) (*128*) and its 3-deoxy derivative **VI-73** (*84*) (Fig. 32). For example, compound **VI-72a** is essentially equipotent with Epo B (which is the most potent natural epothilone) against drug-sensitive human cancer cell lines (IC_{50} values against the human cervix carcinoma cell line KB-31 are 0.25 n*M* and 0.29 n*M* for **VI-72a** and Epo B (*128*)). Likewise, the modification can almost completely revert the loss in cellular potency caused by the removal of the 3-hydroxyl function in Epo B (compound **VI-71**: IC_{50} (KB-31) = 0.58 n*M* *vs.* 0.29 n*M* for Epo B (*84*)). Lastly, the antiproliferative activity of compound **VI-73**, in spite of significant structural deviations from the original natural product leads, is still comparable with that of Epo A, Epo D, or taxol (IC_{50} (KB-31) = 3.16 n*M*) (*84*). This latter analog might in fact be considered as the first representative of a new structural class of microtubule stabilizers, whose overall pharmacological profile may be distinct from that of Epo A/B and more closely related analogs.

While dimethyl-benzimidazole-based epothilone analogs are generally more active against drug-sensitive human cancer cell lines than the corresponding derivatives incorporating the natural epothilone side chain and also tend to be more water-soluble, they exhibit slightly increased susceptibility to P-gp-mediated drug efflux. For example, the IC_{50} values of compounds **VI-71** and **VI-72a** against the P-gp-overexpressing human cervix carcinoma cell line KB-8511 are 1.89 n*M* and 1.36 n*M* *vs.* 0.58 n*M* and 0.25 n*M*, respectively, for the drug-sensitive KB-31 line (*128*). However, it was recently demonstrated that the P-gp-susceptibility of benzimidazole-derived epothilone analogs can be overcome through the replacement of the epoxide oxygen by a more lipophilic methylene group, thus leading to side-chain-modified cyclopropane derivatives, such as **VI-72b** (*154*).

The potency-enhancing properties of a dimethyl-benzimidazole side chain have also been exploited by *Altmann* and co-workers to enhance the activity of 12-aza-macrolide-based epothilone analogs (azathilones,

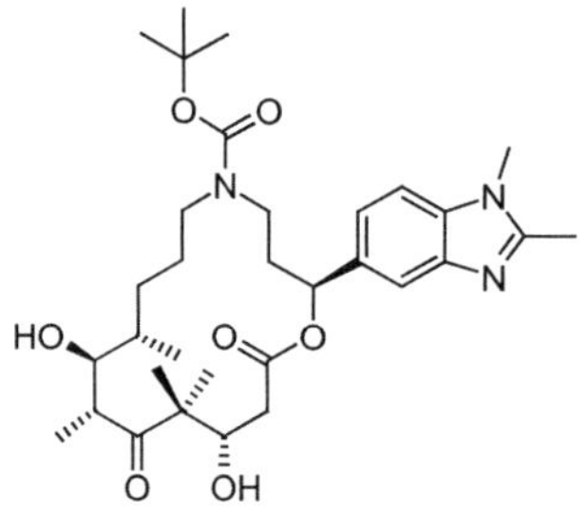

VI-74

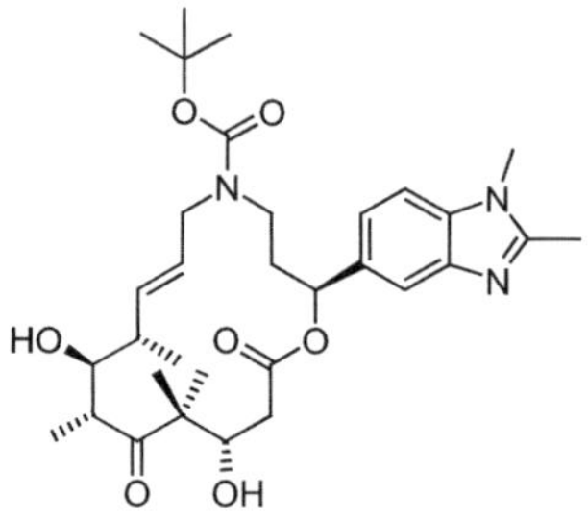

VI-75

Fig. 33. Structures of azathilones

vide supra) (*127*). Based on the available SAR data for this family of compounds, the most promising target for synthesis appeared to be analog **VI-74** (Fig. 33), which is derived from azathilone **VI-34a** (as the most potent "natural" derivative investigated so far, see Fig. 18).

Azathilone **VI-74** was found to be a highly active antiproliferative agent that inhibits the growth of different types of drug-sensitive human cancer cell lines with similar potency to Epo A. (*E.g.*, IC_{50}'s of **VI-74** against the human cancer cell lines A549 (lung), HCT-116 (colon), PC-3M (prostate), and KB-31 (cervix) are 1.9 n*M*, 1.6 n*M*, 2.3 n*M*, and 0.34 n*M* *vs.* 3.2 n*M*, 2.2 n*M*, 3.4 n*M*, and 2.15 n*M* for Epo A (*127*).) Compound **VI-74** also promotes microtubule assembly from soluble tubulin with the same efficiency as Epo A (*127*). The replacement of the natural thiazolyl-vinyl side chain in **VI-34a** by a dimethyl-benzimidazole moiety is associated with a >60-fold increase in cellular potency, which is significantly higher than what had been observed previously in the context of polyketide-based macrocycles (where incorporation of a dimethyl-benzimidazole side chain leads to a potency increase of 2–15-fold). In light of the profound structural differences between **VI-74** and the original natural product lead Epo A, the activity of **VI-74** is quite remarkable and the compound has been suggested to represent a new structural class of microtubule-stabilizing agents. While epothilone analogs based on a 12-aza-macrolide core still retain many of the (two-dimensional) structural features of natural epothilones, their macrolactone ring is no longer based on a contiguous polyketide backbone. Analogs of this type have therefore been classified as "non-natural natural products" by *Altmann* and co-workers (*127*).

The importance of the *tert*-butyl group in **VI-74** for biological activity was probed by its replacement with a less bulky and less lipophilic ethyl group. The resulting analog proved to be less potent than **VI-74**, but IC_{50} values for growth inhibition of drug-sensitive human cancer cell lines are still in the 10–20 n*M* range. Intriguingly, the presence of an (*E*)-configured double bond between C9 and C10 (compound **VI-75**, Fig. 33) produced a dramatic loss in biological potency (*127*), which is contrary to the effects of this modification within the structural framework of natural epothilones (*vide supra*). This finding may point to differences in the tubulin-bound conformation between natural epothilones and the aza-macrolide-based azathilones, but further experimental work will be necessary to either confirm or refute this hypothesis. Unfortunately, and in marked contrast to natural epothilones, the activity of **VI-74** (as well as the corresponding ethyl carbamate) is significantly reduced against the multidrug-resistant KB-8511 human cervix carcinoma cell line, which indicates that the compound is highly susceptible to P-gp-mediated drug efflux.

Epothilone analogs incorporating conformationally constrained side chains of the type found in **VI-67–VI-69** have also been used as templates for a more detailed investigation of the significance of the position of the side chain N-atom (*155*). The picture that has emerged from these studies with regard to the relationship between biological activity and N-positioning in the epothilone side chain is rather complex and cannot be fully rationalized at this point.

Thus, while both isomers within compound pairs **VI-67a/VI-75** and **VI-68a/VI-76** (Fig. 34) have comparable effects on tubulin polymerization, they show very different antiproliferative activities, with the

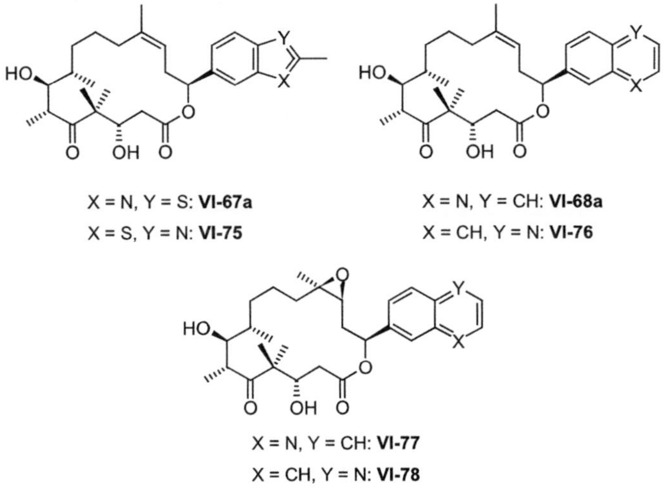

Fig. 34. Epothilone analogs with conformationally constrained side chains: Isomeric heterocycles

"natural" isomer (*i.e.*, the compound with the N-atom *meta* to the bond between the phenyl moiety and the macrolactone ring) being significantly more potent (>50-fold) (*155*). Based on these observations the position of the side chain N-atom is highly relevant for cellular potency, but not for the interactions of these analogs with the tubulin/microtubule system. On the other hand, virtually no effect of the location of the N-atom on biological activity was observed between epoxide-containing analogs **VI-77** and **VI-78** either at the level of tubulin polymerization *in vitro* or at the level of cancer cell growth inhibition (*155*). Apparently, factors other than tubulin-polymerizing activity *in vitro* must have a pronounced impact on the cellular activity of epothilone analogs **VI-67a**, **VI-68a**, **VI-77**, and **VI-78**. However, further experiments will be required in order to establish a mechanistic understanding of the observed differences in cellular potency and of their virtual disappearance upon conversion of **VI-68a** and **VI-75** to the corresponding epoxides.

A different family of side-chain-modified epothilone analogs, which contain a (2-substituted) thiazol-4-yl moiety directly attached to C15 of the (*E*)-9,10-didehydro-Epo D (**VI-22**, Fig. 13) macrocyclic core structure, has recently been reported by the group at Kosan (Fig. 35) (*156*). The design of these structures was inspired by the potency enhancement associated with either the incorporation of an (*E*)-configured double bond between C9 and C10 of the epothilone macrocycle (*vide supra*) or the presence of conformationally constrained side chains of the type present in **VI-67**–**VI-69**, which have an aryl moiety directly attached to C15. Not unexpectedly, the simple removal of the vinyl linker between the thiazole moiety and the macrolactone ring in (*E*)-9,10-didehydro-Epo D, leading to **VI-79**, results in a significant loss in activity (IC_{50} of 380 n*M* against the human breast carcinoma cell line MCF-7 *vs.* 0.6 n*M* for **VI-22**). This finding confirms earlier data from the *Danishefsky* lab, which

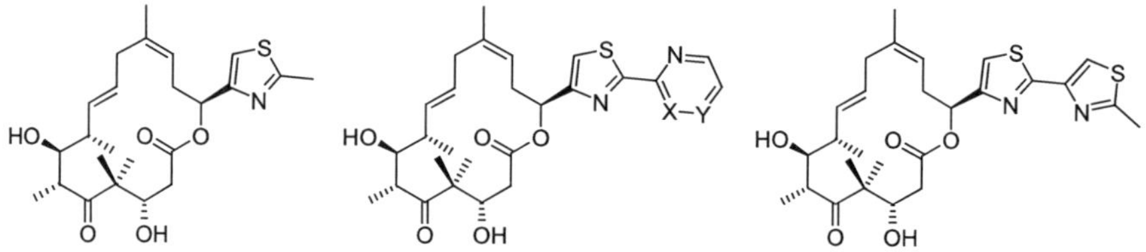

VI-79

X = Y = CH: **VI-80a**
X, Y = N, CH: **VI-80b**
X, Y = CH, N: **VI-80c**

VI-81

Fig. 35. Epothilone analogs with C15 heteroaryl moieties

had shown the corresponding Epo D analog to be substantially less potent than Epo D (*64*). However, the activity loss resulting from the absence of a linker element between the thiazole moiety and the macrocycle could be compensated for, at least to some extent, by the replacement of the methyl group at the 2-position of the thiazole ring with different aryl moieties, such as in **VI-80** and **VI-81** (Fig. 35). Analogs **VI-80a**, **VI-80b**, **VI-80c**, and **VI-81** exhibit IC_{50} values against the MCF-7 human breast cancer cell line of 40 n*M*, 52 n*M*, 34 n*M*, and 38 n*M* (*156*). Thus, the potency of these compounds is still more than 50-fold lower than that of **VI-22**, but there is significant room for additional modifications of both hetero-aryl moieties, which may well lead to further improvements in potency.

6.4.3.2. C14 Modifications

Analogs of Epo B and D with a methyl group attached to C14 have been reported by *Taylor* and co-workers (*114*, *157*) (**VI-82–VI-84**, Fig. 36). These studies have demonstrated that the incorporation of a methyl group attached to C14 is reasonably well tolerated for an (*S*)-configuration of the newly created chiral center (*e.g.*, the IC_{50} of (*S*)-C14-methyl-Epo D (**82a**, Fig. 36) against the human breast cancer cell line MCF-7 is 35 n*M* *vs.* 5 n*M* for Epo D (*157*)), whereas the corresponding C14-(*R*)-isomers are virtually inactive (IC_{50}'s of >1 μM in cell proliferation assays).

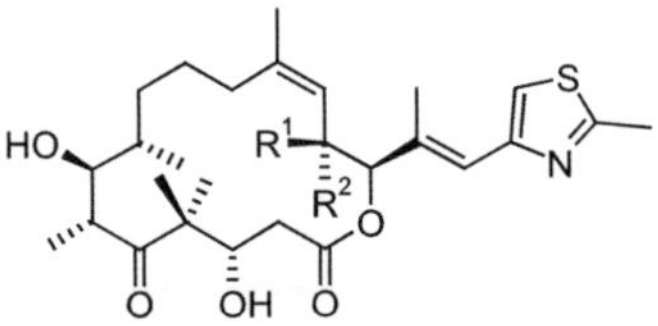

R1 = H, R2 = CH3: **VI-82a**
R1 = CH3, R2 = H: **VI-82b**

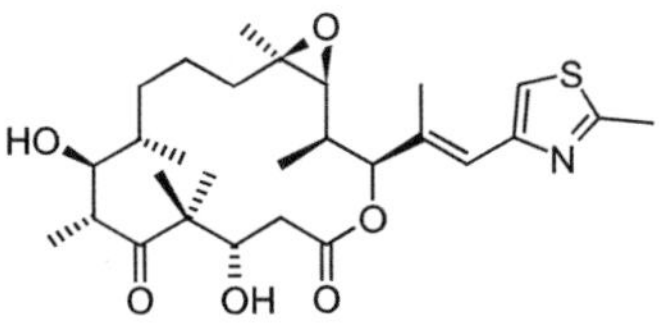

VI-83

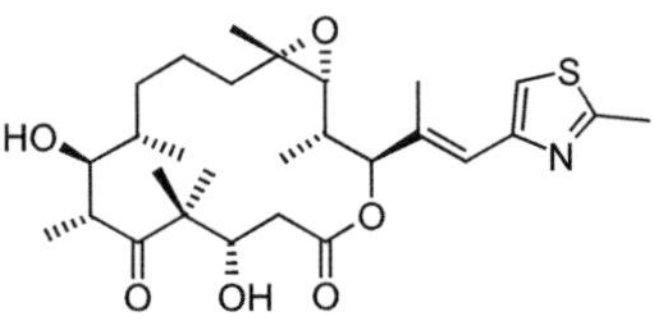

VI-84

Fig. 36. Modifications at C14

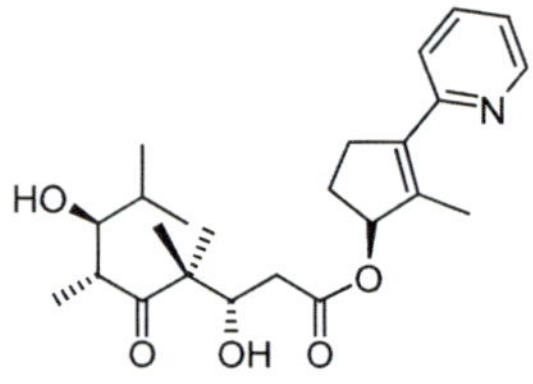

VI-85

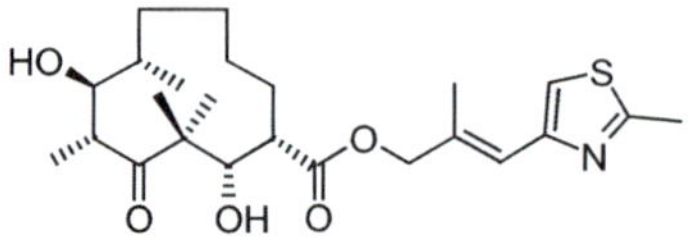

VI-86

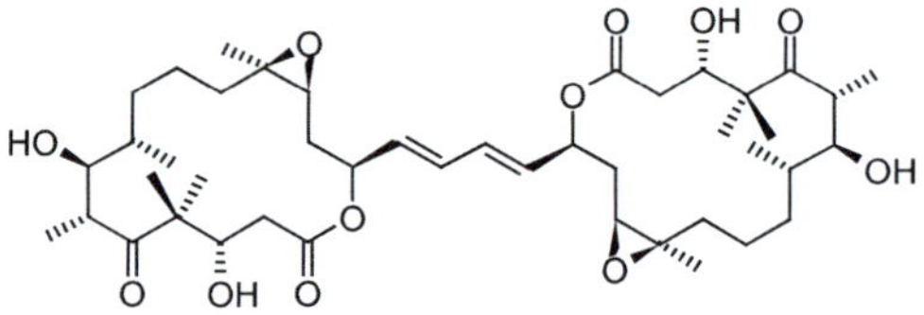

VI-87

Fig. 37. Structures of simplified epothilone analogs and "dimeric" compound

6.4.4. Miscellaneous Modifications

Attempts have been made to design grossly simplified epothilone analogs, including, *e.g.*, analogs **VI-85** (*158*) and **VI-86** (*159*) (Fig. 37). All of these compounds proved to be virtually inactive. Of significantly higher interest is the observation by *Nicolaou et al.* that "dimeric" compound **VI-87** (Fig. 37), which was first detected as a side product in the synthesis of pyridine-based Epo B analogs, exhibits very potent biological activity (no specific data are available) (*149*).

In an attempt to capitalize on this finding, *Nicolaou et al.* have prepared a number of related analogs, either with dimeric or monomeric structures, all of which are characterized by the absence of a heterocyclic appendage or side chain. Unfortunately, none of these compounds was biologically active (*149*).

6.5. Structural Studies and Pharmacophore Modeling

Ever since the discovery of the microtubule-stabilizing properties of epothilones in 1995 efforts have been made to develop a predictive pharmacophore model for this class of compounds, in order to guide the design of improved and perhaps structurally simplified analogs. In the absence of any structural information on complexes between tubulin/microtubules and an epothilone-type molecule, early attempts on the development of an epothilone pharmacophore model (*59*, *160–163*) were generally based on the assumption of a common tubulin binding site for epothilones and taxol in conjunction with electron-crystallography-derived structural data (at 3.7 Å resolution) on a complex between two-dimensional tubulin polymer sheets and taxol (*58*). Based on this premise *Giannakakou et al.* (*59*) have proposed a pharmacophore model, which places the epoxide oxygen of epothilones in the same position as the oxetane oxygen in taxol, while the epothilone side chain is located in the same region as either the C3′-phenyl group or, alternatively, the C2-benzoyloxy moiety. The model also suggests that the methyl group attached to C12 in Epo B is involved in hydrophobic interactions with the side chains of Leu-β273, Leu-β215, Leu-β228, and Phe-β270, and that this may account for the higher activity of Epo B *vs.* Epo A. Different conclusions with regard to the relative positioning of taxol and epothilones within the microtubule binding site have been reached by *Wang et al.* (*161*). In their model, the position of the thiazole moiety in epothilones within the microtubule binding pocket matches the position of the phenyl group of the C-3′-benzamido substituent in taxol. Furthermore, the epoxide oxygen is concluded not to be involved in interactions with the protein, which is in line with the experimental data discussed above for cyclopropane-based epothilone analogs. A model similar to that of *Wang et al.* has recently been proposed by *Botta* and coworkers (*162*). Although these computational models by their very nature are of limited accuracy, some of them are able to reproduce at least part of the published epothilone SAR to a reasonable degree (*161*, *162*). However, all of these models may need to be revisited in light of more recent structural studies on the tubulin-bound conformation of Epo A either by NMR spectroscopy on a soluble β-tubulin/Epo A complex (*88*) or by a combination of electron crystallography, NMR spectroscopic conformational analysis, and molecular modeling of a complex between Epo A and a Zn^{2+}-stabilized two-dimensional α,β-tubulin sheet (solved at 2.89 Å resolution) (*45*). These studies have provided completely new insights into the bioactive conformation of the epothilone-class of microtubule inhibitors. In particular, the solid-state study by *Nettles et al.* (*45*) seriously calls into question the notion of a common pharmacophore between taxol and epothilones, which has been a central provision in all earlier modeling studies. According to the data presented by *Nettles et al.*, Epo A and taxol do indeed occupy the same gross binding pocket, but the actual binding is mediated through different sets of hydrogen bonding and hydrophobic interactions for the two compounds. Interestingly, the tubulin-bound conformation of Epo A derived from the electron crystallographic data at 2.89 Å is different from any of the above computational models, but also from the recent NMR-derived structure of

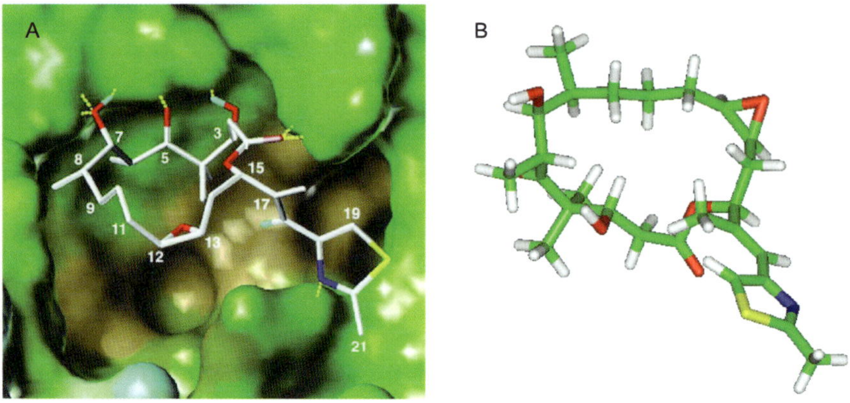

Fig. 38. Putative bioactive (tubulin-bound) conformations of Epo A as determined by electron-crystallography (A) or solution-state NMR spectroscopy (B). (A): From *J. H. Nettles, H. Li, B. Cornett, J. M. Krahn, J. P. Snyder, K. H. Downing*, The Binding Mode of Epothilone A on alpha,beta-Tubulin by Electron Crystallography, Science 2004, 305, 866–869. Reprinted with permission from AAAS. (B) Courtesy of Dr. *T. Carlomagno*

(soluble) tubulin-bound Epo A as determined by *Carlomagno et al.* (*88*) by means of magnetization transfer NMR techniques (Fig. 38).

The relative validity of the two models is difficult to assess, but it should be noted that one important SAR feature of epothilones is not well explained by the structure of *Nettles et al.*, namely, the fact that formal dehydration across the C2–C3-bond (producing an (*E*)-configured double bond) is not associated with any major loss in tubulin-polymerizing or antiproliferative activity (*vide supra*). This finding suggests the C2–C3 torsion angle in the bioactive conformation of epothilones to be in the vicinity of 180°, which corresponds to the conformation in the NMR-derived structure, but is not the conformation suggested by the *Nettles* structure. On the other hand, a 180° torsion angle about the C17–C18 bond is inferred by both structures, thus locating the 27-methyl group and the thiazole nitrogen on opposite sides of the C16–C17 double bond. In fact, *Nettles et al.* have proposed this nitrogen atom to be involved in a H-bond with the protonated form of β-His-227. These conclusions are consistent with the observation that the antiproliferative activity of pyridine-based analogs of Epo B (thiazole ring replaced by a 2-, 3-, or 4-pyridyl moiety) decreases in the order 2-pyridyl $\gg$ 3-pyridyl $\approx$ 4-pyridyl, thus indicating a requirement for the N-atom to be located in the position *ortho* to the attachment point of the linker between the heterocycle and the macrocyclic core (*vide supra*). However, as highlighted above in the discussion of Epo B and D analogs containing isomeric benzothiazole-

or quinoline-based side chains, cellular effects may not be entirely reflective of the strength of ligand/tubulin interactions and the interpretation of changes in cellular activity in terms of structural features of a ligand/tubulin complex could in fact be misleading.

It remains to be seen how the various pharmacophore models and the structural proposals made on the basis of NMR or electron crystallography data will actually relate to the true (*i.e.*, derived from atomic resolution structural data, if they become available) bioactive conformation of epothilones (should there even be a single unique bioactive conformation). In the meantime, both the experimentally derived as well as the purely computational structures may serve as a useful basis for the design of new analogs.

6.6. Conclusions

The discovery that epothilones A and B inhibit human cancer cell growth by a "taxol-like" mechanism of action without being subject to significant P-gp-mediated drug efflux has established these compounds as important lead structures for anticancer drug discovery. Although other natural products have been recognized to act as microtubule stabilizers since the Merck group's discovery, epothilones (together with taxanes) today represent the most widely explored leads for tubulin-directed anticancer drug discovery. The ultimate objective of this research, be it basic or more applied, has been (and continues to be) the transformation of a promising family of natural product leads into a viable, clinically useful anticancer drug (or perhaps even a series of pharmacologically distinct clinical agents). On the way to this goal hundreds of epothilone analogs and derivatives have been prepared over the past decade and their pharmacological properties evaluated *in vitro*, but also in animal models. This has led to a comprehensive map of the structural elements associated with potent biological activity, but also those that are not well tolerated and should be avoided in the design of potent analogs. Most of the structure-activity work was conducted in the absence of structural information on tubulin/epothilone complexes and in an iterative process had to be guided by the activity of previously synthesized analogs. Progress in the medicinal chemistry of epothilones was thus strongly linked to the rapid evaluation of newly synthesized compounds by biochemists, cell biologists, and pharmacologists. In this context it is also worth noting that most of the SAR information available for epothilones today is derived from the investigation of fully synthetic analogs that would not have been accessible through semi-synthesis.

More recently, structural proposals for the tubulin-bound bioactive conformation of Epo A have been advanced on the basis of solution NMR and electron crystallography data. Unfortunately, the conclusions derived from these structural studies have not produced a consistent and unique structural picture for the bioactive conformation of Epo A and it remains to be seen to what extent this information will aid the future design of new and improved analogs.

Epothilone-based drug discovery research so far has delivered seven (publicly known) compounds that have entered clinical evaluation in humans. While the clinical development of epothilones will be discussed in more detail in Chapter 7, it should already be noted that one of these compounds (BMS-247550, ixabepilone) has been approved by the US FDA for the treatment of taxol-resistant metastatic breast cancer in the fall of last year. Thus, the initial promise held by the natural product leads in one case has now been transformed into a new anticancer drug and become manifest in clinical practice. Most of the analogs currently undergoing clinical trials are either directly produced through fermentation or through semi-synthesis from Epo B; however, the compound developed by Bayer-Schering AG is entirely synthetic in nature and is produced through total chemical synthesis.

This well-filled clinical development pipeline is complemented by a number of more recent analogs, which are currently at various stages of preclinical profiling or development. Of particular note among these agents is fludelone, which exhibits a superb preclinical profile and which will hopefully enter clinical studies in humans in the very near future. On the other hand, highly potent new analogs with modified side chain structures have been discovered very recently as well as compounds, such as azathilones, whose overall structural framework deviates significantly from that of the natural product leads and which may therefore be associated with distinctly different overall pharmacological profiles. Clearly, more work is required to establish the development potential of these latter, most recent additions to the extensive collection of epothilone-derived structures. However, it is more than likely that the current group of clinical development compounds is not the end of the line in epothilone-based drug discovery and development.

References

1. Lu MC (1995) Antimitotic Agents. In: WO Foye (ed.) Cancer Chemotherapeutic Agents, p. 345. American Chemical Society, Washington, DC
2. Mekhail TM, Markman M (2002) Paclitaxel in Cancer Therapy. Exp Opin Pharmacother **3**: 755

3. Obasaju C, Hudes GR (2001) Paclitaxel and Docetaxel in Prostate Cancer. Hematol Oncol Clin N Am **15**: 525

4. Hamel E (1996) Antimitotic Natural Products and Their Interaction with Tubulin. Med Res Rev **16**: 207

5. Rowinsky EK (1997) The Development and Clinical Utility of the Taxane Class of Antimicrotubule Chemotherapy Agents. Ann Rev Med **48**: 35

6. Prinz H (2002) Recent Advances in the Field of Tubulin Polymerization Inhibitors. Exp Rev Anticancer Ther **2**: 695

7. Altmann K-H (2001) Microtubule-stabilizing Agents: A Growing Class of Important Anticancer Drugs. Curr Opin Chem Biol **5**: 424

8. Miles DC (2002) Emerging Microtubule Stabilizing Agents for Cancer Chemotherapy. Ann Rep Med Chem **37**: 125

9. Altmann K-H, Gertsch J (2007) Anticancer Drugs from Nature – Natural Products as a Unique Source of New Microtubule-Stabilizing Agents. Nat Prod Rep **24**: 327

10. Schiff PB, Fant J, Horwitz SB (1979) Promotion of Microtubule Assembly in Vitro by Taxol. Nature **277**: 665

11. Höfle G, Reichenbach H (2005) Epothilone, a Myxobacterial Metabolite with Promising Antitumor Activity. In: Cragg GM, Kingston DGI, Newman DJ (eds.) Anticancer Agents from Natural Products, p. 413. Taylor & Francis, Boca Raton, FL

12. Gerth K, Bedorf N, Höfle G, Irschik H, Reichenbach H (1996) Epothilones A and B: Antifungal and Cytotoxic Compounds from *Sorangium cellulosum* (Myxobacteria). Production, Physico-chemical and Biological Properties. J Antibiotics **49**: 560; see also: Höfle G, Bedorf N, Gerth K, Reichenbach H (1993) German Patent Disclosure, DE 4138042, 1993; Chem Abstr **120**: 52841

13. Bollag DM, McQuenay PA, Zhu J, Hensens O, Koupal L, Liesch J, Goetz M, Lazarides E, Woods CM (1995) Epothilones, a New Class of Microtubule-stabilizing Agents with a Taxol-like Mechanism of Action. Cancer Res **55**: 2325

14. Hardt IH, Steinmetz H, Gerth K, Sasse F, Reichenbach H, Höfle G (2001) New Natural Epothilones from *Sorangium cellulosum*, Strains So ce90/B2 and So ce90/D13: Isolation, Structure Elucidation, and SAR Studies. J Nat Prod **64**: 847

15. Höfle G, Bedorf N, Steinmetz H, Schomburg D, Gerth K, Reichenbach, H (1996) Antibiotics from Gliding Bacteria. 77. Epothilone A and B – Novel 16-Membered Macrolides with Cytotoxic Activity: Isolation, Crystal Structure, and Conformation in Solution. Angew Chem Int Ed **35**: 1567

16. Kowalski RJ, Giannakakou P, Hamel E (1997) Activities of the Microtubule-stabilizing Agents Epothilones A and B with Purified Tubulin and in Cells Resistant to Paclitaxel (Taxol). J Biol Chem **272**: 2534

17. Altmann K-H, Wartmann M, O'Reilly T (2000) Epothilones and Related Structures – A New Class of Microtubule Inhibitors with Potent in Vivo Antitumor Activity. Biochim Biophys Acta **1470**: M79

18. Wolff A, Technau A, Brandner G (1997) Epothilone A Induces Apoptosis in Neuroblastoma Cells with Multiple Mechanisms of Drug Resistance. Int J Onc **11**: 123

19. Giannakakou P, Sackett DL, Kang YK, Zhan Z, Buters JT, Fojo T, Poruchynsky MS (1997) Paclitaxel-resistant Human Ovarian Cancer Cells Have Mutant β-Tubulins That Exhibit Impaired Paclitaxel-driven Polymerization. J Biol Chem **272**: 17118

20. Nicolaou KC, Roschangar F, Vourloumis D (1998) Chemical Biology of Epothilones. Angew Chem Int Ed **37**: 2014

21. Harris CR, Danishefsky SJ (1999) Complex Target-Oriented Synthesis in the Drug Discovery Process: A Case History in the dEpoB Series. J Org Chem **64**: 8434

22. Mulzer J, Martin HJ, Berger M (1999) Progress in the Synthesis of Chiral Heterocyclic Natural Products: Epothilone B and Tartrolon B. J Heterocycl Chem **36**: 1421
23. Nicolaou KC, Ritzen A, Namoto K (2001) Recent Developments in the Chemistry, Biology and Medicine of the Epothilones. J Chem Soc Chem Commun: 1523
24. Altmann K-H (2004) The Merger of Natural Product Synthesis and Medicinal Chemistry: On the Chemistry and Chemical Biology of Epothilones. Org Biomol Chem **2**: 2137
25. Watkins EB, Chittiboyina AG, Jung JC, Avery MA (2005) The Epothilones and Related Analogues – A Review of Their Syntheses and Anti-cancer Activities. Curr Pharm Des **11**: 1615
26. Watkins EB, Chittiboyina AG, Jung JC, Avery MA (2006) Recent Developments in the Syntheses of the Epothilones and Related Analogues. Eur J Org Chem **18**: 4071
27. Wartmann M, Altmann K-H (2002) The Biology and Medicinal Chemistry of Epothilones. Curr Med Chem Anti-Cancer Agents **2**: 123
28. Altmann K-H (2003) Epothilone B and Its Analogs – A New Family of Anticancer Agents. Mini-Rev Med Chem **3**: 149
29. Borzilleri RM, Vite GD (2003) Epothilones: New Tubulin Polymerization Agents in Preclinical and Clinical Development. Drugs of the Future **27**: 1149
30. Altmann K-H (2005) Recent Developments in the Chemistry and Biology of Epothilones. Curr Pharm Des **11**: 1595
31. Altmann K-H, Pfeiffer B, Arseniyadis S, Pratt BA, Nicolaou KC (2007) The Chemistry and Biology of Epothilones – The Wheel Keeps Turning. ChemMedChem **2**: 397
32. Chappell MD, Stachel SJ, Lee CB, Danishefsky SJ (2000) En Route to a Plant Scale Synthesis of the Promising Antitumor Agent 12,13-Desoxyepothilone B. Org Lett **2**: 1633
33. He L, Orr GA, Horwitz SB (2001) Novel Molecules That Interact with Microtubules and Have Functional Activity Similar to Taxol. Drug Discov Today **6**: 1153
34. Altaha R, Fojo T, Reed E, Abraham J (2002) Epothilones: A Novel Class of Non-taxane Microtubule-stabilizing Agents. Curr Pharm Des **8**: 1707
35. Jordan MA, Wilson L (2004) Microtubules as a Target for Anticancer Drugs. Nature Rev, Cancer **4**: 253
36. Nogales E (2001) Structural Insights into Microtubule Function. Annu Rev Biophys Biomol Struct **30**: 397
37. Valiron O, Caudron N, Job D (2001) Microtubule Dynamics. Cell Mol Life Sci **58**: 2069
38. Amos L (2004) Microtubule Structure and Its Stabilization. Org Bioorg Chem **2**: 2153
39. Jordan MA, Wilson L (1998) Microtubules and Actin Filaments: Dynamic Targets for Cancer Chemotherapy. Curr Opin Cell Biol **10**: 123
40. Jordan MA (2002) Mechanism of Action of Antitumor Drugs that Interact with Microtubules and Tubulin. Curr Med Chem: Anti-Cancer Agents **2**: 1
41. Zhai Y, Kronebusch PJ, Simon PM, Borisy GG (1996) Microtubule Dynamics at the G2/M Transition: Abrupt Breakdown of Cytoplasmic Microtubules at Nuclear Envelope Breakdown and Implications for Spindle Morphogenesis. J Cell Biol **135**: 201
42. Rusan NM, Fagerstrom CJ, Yvon AMC, Wadsworth P (2001) Cell Cycle-dependent Changes in Microtubule Dynamics in Living Cells Expressing Green Fluorescent Protein-α Tubulin. Mol Biol Cell **12**: 971
43. Altmann K-H, Bold G, Caravatti G, End N, Flörsheimer A, Guagnano V, O'Reilly T, Wartmann M (2000) Epothilones and Their Analogs – Potential New Weapons in the Fight Against Cancer. Chimia **54**: 612

44. Hamel E, Sackett DL, Vourloumis D, Nicolaou KC (1999) The Coral-Derived Natural Products Eleutherobin and Sarcodictyins A and B: Effects on the Assembly of Purified Tubulin with and without Microtubule-Associated Proteins and Binding at the Polymer Taxoid Site. Biochemistry **38**: 5490
45. Nettles JH, Li HL, Cornett B, Krahn JM, Snyder JP, Downing KH (2004) The Binding Mode of Epothilone A on α,β-Tubulin by Electron Crystallography. Science **305**: 866
46. Buey RM, Diaz JF, Andreu JM, O'Brate A, Giannakakou P, Nicolaou KC, Sasmal PK, Ritzen A, Namoto K (2004) Interactions of Epothilone Analogs with the Paclitaxel Binding Site: Relationship Between Binding Affinity, Microtubule Stabilization, and Cytotoxicity. Chem Biol **11**: 225
47. Jordan MA, Wendell K, Gardiner S, Derry WB, Copp H, Wilson L (1996) Mitotic Block Induced in HeLa Cells by Low Concentrations of Paclitaxel (Taxol) Results in Abnormal Mitotic Exit and Apoptotic Cell Death. Cancer Res **56**: 816
48. Lichtner RB, Rotgeri A, Bunte T, Buchmann B, Hoffmann J, Schwede W, Skuballa W, Klar U (2001) Subcellular Distribution of Epothilones in Human Tumor Cells. Proc Natl Acad Sci USA **98**: 11743
49. Kamath K, Jordan MA (2003) Suppression of Microtubule Dynamics by Epothilone B is Associated with Mitotic Arrest. Cancer Res **63**: 6026
50. Chen J-GC, Horwitz SB (2002) Differential Mitotic Responses to Microtubule-stabilizing and -destabilizing Drugs. Cancer Res **14**: 1935
51. Chen J-GC, Yang C-PHY, Cammer M, Horwitz SB (2003) Gene Expression and Mitotic Exit Induced by Microtubule-Stabilizing Drugs. Cancer Res **15**: 7891
52. McDaid HM, Horwitz SB (2001) Selective Potentiation of Paclitaxel (Taxol)-Induced Cell Death by Mitogen-Activated Protein Kinase Kinase Inhibition in Human Cancer Cell Lines. Mol Pharmacol **60**: 290
53. Wittmann S, Bali P, Donapaty S, Nimmanapalli R, Guo F, Yamaguchi H, Huang M, Jove R, Wang HG, Bhalla K (2003) Flavopiridol Down-Regulates Antiapoptotic Proteins and Sensitizes Human Breast Cancer Cells to Epothilone B-induced Apoptosis. Cancer Res **63**: 93
54. Giannakakou P, Nakano M, Nicolaou KC, O'Brate A, Yu J, Blagosklonny MV, Greber UF, Fojo T (2002) Enhanced Microtubule-dependent Trafficking and p53 Nuclear Accumulation by Suppression of Microtubule Dynamics. Proc Natl Acad Sci USA **99**: 10855
55. Bröker LE, Huisman C, Ferreira CG, Rodriguez JA, Kruyt FAE, Giaccone G (2002) Late Activation of Apoptotic Pathways Plays a Negligible Role in Mediating the Cytotoxic Effects of Discodermolide and Epothilone B in Non-small Cell Lung Cancer Cells. Cancer Res **62**: 4081
56. Bröker LE, Huisman C, Span SW, Rodriguez JA, Kruyt FAE, Giaccone G (2004) Cathepsin B Mediates Caspase-Independent Cell Death Induced by Microtubule Stabilizing Agents in Non-Small Cell Lung Cancer Cells. Cancer Res **64**: 27
57. Garraway LA, Chabner B (2002) MDR1 Inhibition: Less Resistance or Less Relevance? Eur J Cancer **38**: 2337
58. Nogales E, Wolf SG, Downing KH (1998) Structure of the α,β Tubulin Dimer by Electron Crystallography. Nature **391**: 199
59. Giannakakou P, Gussio R, Nogales E, Downing KH, Zaharevitz D, Bollbuck B, Poy G, Sackett D, Nicolaou KC, Fojo T (2000) A Common Pharmacophore for Epothilone and Taxanes: Molecular Basis for Drug Resistance Conferred by Tubulin Mutations in Human Cancer Cells. Proc Natl Acad Sci USA **97**: 2904

60. He L, Yang C-P, Horwitz SB (2001) Mutations in β-Tubulin Map to Domains Involved in Regulation of Microtubule Stability in Epothilone-resistant Cell Lines. Mol Cancer Ther **1**: 3
61. Yang CPH, Verdier-Pinard P, Wang F, Lippaine-Horvath E, He LF, Li DS, Höfle G, Ojima I, Orr GA, Horwitz SB (2005) A Highly Epothilone B-Resistant A549 Cell Line with Mutations in Tubulin that Confer Drug Dependence. Mol Cancer Ther **4**: 987
62. Verrills NM, Flemming CL, Liu M, Ivery MT, Cobon GS, Norris MD, Haber M, Kavallaris M (2003) Microtubule Alterations and Mutations Induced by Desoxyepothilone B: Implications for Drug-Target Interactions. Chem Biol **10**: 597
63. Cabral F, Barlow SB (1991) Resistance to Antimitotic Agents as Genetic Probes of Microtubule Structure and Function. Pharmacol Ther **52**: 159
64. Su DS, Balog A, Meng D, Bertinato P, Danishefsky SJ, Zheng YH, Chou TC, He L, Horwitz SB (1997) Structure-activity Relationships of the Epothilones and the First in Vivo Comparison with Paclitaxel. Angew Chem Int Ed **36**: 2093
65. Chou TC, Zhang XG, Balog A, Su DS, Meng D, Savin K, Bertino JR, Danishefsky SJ (1998) Desoxyepothilone B: An Efficacious Microtubule-targeted Antitumor Agent with a Promising in Vivo Profile Relative to Epothilone B. Proc Natl Acad Sci USA **95**: 9642
66. O'Reilly T, McSheehy PMJ, Wenger F, Hattenberger M, Muller M, Vaxelaire J, Altmann K-H, Wartmann M (2005) EPO906 (Epothilone B) is Active in Vivo Against Experimental Prostate Tumors. Prostate **65**: 231
67. Rothermel J, Wartmann M, Chen T, Hohneker T (2003) EPO906 (Epothilone B): A Promising Novel Microtubule Stabilizer. Semin Oncol **30** (Suppl 6): 51
68. Pietras K, Stumm M, Hubert M, Buchdunger E, Rubin K, Heldin C-K, McSheehy P, Wartmann M, Oestman A (2003) STI571 Enhances the Therapeutic Index of Epothilone B by a Tumor-selective Increase of Drug Uptake. Clin Cancer Res **9**: 3779
69. Altmann K-H, Blommers MJJ, Caravatti G, Flörsheimer A, Nicolaou KC, O'Reilly T, Schmidt A, Schinzer D, Wartmann M (2001) Synthetic and Semisynthetic Analogs of Epothilones: Chemistry and Biological Activity. In: Ojima I, Vite GD, Altmann K-H (eds.) Anticancer Agents – Frontiers in Cancer Chemotherapy. ACS Symposium Series 796, p. 112. American Chemical Society, Washington, DC
70. Klar U, Buchmann B, Schwede W, Skuballa W, Hoffmann J, Lichtner RB (2006) Total Synthesis and Antitumor Activity of ZK-EPO (SAGOPILONE): The First Fully Synthetic Epothilone in Clinical Development. Angew Chem Int Ed **45**: 7942
71. Klar U, Skuballa W, Buchmann B, Schwede W, Bunte T, Hoffmann J, Lichtner R (2001) Synthesis and Biological Activity of Epothilones. In: Ojima I, Vite GD, Altmann K-H (eds.) Anticancer Agents – Frontiers in Cancer Chemotherapy. ACS Symposium Series 796, p. 131. American Chemical Society, Washington, DC
72. Sefkow M, Kiffe M, Schummer D, Höfle G (1998) Oxidative and Reductive Transformations of Epothilone A. Bioorg Med Chem Lett **8**: 3025
73. Sefkow M, Kiffe M, Höfle G (1998) Derivatization of the C12–C13 Functional Groups of Epothilones A, B and C. Bioorg Med Chem Lett **8**: 3031
74. Vite GD, Borzilleri RM, Kim SH, Regueiro-Ren A, Humphreys WG, Lee FYF (2001) Epothilones A and B: Springboards for Semisynthesis of Promising Antimitotic Agents. In: Ojima I, Vite GD, Altmann K-H (eds.) Anticancer Agents – Frontiers in Cancer Chemotherapy. ACS Symposium Series 796, p. 148. American Chemical Society, Washington, DC
75. Borzilleri RM, Zheng X, Schmidt RJ, Johnson JA, Kim SH, DiMarco JD, Fairchild CR, Gougoutas JZ, Lee FYF, Long BH, Vite GD (2000) A Novel Application of a

Pd(0)-Catalyzed Nucleophilic Substitution Reaction to the Regio- and Stereoselective Synthesis of Lactam Analogues of the Epothilone Natural Products. J Am Chem Soc **122**: 8890

76. Lee FY, Borzilleri R, Fairchild CR, Kim SH, Long BH, Reventos-Suarez C, Vite GD, Rose WC, Kramer RA (2001) BMS-247550: A Novel Epothilone Analog with a Mode of Action Similar to Paclitaxel but Possessing Superior Antitumor Efficacy. Clin Cancer Res **7**: 1429
77. Chou TC, O'Connor OA, Tong WP, Guan Y, Zhang ZG, Stachel SJ, Lee C, Danishefsky SJ (2001) The Synthesis, Discovery, and Development of a Highly Promising Class of Microtubule Stabilization Agents: Curative Effects of Desoxyepothilones B and F Against Human Tumor Xenografts in Nude Mice. Proc Natl Acad Sci USA **98**: 8113
78. Schinzer D, Altmann K-H, Stuhlmann F, Bauer A, Wartmann M (2000) Synthesis and Biological Evaluation of aza-Epothilones. ChemBioChem **1**: 67
79. Stachel SJ, Lee CB, Spassova M, Chappell MD, Bornmann WG, Danishefsky SJ, Chou T-C, Guan Y (2001) On the Interactivity of Complex Synthesis and Tumor Pharmacology in the Drug Discovery Process: Total Synthesis and Comparative in Vivo Evaluations of the 15-Aza Epothilones. J Org Chem **66**: 4369
80. Yamaguchi H, Paranawithana SR, Lee MW, Huang Z, Bhalla KN, Wang HG (2002) Epothilone B Analogue (BMS-247550)-mediated Cytotoxicity Through Induction of Bax Conformational Change in Human Breast Cancer Cells. Cancer Res **62**: 466
81. Griffin D, Wittmann S, Guo F, Nimmanapalli R, Bali P, Wang HG, Bhalla K (2003) Molecular Determinants of Epothilone B Derivative (BMS 247550) and Apo-2L/TRAIL-induced Apoptosis of Human Ovarian Cancer Cells. Gynecol Oncol **89**: 37
82. Kim J-C, Kim J-S, Saha D, Cao Q, Shyr Y, Choy H (2003) Potential Radiation-sensitizing Effect of Semisynthetic Epothilone B in Human Lung Cancer Cells. Radiother Oncol **68**: 305
83. Regueiro-Ren A, Leavitt K, Kim SH, Höfle G, Kiffe M, Gougoutas JZ, DiMarco JD, Lee FYF, Fairchild CR, Long BH, Vite GD (2002) SAR and pH Stability of Cyano-Substituted Epothilones. Org Lett **4**: 3815
84. Cachoux F, Isarno T, Wartmann M, Altmann K-H (2005) Scaffolds for Microtubule Inhibition Through Extensive Modification of the Epothilone Template. Angew Chem Int Ed **44**: 7469
85. Nicolaou KC, Vourloumis D, Li T, Pastor J, Winssinger N, He Y, Ninkovic S, Sarabia F, Vallberg H, Roschangar F, King NP, Finlay MR, Giannakakou P, Verdier-Pinard P, Hamel E (1997) Designed Epothilones: Combinatorial Synthesis, Tubulin Assembly Properties, and Cytotoxic Action Against Taxol-resistant Tumor Cells. Angew Chem Int Ed **36**: 2097
86. Hardt IH, Steinmetz H, Gerth K, Sasse F, Reichenbach H, Höfle G (2001) New Natural Epothilones from *Sorangium cellulosum*, Strains So ce90/B2 and So ce90/D13: Isolation, Structure Elucidation, and SAR Studies. J Nat Prod **64**: 847
87. Cachoux F, Schaal F, Teichert A, Wagner T, Altmann K-H (2004) Synthesis of 4-aza Epothilone D Analogs. Synlett **14**: 2709
88. Carlomagno T, Blommers MJJ, Meiler J, Jahnke W, Schupp T, Petersen F, Schinzer D, Altmann K-H, Griesinger C (2003) The High-Resolution Solution Structure of Epothilone A Bound to Tubulin: An Understanding of the Structure-activity Relationships for a Powerful Class of Antitumor Agents. Angew Chem Int Ed **42**: 2511
89. Klar U, Schwede W, Skuballa W, Buchmann B, Hoffmann J, Lichtner R (2000) 6-Alkenyl, 6-Alkynyl-, and 6-Epoxy-epothilone Derivatives, Process for Their

Production and Their Use in Pharmaceutical Preparations. PCT Int Appl WO 200066589 A1 20001109

90. Martin HJ, Pojarliev P, Kahlig H, Mulzer J (2001) The 12,13-Diol Cyclization Approach for a Truly Stereocontrolled Total Synthesis of Epothilone B and the Synthesis of a Conformationally Restrained Analog. Chemistry **7**: 2261

91. Balog A, Bertinato P, Su DS, Meng D, Sorensen E, Danishefsky SJ, Zheng YH, Chou TC, He L, Horwitz SB (1997) Stereoselective Syntheses and Evaluation of Compounds in the 8-Desmethylepothilone A Series: Some Surprising Observations Regarding Their Chemical and Biological Properties. Tetrahedron Lett **38**: 4529

92. Sefkow M, Kiffe M, Schummer D, Höfle G (1998) Oxidative and Reductive Transformations of Epothilone A. Bioorg Med Chem Lett **8**: 3025

93. Nicolaou KC, Sarabia F, Ninkovic S, Finlay MR, Boddy CNC (1998) Probing the Ring Size of Epothilones: Total Synthesis of [14]-, [15]-, [17]-, and [18]-Epothilones A. Angew Chem Int Ed **37**: 81

94. End N, Furet P, van Campenhout N, Wartmann M, Altmann K-H (2004) Total Synthesis and Biological Evaluation of a C(10)/C(12)-Phenylene-bridged Analog of Epothilone D. Chem Biodiversity **1**: 1771

95. For other epothilone pharmacophore models cf. (*59*), (*160–163*)

96. Arslanian RL, Tang L, Blough S, Ma W, Qiu R-G, Katz L, Carney JR (2002) A New Cytotoxic Epothilone from Modified Polyketide Synthases Heterologously Expressed in *Myxococcus xanthus*. J Nat Prod **65**: 1061

97. Biswas K, Lin H, Njardson JT, Chappell MD, Chou T-C, Guan Y, Tong WP, He L, Horwitz SB, Danishefsky SJ (2002) Highly Concise Routes to Epothilones: The Total Synthesis and Evaluation of Epothilone 490. J Am Chem Soc **124**: 9825

98. Rivkin A, Njardarson JT, Biswas K, Chou T-C, Danishefsky SJ (2000) Total Syntheses of [17]- and [18]-Dehydrodesoxyepothilones B via a Concise Ring-Closing Metathesis-Based Strategy: Correlation of Ring Size with Biological Activity in the Epothilone Series. J Org Chem **67**: 7737

99. Rivkin A, Biswas K, Chou T-C, Danishefsky SJ (2002) On the Introduction of a Trifluoromethyl Substituent in the Epothilone Setting: Chemical Issues Related to Ring Forming Olefin Metathesis and Earliest Biological Findings. Org Lett **4**: 4081

100. Rivkin A, Yoshimura F, Gabarda AE, Chou T-C, Dong H, Tong WP, Danishefsky SJ (2003) Complex Target-Oriented Total Synthesis in the Drug Discovery Process: The Discovery of a Highly Promising Family of Second Generation Epothilones. J Am Chem Soc **125**: 2899

101. Yoshimura F, Rivkin A, Gabarda AE, Chou T-C, Dong H, Sukenick G, Morel FF, Taylor RE, Danishefsky SJ (2003) Synthesis and Conformational Analysis of (*E*)-9,10-Dehydroepothilone B: A Suggestive Link Between the Chemistry and Biology of Epothilones. Angew Chem Int Ed **42**: 2518

102. Chou T-C, Dong H, Rivkin A, Yoshimura F, Gabarda AE, Cho YS, Tong WP, Danishefsky SJ (2003) Design and Total Synthesis of a Superior Family of Epothilone Analogues, Which Eliminate Xenograft Tumors to a Nonrelapsable State. Angew Chem Int Ed **42**: 4762

103. Rivkin A, Yoshimura F, Gabarda AE, Cho YS, Chou TC, Dong HJ, Danishefsky SJ (2004) Discovery of (*E*)-9,10-Dehydroepothilones Through Chemical Synthesis: On the Emergence of 26-Trifluoro-(*E*)-9,10-dehydro-12,13-desoxyepothilone B as a Promising Anticancer Drug Candidate. J Am Chem Soc **126**: 10913

104. Rivkin A, Cho YS, Gabarda AE, Yoshimura F, Danishefsky SJ (2004) Application of Ring-Closing Metathesis Reactions in the Synthesis of Epothilones. J Nat Prod **67**: 139

105. Rivkin A, Chou TC, Danishefsky SJ (2005) On the Remarkable Antitumor Properties of Fludelone: How we Got There. Angew Chem Int Ed **44**: 2838

106. Tang L, Li R-G, Yong L, Katz L (2003) Generation of Novel Epothilone Analogs with Cytotoxic Activity by Biotransformation. J Antibiotics **56**: 16

107. Starks CM, Zhou Y, Liu F, Licari P (2003) Isolation and Characterization of New Epothilone Analogues from Recombinant *Myxococcus xanthus* Fermentations. J Nat Prod **66**: 1313

108. White JD, Carter RG, Sundermann KF, Wartmann M (2001) Total Synthesis of Epothilone B, Epothilone D, and *cis*- and *trans*-9,10-Dehydroepothilone D. J Am Chem Soc **123**: 5407

109. White JD, Carter RG, Sundermann KF, Wartmann M (2003) Total Synthesis of Epothilone B, Epothilone D, and *cis*- and *trans*-9,10-Dehydroepothilone D [Erratum for J Am Chem Soc 2001, **123**: 5407]. J Am Chem Soc **125**: 3190

110. Chou T-C, Dong HJ, Zhang XG, Tong WP, Danishefsky SJ (2005) Therapeutic Cure Against Human Tumor Xenografts in Nude Mice by a Microtubule Stabilization Agent, Fludelone, via Parenteral or Oral Route. Cancer Res **65**: 9445

111. Wu KD, Cho YS, Katz J, Ponomarev V, Chen-Kiang S, Danishefsky SJ, Moore MAS (2005) Investigation of Antitumor Effects of Synthetic Epothilone Analogs in Human Myeloma Models *in Vitro* and *in Vivo*. Proc Natl Acad Sci USA **102**: 10640

112. Qintard D, Bertrand P, Vielle S, Raimbaud E, Renard P, Pfeiffer B, Gesson J-P (2003) Enantioselective Synthesis of 2,3-Dehydro-3-desoxy-10-oxa Epothilone D. Synlett **13**: 2033

113. Quintard D, Bertrand P, Bachmann C, Gesson J-P (2004) Synthesis and Conformational Analysis of Macrocycles Related to 10-oxa-Epothilone. Eur J Org Chem **23**: 4762

114. Taylor RE, Chen Y, Galvin GM, Pabba PK (2004) Conformation-activity Relationships in Polyketide Natural Products. Towards the Biologically Active Conformation of Epothilone. Org Biomol Chem **2**: 127

115. Schinzer D, Böhm OM, Altmann K-H, Wartmann M (2004) Synthesis and Biological Evaluation of Furano-Epothilone C. Synlett **14**: 1375

116. Meng D, Su DS, Balog A, Bertinato P, Sorensen EJ, Danishefsky SJ, Zheng YH, Chou T-C, He L, Horwitz SB (1997) Remote Effects in Macrolide Formation Through Ring-Forming Olefin Metathesis: An Application to the Synthesis of Fully Active Epothilone Congeners. J Am Chem Soc **119**: 2733

117. Su DS, Meng D, Bertinato P, Balog A, Sorensen EJ, Danishefsky SJ, Zheng YH, Chou T-C, He L, Horwitz SB (1997) Total Synthesis of (–)-Epothilone B: An Extension of the Suzuki Coupling Method and Insights into Structure-activity Relationships of the Epothilones. Angew Chem Int Ed **36**: 757

118. Nicolaou KC, Winssinger N, Pastor J, Ninkovic S, Sarabia F, He Y, Vourloumis D, Yang Z, Li T, Giannakakou P, Hamel E (1997) Synthesis of Epothilones A and B in Solid and Solution Phase. Nature **387**: 268

119. Chou T-C, Zhang XG, Harris CR, Kuduk SD, Balog A, Savin KA, Bertino JR, Danishefsky SJ (1998) Desoxyepothilone B is Curative Against Human Tumor Xenografts that are Refractory to Paclitaxel. Proc Natl Acad Sci USA **95**: 15798

120. Sinha SC, Sun J, Wartmann M, Lerner RA (2001) Synthesis of Epothilone Analogues by Antibody-Catalyzed Resolution of Thiazole Aldol Synthons on a Multigram Scale. Biological consequences of C-13 alkylation of epothilones. ChemBioChem **2**: 656

121. Buchmann B, Klar U, Skuballa W, Schwede W, Schirner M, Menrad A (2000) The Preparation Process, Intermediate Products and Pharmaceutical Use of Epothilone Derivatives. PCT Int Appl WO 2000000485 A1 20000106

122. Altmann K-H, Bold G, Caravatti G, Denni D, Flörsheimer A, Schmidt A, Rihs G, Wartmann M (2002) The Total Synthesis and Biological Assessment of *trans*-Epothilone A. Helv Chim Acta **85**: 4086
123. Nicolaou KC, Namoto K, Ritzen A, Ulven T, Shoji M, Li J, D'Amico G, Liotta D, French CT, Wartmann M, Altmann K-H, Giannakakou P (2001) Chemical Synthesis and Biological Evaluation of *cis*- and *trans*-12,13-Cyclopropyl and 12,13-Cyclobutyl Epothilones and Related Pyridine Side Chain Analogues. J Am Chem Soc **123**: 9313
124. Nicolaou KC, Ritzen A, Namoto K, Ruben MB, Diaz F, Andreu JM, Wartmann M, Altmann K-H, O'Brate A, Giannakakou P (2002) Chemical Synthesis and Biological Evaluation of Novel Epothilone B and *trans*-12,13-Cyclopropyl Epothilone B Analogues. Tetrahedron **58**: 6413
125. Altmann K-H, Flörsheimer A, Bold G, Caravatti G, Wartmann M (2004) Natural Product-Based Drug Discovery – Epothilones as Lead Structures for the Development of New Anticancer Agents. Chimia **58**: 686
126. Glunz PW, He L, Horwitz SB, Chakravarty S, Ojima I, Chou TC, Danishefsky SJ (1999) The Synthesis and Evaluation of 12,13-Benzodesoxyepothilone B: A Highly Convergent Route. Tetrahedron Lett **40**: 6895
127. Feyen F, Gertsch J, Wartmann M, Altmann K-H (2006) Design and Synthesis of 12-aza-Epothilones (Azathilones) – Non-natural Natural Products with Potent Anticancer Activity. Angew Chem Int Ed **45**: 5880
128. Cachoux F, Isarno T, Wartmann M, Altmann K-H (2006) Total Synthesis and Biological Assessment of Benzimidazole-based Analogues of Epothilone A: Ambivalent Effects on Cancer Cell Growth Inhibition. ChemBioChem **7**: 54
129. Johnson J, Kim SH, Bifano M, DiMarco J, Fairchild C, Gougoutas J, Lee F, Long B, Tokarski J, Vite GD (2000) Synthesis, Structure Proof, and Biological Activity of Epothilone Cyclopropanes. Org Lett **2**: 1537
130. Nicolaou KC, Namoto K, Li J, Ritzen A, Ulven T, Shoji M, Zaharevitz D, Gussio R, Sackett DL, Ward RD, Hensler A, Fojo T, Giannakakou P (2000) Synthesis and Biological Evaluation of 12,13-Cyclopropyl and 12,13-Cyclobutyl Epothilones. Chem BioChem **2**: 69
131. Regueiro-Ren A, Borzilleri RM, Zheng X, Kim SH, Johnson JA, Fairchild CR, Lee FY, Long BH, Vite GD (2001) Synthesis and Biological Activity of Novel Epothilone Aziridines. Org Lett **3**: 2693
132. Nicolaou KC, Ninkovic S, Finlay MR, Sarabia F, Li T (1997) Total Synthesis of 26-Hydroxyepothilone B and Related Analogs. J Chem Soc Chem Commun: 2343
133. Chappell MD, Harris CR, Kuduk SD, Balog A, Wu Z, Zhang F, Lee CB, Stachel SJ, Danishefsky SJ, Chou TC, Guan Y (2002) Probing the SAR of dEpoB via Chemical Synthesis: A Total Synthesis Evaluation of C26-(1,3-Dioxolanyl)-12,13-desoxyepothilone B. J Org Chem **67**: 7730
134. Newman RA, Yang J, Finlay MRV, Cabral F, Vourloumis D, Stevens LC, Troncoso LP, Wu X, Logothetis CJ, Nicolaou KC, Navone NM (2001) Antitumor Efficacy of 26-Fluoroepothilone B Against Human Prostate Cancer Xenografts. Cancer Chemother Pharmacol **48**: 319
135. Altmann K-H, Nicolaou KC, Wartmann M, O'Reilly T (2001) Proc Am Assoc Cancer Res **42**: Abstract #1979
136. Nicolaou KC, Finlay MR, Ninkovic S, King NP, He Y, Li T, Sarabia F, Vourloumis D (1998) Synthesis and Biological Properties of C12,13-Cyclopropyl-Epothilone A and Related Epothilones. Chem Biol **5**: 365

137. Höfle G, Glaser N, Kiffe M, Hecht H-J, Sasse F, Reichenbach H (1999) N-Oxidation of Epothilone A–C and O-Acyl Rearrangement to C-19- and C21-Substituted Epothilones. Angew Chem Int Ed **38**: 1971
138. Sefkow M, Höfle G (1998) Substitutions at the Thiazole Moiety of Epothilone. Heterocycles **48**: 2485
139. Nicolaou KC, King NP, Finlay MRV, He Y, Roschangar F, Vourloumis D, Vallberg H, Sarabia F, Ninkovich S, Hepworth D (1999) Total Synthesis of Epothilone E and Related Side-chain Modified Analogues via a Stille Coupling Based Strategy. Bioorg Med Chem **7**: 665
140. Höfle G, Glaser N, Leibold T, Vite G, Kim S (2000) Preparation of C-21 Modified Epothilones for Use as Anticancer Agents. PCT Int Appl WO 2000050423 A1 20000831
141. Lee CB, Chou T-C, Zhang X-G, Wang Z-G, Kuduk SD, Chappell MD, Stachel SJ, Danishefsky SJ (2000) Total Synthesis and Antitumor Activity of 12,13-Desoxyepothilone F: An Unexpected Solvolysis Problem at C15, Mediated by Remote Substitution at C21. J Org Chem **65**: 6525
142. Kolman A (2004) BMS-310705 Bristol-Myers Squibb/GBF. Curr Opin Investig Drugs **5**: 1292
143. Kamath AV, Chang M, Lee F, Zhang YP, Marathe PH (2005) Preclinical Pharmacokinetics and Oral Bioavailability of BMS-310705, a Novel Epothilone B Analog. Cancer Chemother Pharmacol **56**: 145
144. Höfle G, Glaser N, Leibold T (2000) Synthesis and Cytotoxicity of C-21 Modified Epothilones. Ger Offen DE 19907588
145. Uyar D, Takigawa N, Mekhail T, Grabowski D, Markman M, Lee F, Canetta R, Peck R, Bukowski R, Ganapathi R (2003) Apoptotic Pathways of Epothilone BMS 310705. Gynecol Oncol **91**: 173
146. Wartmann M, Loretan J, Reuter R, Hattenberger M, Muller M, Vaxelaire J, Maira S-M, Flörsheimer A, O'Reilly T, Nicolaou KC, Altmann K-H (2004) Preclinical Pharmacological Profile of ABJ879, a Novel Epothilone B Analog with Potent and Protracted Anti-tumor Activity. Proc Am Assoc Cancer Res **45**: Abstract #5440
147. Nicolaou KC, Sasmal PK, Rassias G, Reddy MV, Altmann K-H, Wartmann M, O'Brate A, Giannakakou P (2003) Design, Synthesis, and Biological Properties of Highly Potent Epothilone B Analogues. Angew Chem Int Ed **42**: 3515
148. Nicolaou KC, Scarpelli R, Bollbuck B, Werschkun B, Pereira MM, Wartmann M, Altmann K-H, Zaharevitz D, Gussio R, Giannakakou P (2000) Chemical Synthesis and Biological Properties of Pyridine Epothilones. Chem Biol **7**: 593
149. Nicolaou KC, Hepworth D, King NP, Finlay MR, Scarpelli R, Pereira MM, Bollbuck B, Bigot A, Werschkun B (2000) Total Synthesis of 16-Desmethylepothilone B, Epothilone B10, Epothilone F, and Related Side Chain Modified Epothilone B Analogues. Chem Eur J **6**: 2783
150. Nicolaou KC, Pratt BA, Arseniyadis S, Wartmann M, O'Brate A (2006) Molecular Design and Chemical Synthesis of a Highly Potent Epothilone. ChemMedChem **1**: 41
151. Nicolaou KC, Finlay MRV, Ninkovic S, King NP, He Y, Li TH, Sarabia F, Vourloumis D (1998) Synthesis and Biological Properties of C12,13-Cyclopropyl-epothilone A and Related Epothilones. Chem Biol **5**: 365
152. Altmann K-H, Bold G, Caravatti G, Flörsheimer A, Guagnano V, Wartmann M (2000) Synthesis and Biological Evaluation of Highly Potent Analogues of Epothilones B and D. Bioorg Med Chem Lett **10**: 2765

153. Dong SD, Sundermann K, Smith KMJ, Petryka J, Liu FH, Myles DC (2004) Rapid Access to Epothilone Analogs via Semisynthetic Degradation and Reconstruction of Epothilone D. Tetrahedron Lett **45**: 1945

154. Cachoux F, Isarno T, Wartmann M, Altmann K-H (2006) Total Synthesis and Biological Assessment of Cyclopropane-based Epothilone Analogues – Modulation of Drug Efflux Through Polarity Adjustments. Synlett **16**: 1384

155. Bold G, Wojeik S, Caravatti G, Lindauer R, Stierlin C, Gertsch J, Wartmann M, Altmann K-H (2006) Structure-activity Relationships in Side-chain-modified Epothilone Analogues. How Important is the Position of the Nitrogen Atom? ChemMedChem **1**: 37

156. Hearn BR, Zhang D, Li Y, Myles DC (2006) C-15 Thiazol-4-yl Analogues of (*E*)-9,10-Didehydroepothilone D: Synthesis and Cytotoxicity. Org Lett **8**: 3057

157. Taylor RE, Chen Y, Beatty A (2003) Conformation-Activity Relationships in Polyketide Natural Products: A New Perspective on the Rational Design of Epothilone Analogues. J Am Chem Soc **125**: 26

158. Winkler JD, Holland JM, Kasparec J, Axelsen PH (1999) Design and Synthesis of Constrained Epothilone Analogs: The Efficient Synthesis of Eleven-membered Rings by Olefin Metathesis. Tetrahedron **55**: 8199

159. Alhamadsheh MM, Hudson RA, Tillekeratne LMV (2006) Design, Total Synthesis, and Evaluation of Novel Open-Chain Epothilone Analogues. Org Lett **8**: 685

160. Ojima I, Chakravarty S, Inoue T, Lin S, He L, Horwitz SB, Kuduk SD, Danishefsky SJ (1999) A Common Pharmacophore for Cytotoxic Natural Products that Stabilize Microtubules. Proc Natl Acad Sci USA **96**: 4256

161. Wang M, Xia X, Kim Y, Hwang D, Jansen JM, Botta M, Liotta DC, Snyder JP (1999) A Unified and Quantitative Receptor Model for the Microtubule Binding of Paclitaxel and Epothilone. Org Lett **1**: 43

162. Manetti F, Forli S, Maccari L, Corelli F, Botta M (2003) 3D QSAR Studies of the Interaction Between β-Tubulin and Microtubule Stabilizing Antimitotic Agents (MSAA). A Combined Pharmacophore Generation and Pseudoreceptor Modeling Approach Applied to Taxanes and Epothilones. Il Farmaco **58**: 357

163. Manetti F, Maccari L, Corelli F, Botta M (2004) 3D QSAR Models of Interactions Between β-Tubulin and Microtubule Stabilizing Antimitotic Agents (MSAA): A Survey on Taxanes and Epothilones. Curr Topics Med Chem **4**: 203

7. Clinical Studies with Epothilones

Karl-Heinz Altmann

Department of Chemistry and Applied Biosciences, Institute of Pharmaceutical Sciences, Swiss Federal Institute of Technology (ETH), Zürich, Switzerland

Contents

7.1. Introduction

As indicated in previous chapters, epothilone research so far has delivered seven new chemical entities that have been advanced to clinical trials in humans (Fig. 1). However, the amount of clinical data publicly available at this time strongly varies between individual compounds, depending on their development stage, but also on the general publication policy of the developing company. The compound that has been most comprehensively characterized in the clinical literature is ixabepilone (BMS-247550), for which trial results have been described in a number of articles in peer-reviewed journals and which has been granted FDA approval for two clinical indications on Oct. 16, 2007. For all other compounds, most of the information on clinical trials is available only in abstract form. In all these cases it remains uncertain, whether the content of these abstracts fully reflects the content of the

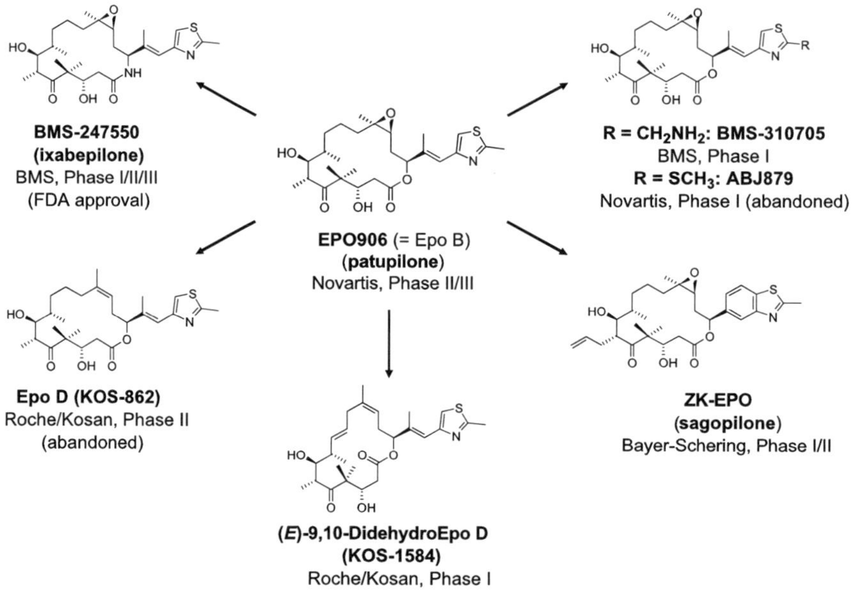

Fig. 1. Molecular structures of epothilone-derived clinical development compounds. All compounds are derivatives or analogs of Epo B/D

subsequent (poster or oral) presentations at the corresponding meeting; in fact, it seems likely that additional data will have been included in the actual meeting presentations that may not have been available at the time of abstract submission. As this is unknown to the author, such additional information cannot be considered in this chapter, which is solely based on information documented in accessible abstracts or journal publications. It should also be kept in mind that the interpretation of data from ongoing clinical trials or forward looking statements based on data from completed trials are always preliminary in character. Compounds that may appear promising at the time of writing of this article may still fail in further development and have been abandoned by the time of publication of this chapter.

The most advanced epothilone in clinical development at this point is the Epo B lactam ixabepilone, which has recently received FDA approval for the treatment of metastatic or advanced breast cancer, either as single agent or in combination with capecitabine, and will be marketed in the US by Bristol-Myers Squibb (BMS) under the trade name Ixempra® (*1*). Based on information from the Novartis webpage, Phase III trials are also ongoing with patupilone in ovarian cancer (*2*). Data from Phase II studies (as the highest phase investigated) have been reported (in abstract form) for KOS-862 (*3*, *4*); however, according to a press a release of February 2007, Roche and Kosan have now discontinued the clinical development of the compound, allegedly, in order to focus their resources on the development of KOS-1584 and in spite of encouraging Phase II data (*5*). Phase II studies are ongoing with ZK-EPO (sagopilone) (*6*), while only Phase I data have been disclosed so far for KOS-1584 and BMS-310705 (*vide infra*). No clinical data have been reported for Novartis' ABJ879, which had been announced to have entered Phase I trials in 2004 (*7*).

In the following, the existing clinical data and the current status of development will be summarized for each of the seven epothilone-based agents that have entered clinical trials so far. However, rather than trying to provide a detailed recapitulation of each individual trial, the major objective of these accounts is to provide a high level overview on important results and conclusions derived from these clinical studies for a non-specialist reader. A number of excellent and detailed reviews on clinical trials with epothilones are available in the recent literature (*8–10*).

7.2. Patupilone (EPO906, Epo B)

Patupilone was the first epothilone to enter clinical trials and Phase I studies with three different treatment schedules have been reported for the compound as single agent (*11*, *12*). In all cases patupilone was formulated as a solution in polyethylene glycol 300 (PEG 300)/0.9% saline and administration occurred by *i.v.* bolus injection over 5 or 15 min. In two studies, patients were treated with the drug on a weekly basis, either for three weeks followed by one week without treatment (3 weeks on/one week off schedule), or for 6 weeks followed by a treatment-free interval of 3 weeks (6 weeks on/3 weeks off schedule) (*11*). Most patients could be treated on either of these schedules for at least 18 weeks. For both treatment regimens the maximum tolerated dose was determined as 2.5 mg/m^2, with diarrhea as the most common dose-limiting toxicity (DLT). Other common non-hematological toxicities included nausea, fatigue and vomiting, but not alopecia. In contrast to Taxol®, severe hematological toxicities were rare, thus suggesting that weekly patupilone was not associated with significant myelosuppression. Pharmacokinetic analysis showed that patupilone blood concentrations after a short infusion declined in a biphasic manner with a terminal half-life of 4 days, which clearly indicates that plasma stability, in contrast to rodent species, is not critical in humans. Partial responses to treatment with patupilone were observed for 3 out of a total of 60 assessable patients (5%), two of which had received prior taxane therapy. Altogether, 31 patients experienced stable disease with a median duration of 16.3 weeks. Similar conclusions as for the weekly treatment schedules have been derived for a once-every-3-weeks regimen with regard to the nature of the most common toxicities. The MTD for the 3-weekly regimen was identified as 6 mg/m^2 (*12*). More recently, the 3-weekly patupilone regimen has also been investigated in a number of Phase I and Phase I/II trials that involved proactive diarrhea management (*13–16*) and this has resulted in a significant increase in the MTD with a recommended Phase II dose of 10 mg/m^2 for an ovarian cancer trial (*16*). Interestingly, preliminary data seem to indicate that under these conditions neuropathy becomes dose-limiting (*16*).

Phase II clinical trial results for patupilone have been reported in prostate, colon, gastric, non-small-cell lung (NSCLC), ovarian, neuro-endocrine and renal cancers and, most recently, in advanced hepato-cellular carcinoma (HCC) (*17*) and for the treatment of recurrent or progressive brain metastases in patients with NSCLC (*18*). So far, the most promising data have been obtained in prostate, ovarian, and NSCL cancer patients. Response rates of 25% and 11% were observed for platinum- and taxane-resistant ovarian carcinomas (*16*) and platinum-pretreated NSCLC (*19*), respectively, which are comparable with standard therapy. Superior activity over standard therapy was observed in a study in hormone-refractory prostate cancer (20% objective tumor response, *i.e.*, 4 out of 20 patients with measurable disease showed a partial response *vs.* 12% for docetaxel in combination with prednisone (in a different trial (*20*)) (*21*). It should be kept in mind, however, that all efficacy studies conducted with patupilone to date were single-armed (*i.e.*, they did not include a parallel patient group receiving standard treatment) and thus have to be interpreted very cautiously. In summary, clinical studies with patupilone have shown that the compound can be safely administered to patients and due to non-overlapping toxicity profiles can also be combined with other cytotoxics. Promising efficacy data have been obtained for several tumor types at this point, but more extended studies will be necessary to determine whether patupilone will indeed offer advantages over standard therapy in the treatment of specific types of tumors.

7.3. Ixabepilone

As highlighted above, BMS' ixabepilone is the first (and so far only) epothilone that has been approved by regulatory authorities for clinical

References, pp. 234–237

use in cancer patients (*1*). Early clinical studies with ixabepilone have been reviewed by *Lin et al.* (*22*) and more recent data are summarized in *Pivot et al.* (*23*). Data from a number of Phase I studies with ixabepilone are available from articles in peer reviewed journals, comprising 4 different dosing schedules. In general, the compound was administered as a 1-h or 3-h *i.v.* infusion in Cremophor together with prophylactic antihistamines, in order to avoid complications from hypersensitivity reactions. (Note that no such prophylactic measures are required with the patupilone formulation.) Administration of ixabepilone as a 1-h infusion every three weeks was associated with an MTD of 40 mg/m^2 (*24*, *25*) or 50 mg/m^2 (*26*, *27*), with either of these doses being recommended for Phase II efficacy studies. A Phase II dose of 25 mg/m^2 has been recommended for a weekly 1-h infusion treatment regimen, either contiguous or including a 1-week treatment break after every three doses (*28*). In addition, higher frequency administration schedules have been investigated, including daily 1-h infusions for 3 (*29*) or 5 (*30*) consecutive days, respectively, every 21 days; the respective MTD's were 6 and 8 mg/m^2. The major DLT in all studies was neutropenia, frequent non-hematological toxicities included fatigue, peripheral neuropathy and gastrointestinal discomfort. A number of objective responses to single-agent treatment were observed in Phase I studies with ixabepilone in breast, ovarian, cervical, prostate, colon, lung, and renal cancers as well as in squamous cell cancers of the head and neck, lymphoma, and angiosarcoma. Significantly, the compound was also shown to induce microtubule bundling in peripheral blood monocytes (PBMCs) of treated individuals and a good correlation was observed between the magnitude of this effect and plasma AUCs (*31*). Similar microtubule bundling and subsequent apoptosis could also be detected in a tumor sample, in spite of the expression of MDR1 and MRP1 (*31*). These findings provide mechanistic validation of the *in vitro* pharmacodynamic findings with ixabepilone in the clinical setting.

The efficacy of ixabepilone in Phase II studies has been documented in a number of peer reviewed papers and additional data have been reported in a significant number of meeting abstracts (reviewed in Ref. (*8*)). Tumor types that have been investigated with ixabepilone include breast, prostate, colorectal, non-small-cell lung, gastric, hepatobiliary, gynaecological, and pancreatic cancer. In addition, studies have been conducted for the treatment of sarcoma, melanoma, and non-*Hodgkin*'s lymphoma. Promising activity of ixabepilone has been noted in a Phase II trial in metastatic hormone-refractory prostate cancer (HRPC) employing the 3-weekly treatment schedule at a dose of 40 mg/m^2 (*32*).

Using a prostate-specific antigen (PSA) response as the primary end point, a response rate of 33% was observed (14 out of 42 chemotherapy-naïve patients), with 72% of PSA responders showing a decrease in PSA levels of >80%. Using the same schedule and an ixabepilone dose of 35 mg/m^2 either alone or in combination with estramustine phosphate, response rates of 32% and 48%, respectively, were found in patients with measurable disease in a trial in progressive castrate metastatic prostate cancer (*33*). As observed in the Phase I trials, the major toxicities in both studies consisted of neutropenia, neuropathy, and fatigue (for ixabepilone as single agent). The efficacy of ixabepilone in hormone-refractory prostate cancer, as for patupilone (*vide supra*), compares favorably with a 12% response rate to standard treatment (docetaxel in combination with prednisone), which suggests that epothilones could find applications in the management of the disease, in particular in view of the limited treatment options available in this setting.

Single-agent ixabepilone has also shown clinically relevant activity in a large (123 patients), two-armed Phase II trial in NSCLC, which yielded response rates of 14% (administration of ixabepilone as a 32 mg/m^2 3-h infusion every three weeks; 77 patients) and 11.6% (administration of a 6 mg/m^2 1 h infusion daily for 5 consecutive days in a 3-week cycle, 69 patients) in a patient population that had experienced disease progression after one prior *cis*-platin- or carboplatin-based chemotherapy regimen (*34*). Confirmed response rates of 9% each have been reported for Phase II trials in advanced pancreas cancer (40 mg/m^2 once every 3 weeks, 5 partial responses in 56 patients with measurable disease) (*35*) and in metastatic gastric adenocarcinoma (40 mg/m^2 once every 3 weeks, 2 partial responses in 23 patients) (*36*), which has led the authors of those studies to recommend further testing of ixabepilone in the above diseases. It should be noted, however, that no responses were observed in a second arm of the gastric cancer trial for a treatment regimen of 6 mg/m^2 on days 1–5 every 21 days. No activity was observed with single-agent ixabepilone in a Phase II trial in advanced colorectal cancer (40 mg/m^2 every 3 weeks; 23 patients with disease progression after treatment with irinotecan-5-fluorouracil-leucovorin (IFL)) (*37*).

The most comprehensive clinical profiling of ixabepilone by far has occurred in metastatic breast cancer (MBC), where preliminary results from several Phase II studies have been reported in abstract form or, more recently, as articles in peer-reviewed journals. These studies were conducted with different treatment regimens (all *i.v.*), including 40 mg/m^2 administered as either a 3-h or 1-h *i.v.* infusion every three weeks,

References, pp. 234–237

6 mg/m^2/day given on days 1 through 5 every 3 weeks, and 8 mg/m^2/day applied daily for 3 days for a first 3-week cycle and a dose increase to 10 mg/m^2/day for subsequent cycles, if patients did not show hematological or other toxicities after the first cycle. Major toxicities were those predicted from the Phase I studies (*vide supra*). In one study based on the 3-h infusion 3-weekly dosing regimen, a response rate of 33% (40 mg/m^2 every 3 weeks; 15 out of 44 patients) was observed in a patient population that had been previously treated with an anthracycline in the adjuvant and/or neoadjuvant setting (*38*). A second study in women, who had undergone previous treatment with an anthracycline and a taxane and who had progressed during or within 4 months after their taxane therapy, reported a response rate of 12% (6 out of 49 patients) (*39*). Stable disease was noted in this study in 22 (of 44) patients and 4 of the 6 responders had previously progressed within 1 month of their last taxane dose. A 6 mg/m^2/day on days 1 through 5 every 3 week dosing regimen has been investigated in women with metastatic and locally advanced breast cancer, who had undergone prior neoadjuvant, adjuvant, or metastatic therapy with paclitaxel and/or docetaxel (*40*). A 22% response rate was observed for these treatment conditions (8 out of 37 patients), including one complete response. In this study, levels of glutamate (Glu)-terminated and of acetylated α-tubulin, which are markers of microtubule stabilization, were determined in a subset of matched pre- and post-treatment tumor biopsies by Western blot analysis and immunohistochemistry and found to be increased after ixabepilone therapy. In a more recent study the same dosing regimen (*i.e.* 6 mg/m^2 daily × 5 every 3 weeks) was employed in women with MBC, who had not received previous treatment with a taxane (*41*). A response rate of 57% was observed in this patient group, which raises the prospect for the use of ixabepilone in the first-line metastatic and adjuvant setting (*41*). Obviously, much more extensive trials would be necessary to reinforce this notion.

Interestingly, no responses were noted in a MBC Phase II trial with 12 patients (with prior exposure to taxanes), who were treated with ixabepilone at 8 mg/m^2/day for 3 days for a first 3-week cycle followed by a dose increase to 10 mg/m^2/day for subsequent cycles, if no hematological or other toxicities occurred after the first cycle (*42*). This contrast with the 22% response rate for the 6 mg/m^2 daily dose, but administered for 5 (instead of only 3) consecutive days every 3 weeks (*vide supra*).

The efficacy of ixabepilone in MBC has also been studied in a multinational randomized Phase III trial comprising 752 patients, who were anthracycline pretreated and met predefined resistance criteria to taxanes. In this trial ixabepilone (40 mg/m^2 *i.v.* over 3 h every three weeks) was combined with capecitabine (1000 mg/m^2 p.o. bid q14d) and compared with capecitabine alone (1250 mg/m^2 p.o. bid q14d). The primary endpoint of the study was progression-free survival (PFS) with objective response rate (ORR), safety, and overall survival (available after 2007) as secondary end points. The combination of ixabepilone with capecitabine was found to be superior to capecitabine alone, with "significant benefit [being] consistently maintained across pre-defined subgroups, including HER2-/ER-/PR- and HER2 + " (*43*). As indicated above, BMS-247550 was recently approved by the US FDA for the treatment of metastatic or advanced breast cancer, either as single agent or in combination with capecitabine (*1*). Specifically, approval has been granted for the treatment of (a) anthracycline- and taxane-resistant metastatic or locally advanced breast cancers (in combination with capecitabine) and (b) as monotherapy for the treatment of metastatic or locally advanced breast cancer in patients whose tumors are resistant or refractory to anthracyclines, taxanes, and capecitabine.

7.4. KOS-862

Several Phase I trials have been reported for KOS-862 (R1492, epothilone D, deoxyepothilone B), which have investigated a total of 6 different dosing regimens for single-agent treatment with the compound. In addition, KOS-862 has also been studied in combination with gemcitabine (*44*) and with trastuzumab in HER-2 overexpressing malignancies (*45*). In two of these trials KOS-862 was administered on a 3-weekly schedule either as a single dose of 9–185 mg/m^2 or at 20–50 mg/m^2 daily for three days every three weeks (*46*). Pharmacokinetics (PK) was linear for both schedules and similar. DLTs were gait impairment, chest pain, and cognitive abnormalities. More dose intense schedules for KOS-862 have involved weekly treatment with a single dose for 3 consecutive weeks out of every 4 or for 2 weeks out of every 3 (*47, 48*). Dose levels were between 16–120 mg/m^2, with the 100 mg/m^2 dose being expanded on both schedules. At this dose level mild-to-moderate fatigue and sensory neuropathy were the major toxicities and no cognitive/perceptual changes other than mild dizziness were observed. The compound showed linear PK across the entire dose range investigated with a half-life of 9.1 h. Assessment of microtubule bundle formation in PBMCs gave maximal values at the end-of-infusion and 1-h post-infusion, with bundle formation in *ca.* 40% of PBMCs. After 24 h

References, pp. 234–237

this value had declined to 15%. Tumor shrinkage was observed in these studies in 2 patients with large cell and mediastinal B-cell lymphoma. Based on these studies the Phase II dose for KOS-862 was defined as 100 mg/m^2.

As preclinical findings had demonstrated optimal efficacy in animal models for a 6 h continuous infusion schedule, continuous infusion regimens of KOS-862 have also been investigated in Phase I clinical studies (*49*, *50*). This has included either 24-h or up to 72-h infusions every two weeks at infusion rates between 1 and 6 mg/h, following a loading dose of 75 mg/h for 30 min. Toxicities were similar to those in the short infusion studies and no significant differences in PK parameters were found between the short and long infusion schedules. As for the short infusion regimens, microtubule bundling was detectable in PBMCs and signs of antitumor activity were observed.

Two Phase II trials with KOS-862 have been reported in abstract form, one in MBC (*3*) and one in NSCLC (*4*). In both cases the compound was administered at 100 mg/m^2 as a weekly 90-min *i.v.* infusion for 3 consecutive weeks out of a 4 week treatment cycle. No Phase II trials based on a continuous infusion regimen have been reported. In the breast cancer trial H1/H2 antagonists and corticosteroids were given 30–60 min before administration of KOS-862 as infusion reaction prophylaxis, while no such pretreatment is mentioned in the abstract related to the NSCLC trial. The MBC study included 12 patients, who had been pretreated with either anthracyclines or taxanes and were treated with KOS-862 for a median of 2.6 cycles (range of 1–6 cycles). Two out of 10 evaluable patients showed a confirmed partial response (20%); major toxicities were peripheral neuropathy (one patient) and ataxia (one patient). A response rate of <8% (1 for 35 patients) was found in the first stage of the NSCLC trial, which comprised 55 patients, who had failed platinum-based therapy and of whom 35 were evaluable at the time of reporting. In light of the lack of efficacy of KOS-862 the trial was not advanced to the second stage.

As indicated above Kosan and Roche have recently announced that the clinical development of KOS-862 has been terminated (*5*).

7.5. BMS-310705

Two Phase I studies with BMS-310705 have been reported in abstract form (*51*, *52*); no Phase II data are (publicly) available for the compound at this point. BMS-310705 is more water-soluble than ixabepilone and therefore can be administered as an *i.v.* infusion in

Cremophor-free formulations (*52*, *53*). Two treatment schedules with BMS-310705 have been investigated so far, one involving single dose treatment every three weeks (*51*) and one where the compound was administered weekly for 3 consecutive weeks of a 4 weeks cycle (*52*). In both cases BMS-310705 was given as a 15-min *i.v.* infusion. Doses of 0.6–70 mg/m^2 were investigated for the 3-weekly schedule, with DLTs at the 70 mg/m^2 dose consisting of grade 4 hyponatremia and grade 4 neutropenia; sensory neuropathy was generally observed at doses higher than 40 mg/m^2. The most frequent toxicities at the 40 mg/m^2 dose (which was expanded to 17 patients, out of a total of 59 treated) were neutropenia and sensory neuropathy. In addition, fatigue, vomiting, diarrhea, constipation, dehydration, anorexia, dyspnea, muscle weakness, and anemia were observed, although on a relatively infrequent basis. No DLTs were observed at the 40 mg/m^2 dose level and PK data showed dose-related increases in AUC and C_{max} that were similar for cycles 1 and 2. A complete response to treatment with 40 mg/m^2 of BMS-310705 was observed in one NSCLC patient; in addition, partial responses are reported for one patient with ovarian cancer (at 40 mg/m^2) and one with bladder cancer (at 30 mg/m^2). The investigation of a 3 weeks on/1 week off schedule in the second Phase I study with BMS-310705 was performed at doses between 5 and 30 mg/m^2 (*52*). Diarrhea proved to be the first course DLT at 30 mg/m^2, leading to an expansion of the 20 mg/m^2 dose to 16 patients (out of a total of 21 treated in the study). The major toxicities at this dose level included hypersensitivity reaction that re-occurred under steroid prophylaxis (in one patient), diarrhea, and >G2 sensory neuropathy (for 2/3 patients receiving their 4th course); other toxicities were musculoskeletal pain and asthenia (in 50% of patients), alopecia and ataxia (in one case). Two partial responses were observed in patients with breast cancer and stomach cancer, respectively, in this study. At the time of reporting, the determination of a recommended Phase II dose was still pending. Overall, the toxicities associated with BMS-310705 treatment are similar to those observed for ixabepilone (*8*).

7.6. KOS-1584

At the time of writing of this account three abstracts have been published describing interim results of ongoing Phase I trials with KOS-1584 (R1645) (*54–56*). In the first trial conducted with the compound, KOS-1584 was administered *via* 3-h infusion every 3 weeks at 8 different dose levels between 0.8 and 11.3 mg/m^2 (27 patients) (*54*). Major toxicities (grade 3) were constipation, fatigue and increased aspartate transaminase (AST) levels (1 each), but no Cycle 1 DLT had been seen at the time of reporting. Other toxicities (all grade 1–2, no dose dependence) included gastrointestinal toxicity (diarrhea, constipation, nausea), fatigue, and increases in AST, while no drug-related neurotoxicity was noted. PK assessment after the first and second infusion revealed a half-life of KOS-1584 of 17.7 h, and complete dose linearity of PK parameters was found over the dose range investigated. At the pharmacodynamic level a dose-dependent increase in microtubule bundle formation was observed in PBMCs; at the 8.5 mg/m^2 dose this effect was similar in magnitude to that reported for ixabepilone or KOS-862 at their Phase 2 dose. At the time of reporting, accrual into the trial was still continuing and no recommended Phase II dose had yet been defined.

In a second study KOS-1584 has been administered as a 1-h infusion on days 1, 8, and 15 every 4 weeks at dose levels between 0.8 and 25 mg/m^2 (37 patients enrolled at the time of reporting) (*55*, *56*). DLTs were observed at 20 and 25 mg/m^2, all of which, with one exception, involved diarrhea of increasing severity after successive infusions, despite maximal supportive care. These findings are strongly reminiscent of those discussed above for patupilone. In addition, one patient had typhlitis upon biopsy, and an ovarian cancer patient experienced grade 3 weakness, neutropenia and peripheral sensory neuropathy. (The latter had high plasma drug concentrations that may be related to pre-existing severe hypoalbuminemia and ascites). Common all-grade drug-related toxicities included nausea (51%), diarrhea, fatigue (both 49%), vomiting (32%), anorexia (24%), constipation (24%), peripheral sensory neuropathy (19%), and anemia (16%). Neutropenia/leucopenia (grade 1–2) was only observed at 16–25 mg/m^2 dose levels. Except for the DLT involving peripheral neurotoxicity, all neurotoxicity was mild-to-moderate. PK analysis revealed a half-life of the drug of 28.1 h and dose proportional increases in AUC and C_{max}. Assessment of pharmacodynamic effects was performed through serial sampling of PBMCs for soluble and polymerized tubulin by immunoblot. An increase in the fraction of poylmerized tubulin was observed with a maximal effect at end-of-infusion. The trial so far has yielded one confirmed response in a NSCLC patient. At the time of reporting, accrual into the trial still continued with expansion of the 16 mg/m^2 cohort and using aggressive anti-diarrheal prophylaxis (*56*).

In parallel with the 4-week schedule, a Phase I trial has been initiated with KOS-1584 with a weekly dosing regimen in the first two weeks of a 3-week cycle (*i.e.* with treatment on days 1 and 8 every 3 weeks) (*56*).

At the time of publication of the corresponding abstract one patient had been enrolled at a 16 mg/m^2 dose level.

7.7. Sagopilone (ZK-Epo)

Sagopilone, which is developed by Bayer-Schering, is the only fully synthetic epothilone analog so far that has been advanced to clinical development. At the time of writing this article, data from one Phase I and one Phase II study with the compound have been reported in abstract form (*6*, *57*). In the Phase I study (*57*) sagopilone was administered as a 30 min *i.v.* infusion once every 3 weeks at 12 dose levels ranging from 0.6 up to 29 mg/m^2. At the time of reporting, 37 patients had been treated with the compound. The most common toxicities observed were peripheral sensory neuropathy (mostly grade 1–2) and nausea, while hematological toxicity was mild and infrequent. DLTs occurred at the 16 mg/m^2 and the 29 mg/m^2 dose levels (grade 3 peripheral neuropathy (1 patient) and grade 3 ataxia (1 patient), respectively), but no MTD had yet been reached. Two confirmed responses were observed in patients with taxane-pretreated breast cancer for the 12 mg/m^2 and 16 mg/m^2 dose levels, and stable disease is reported for 13 patients with a median duration of 5.5 months. At the time of reporting, the trial was still ongoing.

According to a company fact-sheet, sagopilone is currently evaluated "in an extensive program of clinical Phase II studies in various oncological indications". Interim data from one of those Phase II trials were recently reported in abstract form (*6*). The trial includes patients with platinum-resistant ovarian cancer and is based on a 3-weekly treatment regimen (*i.e.*, every three week treatment), with the compound being administered at a dose of 16 mg/m^2 either as 3-h or a 30-min infusion. At the time of writing, 63 patients had entered the study, with data being available on 30 patients. Four responses (out of 13 patients) are reported for the 3-h arm in comparison with 1/13 response in the 30-min arm. The most common toxicity for both treatment regimens was grade 2–3 peripheral sensory neuropathy; other noticeable toxicities include fatigue, arthralgia, nausea, and lethargy. The trial was still ongoing when the present chapter has being written.

7.8. ABJ879

ABJ879 had been reported to have entered clinical development in an abstract in 2004 (*7*). No clinical trial data have been published for the compound and according to information from the company, Novartis in the meantime has terminated the development of ABJ879 (*58*).

7.9. Conclusions

At the time of writing of this account, data from more than 20 Phase II trials with 4 different epothilones (patupilone, ixabepilone, KOS-862, sagopilone) have been reported in literature, either in peer-reviewed papers or in meeting abstracts. Measurable antitumor activity has been observed in several of these studies, in particular in metastatic breast cancer, ovarian cancer, and hormone-resistant prostate cancer, while only sporadic activity has been reported in colon cancer. For two of these compounds the results from the Phase II studies have led to the initiation of Phase III trials and, in the case of ixabepilone, to FDA approval for the treatment of metastatic or locally advanced breast cancer (*1*). It remains to be seen, whether antitumor activity sufficient for registration purposes will also be established in other indications and/or for others of the above compounds or for one of those agents currently in Phase I clinical trials. This will require additional and more extensive studies, in particular, in direct comparison with standard therapy (all Phase II studies reported for epothilones to date are single-armed). It should also be noted that the development of one of the above Phase II compounds reportedly has been discontinued, although allegedly this was not a consequence of insufficient activity in the Phase II trials. Only Phase I data have been reported for BMS-310705 and KOS-1584 and the development status of BMS-310705 seems somewhat unclear (based on the fact that no new data have been reported since 2003). Development has definitely been terminated for Novartis' ABJ879, without any clinical data for the compound having been published so far. No reasons have been disclosed by the company for the termination decision.

In summary, the fact that ixabepilone has shown activity superior over standard therapy in metastatic breast cancer and has now been approved by the FDA for clinical use clearly highlights the potential of epothilones to become an important new component of our clinical armamentarium for cancer treatment. Likewise, the advancement of patupilone to Phase III trials in ovarian cancer is a highly encouraging sign (although Novartis expects submission of a registrations dossier for the compound only in 2009 (*58*)). It is hoped that the ongoing trials will reveal clinically useful antitumor activity against different types of cancer so that more than a decade of epothilone-directed research in chemistry, biology, and medicine will ultimately translate into a tangible benefit for cancer patients.

References

1. http://www.cancer.gov/cancertopics/druginfo/fda-ixabepilone
2. http://www.novartisoncology.com/page/patupilone.jsp
3. Overmoyer B, Waintraub S, Kaufman PA, Doyle T, Moore H, Modiano M, Kroener J, Zhang X, Buzdar A, Demario M (2005) Phase II Trial of KOS-862 (Epothilone D) in Anthracycline and Taxane Pretreated Metastatic Breast Cancer. J Clin Oncol **23** (16S): 778
4. Yee L, Lynch T, Villalona-Calero M, Rizvi N, Gabrail N, Sandler A, Cropp G, Palmer G (2005) A Phase II Study of KOS-862 (Epothilone D) as Second-Line Therapy in Non-Small Cell Lung Cancer. J Clin Oncol **23** (16S): 7127
5. http://phx.corporate-ir.net/phoenix.zhtml?c=121014&p=irol-newsArticle&ID=967533&highlight=
6. Rustin GJ, Reed NS, Jayson G, Ledermann JA, Adams M, Stredder C, Wagner A, Giurescu M (2007) Phase II Trial of the Novel Epothilone ZK-EPO (SAGOPILONE) in Patients with Platinum Resistant Ovarian Cancer. J Clin Oncol **25** (18S): 5527
7. Wartmann M, Loretan J, Reuter R, Hattenberger M, Muller M, Vaxelaire J, Maira S-M, Flörsheimer A, O'Reilly T, Nicolaou KC, Altmann K-H (2004) Preclinical Pharmacological Profile of ABJ879, a Novel Epothilone B Analog with Potent and Protracted Anti-tumor Activity. Proc Am Assoc Cancer Res **45**: Abstract 5440
8. Larkin JMG, Kaye SB (2006) Epothilones in the Treatment of Cancer. Exp Opin Invest Drugs **15**: 691
9. Cortes J, Baselga J (2007) Targeting the Microtubules in Breast Cancer Beyond Taxanes: The Epothilones. Oncologist **12**: 271
10. Lee JJ, Swain SM (2005) Development of Novel Chemotherapeutic Agents to Evade the Mechanisms of Multidrug Resistance (MDR). Semin Oncol **32** (6 Suppl 7): S22
11. Rubin EH, Rothermel J, Tesfaye F, Chen T, Hubert M, Ho YY, Hsu C-H, Oza AM (2005) Phase I Dose-Finding Study of Weekly Single-Agent Patupilone in Patients With Advanced Solid Tumors. J Clin Oncol **23**: 9120
12. Rothermel J, Wartmann M, Chen T, Hohneker T (2003) EPO906 (Epothilone B): A Promising Novel Microtubule Stabilizer. Semin Oncol **30** (Suppl 6): 51
13. Casado E, Tabernero J, Melichar B, Bridgewater J, Bennouna J, Delord J, Sopala M, Sklenar I, Cheung W, Johri A (2006) Patupilone in Chemotherapy-pretreated Patients with Advanced Colorectal Cancer (CRC) Receiving Nutritional Support and Intensive Diarrhea Management: A Phase I Multicenter Trial. J Clin Oncol **24** (18S): 3593
14. Melichar B, Tabernero J, Casado E, Bridgewater J, Hamm J, Sklenar I, Holland J, Cheung W, Zaknoen S, Johri A (2005) Phase I Dose Optimization Trial of Patupilone in Previously Treated Patients (pts) with Advanced Colon Cancer (ACC). J Clin Oncol **23** (16S): 3688
15. Østerlind K, Sánchez JM, Zatloukal P, Hamm J, Belani CP, Kim E, Felip E, Johri A, Berton M, Sklenar I (2005) Phase I/II Dose Escalation Trial of Patupilone Every 3 Weeks in Patients with Non-small Cell Lung Cancer. J Clin Oncol **23** (16S): 7110
16. Smit WM, Šufliarsky J, Spanik S, Wagnerová M, Kaye S, Oza AM, Gore M, Williams KJ, Johri A, Ten Bokkel Huinink WW (2005) Phase I/II Dose-Escalation Trial of Patupilone Every 3 Weeks in Patients with Resistant/Refractory Ovarian Cancer. Eur J Cancer Suppl **3**: Poster 909

17. Venook AP, Poon R, Kang YK, Mok TS, Chao Y, de Bedout S, Harbott A, Li J, Johri AR (2007) Evaluation of Patupilone as Monotherapy in Patients with Advanced Hepatocellular Carcinoma (HCC). J Clin Oncol **25** (18S): 15055

18. Abrey L, Wen PY, Govindan R, Reimers HJ, Rigas JR, Robins HI, de Bedout S, O'Connell C, Li X, Johri A (2007) Activity of Patupilone for the Treatment of Recurrent or Progressive Brain Metastases in Patients (pts) with Non-small Cell Lung Cancer (NSCLC): An Open-Label, Multicenter, Phase II Study. J Clin Oncol **25** (18S): 18058

19. Sánchez JM, Mellemgaard A, Perry M, Zatloukal P, Hamm J, Belani CP, Kim E, Felip E, Berton M, Johri A (2006) Efficacy and Safety of Patupilone in Non-small Cell Lung Cancer (NSCLC): A Phase I/II Trial. J Clin Oncol **24** (18S): 7104

20. Tannock IF, de Wit R, Berry WR, Horti J, Pluzanska A, Chi KN, Oudard S, Théodore C, James ND, Turesson I, Rosenthal MA, Eisenberger MA (2004) Docetaxel Plus Prednisone or Mitoxantrone Plus Prednisone for Advanced Prostate Cancer. New Engl J Med **351**: 1502

21. Hussain A, Dipaola RS, Baron AD, Higano CS, Tchekmediyan NS, Miller JA, Rothermel JD (2004) A Phase IIa Trial of Weekly EPO906 in Patients with Hormone-refractory Prostate Cancer (HPRC). J Clin Oncol **22** (14S): 4563

22. Lin N, Brakora K, Seiden M (2003) BMS-247550 (Bristol-Myers Squibb/GBF). Curr Opin Invest Drugs **4**: 746

23. Pivot X, Dufresne A, Villanueva C (2007) Efficacy and Safety of Ixabepilone, a Novel Epothilone Analogue. Clin Breast Cancer **7**: 543

24. Gadgeel SM, Wozniak A, Boinpally RR, Wiegand R, Heilbrun LK, Jain V, Parchment R, Colevas D, Cohen MB, LoRusso PM (2005) Phase I Clinical Trial of BMS-247550, a Derivative of Epothilone B, Using Accelerated Titration 2B Design. Clin Cancer Res **11**: 6233

25. Mani S, McDaid H, Hamilton A, Hochster H, Cohen MB, Khabelle D, Griffin T, Lebwohl DE, Liebes L, Muggia F, Horwitz SB (2004) Phase I Clinical and Pharmacokinetic Study of BMS-247550, a Novel Derivative of Epothilone B, in Solid Tumors. Clin Cancer Res **10**: 1289

26. Aghajanian C, Burris HA III, Jones S, Spriggs DR, Cohen MB, Peck R, Sabbatini P, Hensley M, Greco FA, Dupont J, O'Connor OA (2007) Phase I Study of the Novel Epothilone Analog Ixabepilone (BMS-247550) in Patients with Advanced Solid Tumors and Lymphomas. J Clin Oncol **25**: 1082

27. Shimizu T, Yamamoto N, Yamada Y, Fujisaka Y, Yamada K, Fujiwara Y, Takayama K, Tokudome T, Klimovsky J, Tamura T (2008) Phase I Clinical and Pharmacokinetic Study of 3-Weekly, 3-h Infusion of Ixabepilone (BMS-247550), an Epothilone B Analog, in Japanese Patients with Refractory Solid Tumors. Cancer Chemother Pharmacol **61**: 751

28. Burris HA III, Awada A, Jones S, Calvert S, Crabeels D, Whitaker L, Voi M, Lebwohl D, McCabe S, Hendlisz A, Piccart MJ (2002) Phase I Study of the Novel Epothilone BMS-247550 Administered Weekly in Patients (pts) with Advanced Malignancies. Proc Am Soc Clin Oncol **21**: Abstract 412

29. Zhuang SH, Agrawal M, Edgerly M, Bakke S, Kotz H, Thambi P, Rutt A, Balis FM, Bates S, Fojo T (2005) A Phase I Clinical Trial of Ixabepilone (BMS-247550), an Epothilone B Analog, Administered Intravenously on a Daily Schedule for 3 Days. Cancer **103**: 1932

30. Abraham J, Agrawal M, Bakke S, Rutt A, Edgerly M, Balis FM, Widemann B, Davis L, Damle B, Sonnichsen D, Lebwohl D, Bates S, Kotz H, Fojo T (2003) Phase I Trial and Pharmacokinetic Study of BMS-247550, an Epothilone B Analog, Administered Intravenously on Daily Schedule for Five Days. J Clin Oncol **21**: 1866

31. McDaid HM, Mani S, Shen H-J, Muggia F, Sonnichsen D, Horwitz SB (2002) Validation of the Pharmacodynamics of BMS-247550, an Analogue of Epothilone B, During a Phase I Clinical Study. Clin Cancer Res **8**: 2035
32. Hussain M, Tangen CM, Lara PN Jr, Vaishampayan UN, Petrylak DP, Colevas AD, Sakr WA, Crawford ED (2005) Ixabepilone (Epothilone B Analogue BMS-247550) is Active in Chemotherapy-Naive Patients with Hormone-Refractory Prostate Cancer: A Southwest Oncology Group Trial S0111. J Clin Oncol **23**: 8724
33. Galsky MD, Small EJ, Oh WK, Chen I, Smith DC, Colevas AD, Martone L, Curley T, DeLaCruz A, Scher HI, Kelly WK (2005) Multi-Institutional Randomized Phase II Trial of the Epothilone B Analog Ixabepilone (BMS-247550) with or without Estramustine Phosphate in Patients with Progressive Castrate Metastatic Prostate Cancer. J Clin Oncol **23**: 1439
34. Vansteenkiste J, Lara PN Jr, Le Chevalier T, Breton J-L, Bonomi P, Sandler AB, Socinski MA, Delbaldo C, McHenry B, Lebwohl D, Peck R, Edelman M (2007) Phase II Clinical Trial of the Epothilone B Analog, Ixabepilone, in Patients with Non-Small-Cell Lung Cancer Whose Tumors Have Failed First-Line Platinum-Based Chemotherapy. J Clin Oncol **25**: 3448
35. Whitehead RP, McCoy S, Rivkin SE, Gross HM, Conrad ME, Doolittle GC, Wolff RA, Goodwin JW, Dakhil SR, Abbruzzese JL (2006) A Phase II Trial of Epothilone B Analogue BMS-247550 (NSC #710428) Ixabepilone, in Patients with Advanced Pancreas Cancer: A Southwest Oncology Group Study. Invest New Drugs **24**: 515
36. Ajani JA, Safran H, Bokemeyer C, Shah MA, Lenz H-J, Cutsem E, Burris HA III, Lebwohl D, Mullaney B (2006) A Multi-Center Phase II Study of BMS-247550 (Ixabepilone) by Two Schedules in Patients with Metastatic Gastric Adenocarcinoma Previously Treated with a Taxane. Invest New Drugs **24**: 441
37. Eng C, Kindler HL, Nattam S, Ansari RH, Kasza K, Wade-Oliver K, Vokes EE (2004) A Phase II Trial of the Epothilone B Analog, BMS-247550, in Patients with Previously Treated Advanced Colorectal Cancer. Ann Oncol **15**: 928
38. Roche HH, Cure H, Bunnell C, Cognetti F, Mauriac F, Perez E, Miller K, Sparano J, Peck R, Yelle L (2003) A Phase II Study of Epothilone Analog BMS-247550 in Patients (pts) with Metastatic Breast Cancer (MBC) Previously Treated with an Anthracycline. Proc Am Soc Clin Oncol **22**: Abstract 69
39. Thomas E, Tabernero J, Fornier M, Fumoleau P, Lluch A, Viens P, Vahdat L, Conte P, Peck R, Martin M (2003) A Phase II Study of the Epothilone B Analog BMS-247550 in Patients (pts) with Taxane-resistant Metastatic Breast Cancer (MBC). Proc Am Soc Clin Oncol **22**: Abstract 30
40. Low JA, Wedam SB, Lee JJ, Berman AW, Brufsky A, Yang SX, Poruchynsky MS, Steinberg SM, Mannan N, Fojo T, Swain SM (2005) Phase II Clinical Trial of Ixabepilone (BMS-247550), an Epothilone B Analog, in Metastatic and Locally Advanced Breast Cancer. J Clin Oncol **23**: 2726
41. Denduluri N, Low JA, Lee JJ, Berman AW, Walshe JM, Vatas U, Chow CK, Steinberg SM, Yang SX, Swain SM (2007) Phase II Trial of Ixabepilone, an Epothilone B Analog, in Patients with Metastatic Breast Cancer Previously Untreated with Taxanes. J Clin Oncol **25**: 3421
42. Denduluri N, Lee JJ, Walshe J, Berman AW, Vatas U, Chow CK, Steinberg SM, Cox MC, Low JA, Swain SM (2007) Phase II Trial of Ixabepilone, an Epothilone B Analog, Given Daily for Three Days Every Three Weeks, in Metastatic Breast Cancer. Invest New Drugs **25**: 63
43. Vahdat LT, Thomas E, Li R, Jassem J, Gomez H, Chung H, Peck R, Mukhopadhyay P, Klimovsky J, Roché H (2007) Phase III Trial of Ixabepilone plus Capecitabine

Compared to Capecitabine Alone in Patients with Metastatic Breast Cancer (MBC) Previously Treated or Resistant to an Anthracycline and Resistant to Taxanes. J Clin Oncol **25** (18S): 1006
44. Marshall JL, Ramalingam S, Hwang JJ, Ramanathan R, Malik S, Egorin M, Hannah A, Cropp GF, Johnson RG, Belani C (2005) Phase 1 and Pharmacokinetic (PK) Study of Weekly KOS-862 (Epothilone D) Combined with Gemcitabine (GEM) in Patients (Pts) with Advanced Solid Tumors. J Clin Oncol **23** (16S): 2041
45. Cortes J, Climent MA, Gomez P, Ruiz A, Calvo E, Zhang X, Perez ES, Demario M, Baselga J (2006) A Phase I Trial of Weekly Combination KOS-862 (Epothilone D) and Trastuzumab in HER-2 Overexpressing Malignancies. J Clin Oncol **24** (18S): 2028
46. Piro LD, Rosen LS, Parson M, Cropp GF, McDaid H, Han J, Matela J, Johnson RG, Hannah AL (2003) KOS-862 (Epothilone D): A Comparison of two Schedules in Patients with Advanced Malignancies. Proc Am Soc Clin Oncol **22**: Abstract 539
47. Spriggs D, Dupont J, Pezzulli S, Larkin J, Cropp J, Johnson R, Hannah AL (2003) Phase 1 Dose Escalating and Pharmacokinetic (PK) Study of KOS-862 (Epothilone D): Phase 2 Dose and Schedule Defined. Clin Cancer Res **9** (16S): Abstract A248
48. Spriggs DR, Dupont J, Pezzulli S, Larkin J, Johnson RG, Hannah AL, Cropp G, McDaid H (2003) KOS-862 (Epothilone D): Phase 1 Dose Escalating and Pharmacokinetic (PK) Study in Patients with Advanced Malignancies. Proc Am Soc Clin Oncol **22**: Abstract 894
49. Holen K, Hannah A, Zhou Y, Cropp G, Johnson R, Volkman J, Binger K, Alberti D, Wilding G (2003) A Phase I Study to Determine the Safety and Pharmacokinetics of Continuous Intravenous Administration of KOS-862 (Epothilone D) in Patients with Solid Tumors. Clin Cancer Res **9** (16S): Abstract A261
50. Holen KD, Syed S, Hannah AL, Binger K, Wood L, Zhou Y, Cropp GF, Johnson RG, Rowinsky E, Wilding G (2004) Phase I Study Using Continuous Intravenous (CI) KOS-862 (Epothilone D) in Patients with Solid Tumors. J Clin Oncol **22** (14S): 2024
51. Mekhail T, Chung C, Holden S, Bukowski RM, Eckhardt SG, Cunningham M, Messina M, Cohen M, Peck R, Sikic B (2003) Phase I Trial of Novel Epothilone B Analog BMS-310705 IV q 21 Days. Proc Am Soc Clin Oncol **22**: Abstract 515
52. Sessa C, Perotti A, Malossi A, Capri G, Cresta S, De Braud F, Dall'o E, Voi M, Marsoni S, Gianni L (2003) Phase I and Pharmacokinetic (PK) Study of the Novel Epothilone BMS-310705 in Patients (pts) with Advanced Solid Cancer. Proc Am Soc Clin Oncol **22**: Abstract 519
53. Borzilleri RM, Vite GD (2003) Epothilones: New Tubulin Polymerization Agents in Preclinical and Clinical Development. Drug Future **27**: 1149
54. Villalona-Calero M, Goel S, Schaaf L, McCracken B, Desai K, Cropp G, Zhou Y, Johnson RG, Hannah AL, Mani S (2006) First-in-human Phase I Trial of a Novel Epothilone, KOS-1584. J Clin Oncol **24** (18S): 2003
55. Stopeck A, Thomas E, Jones S, Cohen J, Cropp G, Zhou Y, Feng L, Kersey K, Hannah AL, Burris H (2006) Phase I Trial of a Novel Epothilone, KOS-1584, Using a Weekly Dosing Schedule. J Clin Oncol **24** (18S): 2041
56. Stopeck A, Moulder S, Jones S, Cohen J, McDowell M, Cropp G, Zhong Z, Wells S, Hannah AH, Burris H (2007) Phase I Trial of KOS-1584 (a Novel Epothilone) Using Two Weekly Dosing Schedules. J Clin Oncol **25** (18S): 2571
57. Schmid P, Kiewe P, Kuehnhardt D, Korfel A, Lindemann S, Giurescu M, Reif S, Thiel E, Possinger K (2005) A Phase I Study of the Novel, Third Generation Epothilone ZK-EPO (SAGOPILONE) in Patients with Advanced Solid Tumors. J Clin Oncol **23** (16S): 2051
58. http://sec.edgar-online.com/2007/01/31/0001047469-07-000510/Section5.asp

Author Index

Page numbers printed in *italics* refer to References

Subject Index

1286093R0

Printed in Germany by
Amazon Distribution
GmbH, Leipzig